WITHDRAWN
FROM
COLLECTION

FORDHAM
UNIVERSITY
LIBRARIES

WILLIAM JAMES

William James

THE CENTER OF HIS VISION

DANIEL W. BJORK

COLUMBIA UNIVERSITY PRESS
NEW YORK
1988

Columbia University Press
New York Guildford, Surrey

LIBRARY OF CONGRESS CATALOGING-IN-PUBLICATION DATA

Bjork, Daniel W.
William James: the center of his vision / Daniel W. Bjork.
p. cm.
Bibliography: p.
Includes index.
ISBN 0-231-05674-5
1. James, William, 1842–1910. 2. Philosophers—United States—
Biography. I. Title
B945.J24B43 1988
191—dc19
[B] *87-31798*
CIP

Printed in the United States of America

Hardback editions of Columbia University Press books are Smyth-sewn
and are printed on permanent and durable acid-free paper

BOOK DESIGN BY JENNIFER DOSSIN

William James shortly after the publication of The Principles of Psychology, *in the early 1890s. By permission of the Houghton Library, Harvard University.*

FOR DAVID W. LEVY

Contents

Illustrations

Preface

My inner biography doesn't work either directly or indirectly on my mind, it is *my mind.*

Pure Experience Notebook
ca. 1905

"ANY AUTHOR is easy if you can catch the centre of his vision," remarked William James in 1903.[1] And although he referred to the German philosopher Hegel, James applied the generalization to himself as well. Several years earlier, in a critique of a Harvard student's thesis whose subject was "my poor self," he observed "that the whole . . . industry of building up an author's meaning out of separate texts leads nowhere unless you have grasped his centre of vision by an act of imagination. That, it seems to me, you lack in my case."[2] I have taken James' advice seriously, assuming that one must risk an act of imagination to locate the central biographical truth. In this sense what follows is a quest for the fundamental focus of life and mind.

At first glance one can imagine multiple centers in James. He embraced a pluralistic universe, a reality that refused containment in a monolithic perspective, a world that was multifaceted and flexible, one big enough to harbor endless varieties of experience. James' career also seems multicentered; he successively pursued art, natural science, medicine, psychology, and philosophy. Even his geographical compass seems unfixed, for he traveled to and fro between America and Europe, a cosmopolitan polylingual intellectual who, unlike his novelist brother Henry, could not fully settle in the New World or the Old. Moreover, James' temperament was seldom in repose; his inner self seems a shifting collage of moods, or as he would have preferred, "mental states." He described himself as a "motor," and his sedentary, neurasthenic sister, Alice, called William a "blob of mercury."

All these Jamesianisms have been presented as authentic, and in the five decades and more since Ralph Barton Perry's definitive two-volume biography, *The Thought and Character of William James* (1935), they have calcified into a conventional, little-questioned portrait. The shifting, flexible stance of pragmatism—mistakenly interpreted as James' major philosophical contribution—has also done its part to confirm his multicentered image. William James' perspective and life were perfectly consistent, according to this view; intellectually and personally he was a pragmatic, pluralistic, vocationally unsettled, globetrotting, restless fellow. To search for *the* center of his vision would seem inevitably to make monochromatic the biographical coat of many colors he so brilliantly and naturally wore.

No act of imagination is necessary to envision such a multifaceted James. He seems to be empirically given. Yet as he reminded the student who attempted to recreate the mind of her professor, such easy reconstruction poses the danger of leading nowhere. Although there has been a steady outpouring of writing about James—some of which is excellent—full-scale biographical reconstruction has, with one recent exception, generally been a matter of "covering" his life and thought.[3] In addition there has been very little critical interpretation of James as a person or a thinker. He has a legendary persona. And writers are in such basic sympathy with James' pluralism—an almost reflexive cultural/intellectual value—that to search for the Jamesian center would seem to distort not only James, but all of reality.

Gay Wilson Allen's *William James: A Biography* (1967) is an extensive narrative of James' life that largely avoids discussion of intellectual matters and is mostly uncritical of its subject.[4] Howard Feinstein's *Becoming William James* (1984) takes imaginative psychoanalytic risks and makes important interpretative advances on both Perry and Allen. Feinstein is concerned exclusively with the young man, however, ending his book with William in his mid-thirties.[5] One is left wondering what became of him over the next three and a half decades. Gerald E. Myers' recent *William James: His Life and Thought* (1986) provides the best coverage of James' intellectual contribution since Perry.[6] Myers presents an extensive account of James' psychology and metaphysics that incorporates a half century of scholarship as well as sources unavailable to Perry. But Myers does not search for an irreducible James so much as explicate his intellectual and moral life.

Rather than attempting to get inside the imaginative flow of James' creative process, biographers have more often stood outside and considered his intellectual productions. We learn about what supposedly influenced James, rather than how his mind might have created. The failure to distinguish between James' creative life and his successive careers has made biographical reconstruction more revealing of the biographers themselves than of James. Hence, we learn that Perry was interested in James' metaphysics, or that Allen was interested in James family relationships. That James was indeed interested in metaphysics, and that domestic life concerned him deeply, is obvious and important. But to reconstruct and interpret these historical factors does not make James' mind and emotional life come alive. To breathe biological life into James one must take an imaginative leap into the way he saw reality as his creative life unfolded.[7]

In searching for the Jamesian center, I have made several assumptions which constitute the bias of this biography. First, I take James to be a genius, who regardless of the influences of his mind—and there were, of course, many important ones—forged an original world picture, a novel reality that no American or European intellectual had yet developed. For all of James' contact with other thinkers, he was not essentially derivative of any thinker, including Charles Darwin and Charles Renouvier. Yet paradoxically, nothing of significance that James created was out of historical context; his originality was wonderfully culturally relevant, which is why he has remained so visible.

Second, I am convinced that William's intellectual and emotional indebtedness to his father, the eccentric, one-legged mystic-philosopher, Henry James, Sr., has been considerably overplayed. Here, despite admiration for Howard Feinstein's provocative Freudian reconstruction of three generations of father-son conflict, I find that William not only found little in the elder James' thought to truly stimulate his intellect, but failed to take his father seriously as a role model. Love him he did; derive the deepest springs of his genius from he did not.

On the other hand, James' intellectual-emotional dependence on his wife Alice Howe Gibbens James has been underplayed. Even though Allen alludes to her importance in sustaining William through periods of depression, limited access to their correspondence prevented full disclosure until recently of the importance of their relationship. Since 1980 some 1,400 letters from William to Alice have been available to researchers. This correspondence provides an inner structure upon which to interpret not only their relationship, but connections with other family members as well. Moreover, James expressed to Alice opinions about colleagues at Harvard and thinkers in both America and abroad that he never voiced to others. These letters also shed light on James' continuing health problems, especially those associated with his ailing heart in the period 1898–1910.

Third, I am convinced that the matter of James' psychic or spiritual crisis in the late 1860s and early 1870s has been overemphasized, while the later period of his life has been interpretively slighted. Since Cushing Strout's pathbreaking 1968 article on the oedipal sources of William's melancholia and vocational indecision,[8] there has been an almost exclusive biographical fixation on the late 1860s and early 1870s. This was partly of James' own making since his 1868–1873 diary gives ample evidence of an acute time of troubles. But scholars have worried about James' youthful anxieties to a point well beyond their actual importance in his experience. Their preoccupation says more about the fashion of writing about Victorian neurosis, and about the persistence of self-help mythology, than it does about James. Moreover, such an emphasis has encouraged a de-emphasizing of the intrinsic importance of his ideas. As crucial as the health question was, it never prevented James from being a strikingly original and productive scholar. Indeed, in a curious way, ill health encouraged creative endeavor.

Fourth, I have attempted to find the core of James' vision more in unpublished than in published sources—in correspondence, diaries, and intellectual notebooks. These materials present a less camouflaged James than their published counterparts. No effort has been made to "cover" with anything approaching thoroughness all of his published intellectual productions. For this the reader can more profitably turn to Gerald Myers' book as well as to a multitude of articles and monographs on James' psychological and philosophical contributions, many of which are discussed in the bibiographical essay at the end of this book. One can also turn to the definitive edition of James' major writings, *The Works of William James.*[9]

Finally, I have taken James' own cue to heart. He said that "my inner biography . . . is my mind," and I have assumed that it was so. James' own model of "pure experience" is my conceptual tool for examining his intellectual life. He persistently asked, "Within the world of experience, how do *I* see?" His mind was a creative continuum that focused on art, natural science, medicine, aesthetics, psychology, and philosophy. Rather than assuming that all of these interests, save philosophy, were false starts or mere preparations for metaphysics, I believe that each was a field of mental experience that was intrinsically genuine and yet *related* to another field or fields. As we shall see, there was for example no reason why James could not be a psychologist while considering art objects in Dresden in 1868; nor was there a reason why he could not indulge in psychical speculation while developing a philosophy of radical empiricism at Harvard in the late 1890s. To say that James was essentially a philosopher or a psychologist or an artist is to miss the real import of his mind. Nor does it help to say, although in a superficial sense it was true, that he was all of these. Vocational labels lead away rather than toward the center of his vision.

For James all knowledge was shareable in experience. And in this sense each of his vocations shared something of itself in relation to another field. These relations between fields of knowledge—or, more broadly, fields of experience—became the indispensable catalysts in forging James' perspective: "The essence of my contention is that in a world where connexions are not logically necessary, they may nonetheless adventitiously *come*." Hence, the relational area between the fields—between art, natural science, medicine, aesthetics, psychology, and philosophy—became the great creative ground: "The boundry becomes sensibly alive, is the firing line, the line of action." What follows is an exploration of James' fields of creative experience, areas that were alive and moveable, he suggested, like "a line of fire in burning grass."[10]

THINKING ABOUT James has had one unexpected and most pleasant consequence. Several individuals whose efforts and expertise materially strengthened whatever may be the merit of this study are now counted as new friends. And old friendships—at least from the author's perspective—have been "twice born" through kind sacrifices rendered in the midst of lives already busied with academic chores. Hence, I am not only deeply grateful for criticism, suggestions, and help in preparing the manuscript, but am fortunate to have expanded and revived a circle of friendship.

Professors Clifford Scott, William Baum, David W. Levy, and S. P. Fullinwider read with care and insight all or portions of the manuscript. The book benefits substantially from their collective wisdom. Discussions and correspondence with professors Michael M. Sokal, James Gilbert, and H. Wayne Morgan—now long past but not forgotten by the author—stimulated my thinking about James and his context.

There were two unexpected research pleasures in pursuing James that measurably aided understanding and interpretation.

Mrs. Peggy O'Brien, Librarian of the Keene Valley Library, Keene Valley, New York, acted as guide to the local sights of William's Adirondack summers, most notably to the still-standing Putnam Shanty—the primitive cottage where James and his bride, Alice Howe Gibbens James, honeymooned in 1878. Mrs. O'Brien's Keene Valley knowledge catalyzed my thinking about the large part the area would play in James' imagination.

Mr. and Mrs. Henry Vaux, of Berkeley, California, graciously allowed me free examination of a private collection of James family correspondence in their lovely home. Mr. Vaux is Robertson James' grandson and has a strong interest in the continual biographical reconstruction of the lives of all of his late relatives. This rich source was especially helpful in reconstructing William's relationship with his younger brother Robertson, and in more fully understanding family matters in general.

I have also accumulated a long-standing debt to the librarians and staff of the Houghton Reading Room, Harvard University. For a decade, with uncommon skill and patience, they have assisted in bringing my attention to the considerable mass of James materials housed in this repository. I am particularly thankful for the many services rendered by Houghton curator James J. Lewis and staff member Melanie Wisner.

Mercy College of Detroit generously extended faculty development funding, which facilitated much vital research. Several Mercy College friends have also unselfishly and skillfully assisted in preparing the manuscript. Since 1983 Pauline Yocum has uncomplainingly typed various drafts with admirable professionalism. Elizabeth Szalay graciously made available dozens of publications through a computer-based interlibrary loan, a service that saved many trips to many libraries. And Emmy Yousey expertly prepared photographic reproductions.

Without the support of Columbia University Press, this book would have remained a manuscript. I especially thank former associate editor Susan Koscielniak, assistant executive editor Elizabeth Sutton, and manuscript editor Ann Miller for helping transform this project from idea to hardcover reality.

Finally, there are two people who deserve inexpressible thanks, but thanks which nonetheless must be attempted.

David W. Levy has for twenty years been an inspiration, a wise critic, and a true friend. He has borne with saintly patience and singular wisdom the unending calls for assistance from one who ought to have realized long ago the meaning of academic abuse. I take great pleasure in dedicating this study to Dave, a gesture that is but insufficient and symbolic acknowledgment of his real contribution to my life and mind as well as to my thinking on James.

To my wife, Rhonda, who has been a companion in research and writing now on three books, I again find that verbalizing my appreciation is a trivialization. She spent hundreds of library hours taking notes; she read and criticized with unfailing good humor and sensitivity every word of the manuscript, several times; and she did this while working full time and catering to the author's creature comforts. My gratitude is boundless. If this study has value, credit for it must be shared equally with her. Any shortcomings remain mine alone.

WILLIAM JAMES

William James of Albany, William's paternal grandfather.
By permission of the Houghton Library, Harvard University.

CHAPTER ONE

Repent of the Past

Lose not a moment then, my dear Henry, to convince your father, by every future act of your life, that you repent of the past, and that you determine to act entirely comformably to his advice and wishes.

Advice to Henry James, Sr.,
while a student at Union College
NOVEMBER 1829

ON JULY 28, 1840, the honorable mayor of New York City, Isaac Varian, performed a civil marriage ceremony in the parlor of Mrs. James Walsh at 19 Washington Square. The betrothed were twenty-nine-year-old Henry James, and Mary Walsh, not quite one year his senior. The alliance culminated a courtship which had begun when Mary's brother, Hugh, introduced his sister to Henry when both men were students at Princeton. It was a portentous ceremony, for within three years Mary James would bear two sons who would become intellectual and literary giants in America and Europe.

The James and Walsh families were Irish and Presbyterian; both had been headed by prominent émigré fathers who at the time of Henry and Mary's wedding had long been dead.[1] Family similarities did not, however, produce similar temperaments. Mary was down to earth, shrewd, and not drawn to intellectual life. Henry was mystical, astonishingly impractical, and absorbed by abstractions. These fundamental differences did not prevent the couple from forming a deep attachment. When Mary James died in 1881, Henry found little reason to go on living and followed his beloved a year later. He recalled that "she really did arouse my heart, early in our married life, from its selfish stupor, and so enabled me to become a man."[2]

The "selfish stupor" originated in a troubled childhood and adolescence. Henry was the fourth son of William James of Albany, New York, whose three wives gave him thirteen children. Henry's mother was Catharine Barber, who became William's third wife in 1803 and bore Henry five years later. Catharine remained in the shadow of her aggressive, acquisitive husband whose rise to success in the New World was truly spectacular. Arriving as an eighteen-year-old immigrant in 1789, over the next forty years he built a fortune that was surpassed in New York State only by that of John Jacob Astor. He acquired property in Albany and Syracuse when construction of the Erie Canal sparked a boom that greatly inflated real estate prices. As his fortune increased he diversified into commerce, banking, and public utilities. At his death in 1832 William James of Albany, as he was known locally, had amassed an estate worth three million dollars. William's wealth permeated successive James generations. His success in life and estate in death shaped the character and rebellion of his son Henry.[3]

Born eight years after Emerson and eight years before Melville, the man who would become known as Henry James, Sr. lived during perhaps the most formative period in American history. The age was expansive in economic growth, political action, social experimentation, and intellectual innovation; it was the era of Jacksonian democracy and the first great stirrings of what would become American industrial capitalism.[4] Unfortunately for Henry, his father's timely and imaginative entrepreneurship was not accompanied by intellectual daring. William of Albany was quite satisfied with a conservative Presbyterianism, a Calvinist faith that encouraged the diligence necessary for financial success.

Much later, at the insistence of his own children, Henry sketched a brief, unfinished, thinly disguised autobiography, supposedly the life story of "the late Stephen Dewhurst." In it Dewhurst remembers not having had "much intellectual contact with my father save at family prayers and at meals, for he was always occupied during the day with business. . . . I cannot recollect that he ever questioned me about my out-of-door occupations, or about my companions, or showed any extreme solicitude about my standing in school."[5] Paternal indifference disappears, however, when Dewhurst suffers a gunshot wound in his arm. In real life, Henry at age thirteen suffered an accident that not only reversed his father's apathy, but changed his own life forever.

Playing with some flammable toy hot air balloons, he had kicked at one that ignited and severely burned his leg. The injury never healed properly, and black spots appeared, presaging gangrene. An amputation was performed well above the knee using no anesthetic except whiskey. The leg did not mend, and another operation was performed. His father's new-found compassion could not alter the fact that Henry had been badly disfigured. For two years he remained virtually bedridden. The words of the fictional Dewhurst suggest the fright and misery that possessed young Henry after the accident. God could choose to end Dewhurst's life at any time: "This insane terror pervaded my consciousness . . . made me loath at night to lose myself in sleep, lest this dread hand should clip my thread of life without time for a parting sob or penitence. . . . The conviction of his supernatural being and attributes was burnt into me as with a red-hot iron [as Henry's leg had been burned?], and I am sure no childish sinews were ever more strained than mine were in wrestling with the subtle terror of his name."[6]

Henry's travail, however, was just beginning. During convalescence troubling, persistent questions arose. He could see that his leg was gone, and yet it felt as if it were there. Why was this so? One thing seemed clear: nothing that he had learned from his father or in school had prepared him to cope with the loss of part of himself. The accident initiated Henry's life-long spiritual journey.

His early schooling at Albany Academy had been relatively uneventful. But getting to school was another matter. Years later he recalled that "when I was ten years old I was in the habit of taking a drink of raw gin or brandy on my way to school morning and afternoon. This was not an occasional thing, but as I say, habitual."[7] At Union College Henry continued to drink and became alarmingly irresponsible as well. He entered the Presbyterian school at nearby Schenectady in 1828, four years after losing his leg. Union had acquired a reputation as a way station for troubled youths. William of Albany was the leading parton-trustee, and while his son attended, he saved the college from bankruptcy. To Henry, the place seemed the institutional arm of his father's authority and values. Young James rebelled: he caroused, drank excessively, wrote unauthorized checks off William's account, and eventually ran away.

Years later when his own son, Robertson, was struggling with alchoholism, Henry recalled his protracted bout with intemperance:

> My long sickness previous to the loss of my leg had led my parents, physicians, and nurses to give me all manner of stimulants with a view to keep up my strength. . . . No wonder then that when I emerged from my sick-room, and went to college, I was hopelessly addicted to the vice. In college matters became very much worse with me and by the time I left it I was looked upon as another victim of intemperance. . . . I fell in with professional gamblers, and became demoralized generally. I scarcely ever went to bed sober, and lost my self-respect *almost* utterly. . . . I know what the insatiate craving after the loathsome madness is—how tempting a base enjoyment even for a moment of the deceitful cup is, and I have known . . . how dreary and dismal it is to have lost public confidence and [be] left to struggle alone against your bosom enemy.[8]

Just how far Henry had strayed from William's expectations, and to what degree his aberrant behavior embarrassed his father, may be glimpsed in a letter from William to a friend who was involved with him in shoring up to college's financial fortunes: "It is difficult to conceive of the wounded spirit of a man in my position—my heart pities a poor unfortunate son—who has so perverted the mercies of a kind providence. . . . If you see him, tell him, that when he finds how base he has acted—and when deceiv'd and despised by himself and all others;—to come to me and I shall endeavor to screen him from infamy and as far as possible from reproach—and . . . to ward off the shame and reproach he is creating to his brothers and sisters."[9]

Henry became a mid-nineteenth century dropout, and then not only refused to help carry on his father's business affairs, but failed to find an alternative profession. Later when his own children were asked by friends what their father did for a living, Henry suggested that they reply, "Say I'm a philosopher, say I'm a seeker for truth, say I'm a lover of my kind, say I'm an author of books, if you like; or best of all, just say I'm a student."[10] And at age forty-three he wrote to his mother, "There are two very bad things in this American land of ours, the worship of money and the worship of intellect. Both money and intellect are regarded as good in themselves. . . . Vastly better is it for the truly manly or womanly soul, to be unknown to the public, and so uncommitted to any popular idolatries, and to pursue its heavenly destiny in the path sincerity lit up only by the sincere smile of God."[11] But it would be William of Albany's money that sustained his search for religious idealism.

After dropping out of Union Henry drifted among shady companions in Canandaiqua and Buffalo and ended up in Boston. There he began to pursue the self-education that discontent demanded; he read voraciously, attended Unitarian church services, and felt temporarily free from paternal oppression. In 1830 Henry did return to Union to graduate with his class. But there was no rapprochement between father and son. His flight from Union was symptomatic of a lifelong journey away from William's conventional values. Henry had an almost constitutional aversion to established institutions, be they business corporations, governments, political parties, churches, or schools. Like many Jacksonian Americans, young James nurtured a radically anti-institutional world view. As one biographer aptly put it, Henry "scorned the essence of his father's world and waged open warfare against work, capital and the Calvinist ethos."[12]

In 1832, three years after Henry graduated from Union, Albany was struck by a cholera epidemic. Being a realistic and responsible burgher, William drew up a will. His timing proved auspicious, though he succumbed to a stroke rather than to the epidemic. But Henry was not turned toward Presbyterianism or toward a career in business by his father's death, nor did William's passing free Henry from paternal authority. Trustees were instructed to withhold his share of the estate until he settled into a practical profession. Without steady work he would not share in the final distribution of the estate, scheduled for some twenty years into the future. Henry engaged in protracted but ultimately successful litigation to challenge these stipulations.[13] The legal struggle cast a shadow over his schooling at Princeton, where he enrolled in 1835 to study theology. Although a theological degree would have enabled him to become an ordained Presbyterian minister, Princeton's conservatism proved unacceptable. By this point Henry favored a liberal, antinomian Protestantism, one emphasizing a freely bestowed spirituality. The moral law of his father's Presbyterianism could not compete with the inward joy of simple faith.

In 1837, immediately after the first distribution of the estate, Henry embarked for England. There he discovered Sandermanianism, a creed derived from early Christianity that advocated an uncomplicated relationship with God. Six years later, now a married man with two young children, and again after a favorable announcement on the estate, he returned to Europe. This trip had decisive educational and spiritual consequences. He never forgot the inner panic that seized him one evening in May of 1843, while gazing into the fire in a rented cottage near Windsor Castle:

> To all appearance it was perfectly insane and abject terror without ostensible cause, and only to be accounted for, to my perplexed imagination, by some damned shape squatting invisible to me within the precincts of the room, and raying out from his fetid personality influences fatal to life. The thing had not lasted ten seconds before I felt myself a wreck; that is, reduced from a state of firm and vigorous, joyful manhood to one of almost helpless infancy. The only self-control I was capable of exerting was to keep my seat. I felt the greatest desire to run incontinently to the foot of the stairs and shout for help to my wife—to run to the road-side even, and appeal to the public to protect me; but by an immense effort I controlled these frenzied impulses, and determined not to budge from my chair till I had recovered my lost self-possession.[14]

The "damned shape" which rayed out of "his fetid personality" may have reminded Henry of his tyrannical father. Or it may have evoked memory of the putrification of his own flesh after the accident. But the crucial symbolic element was neither the father nor the amputation. To be reduced "from a state of firm and vigorous, joyful manhood to one of almost helpless infancy" was to lose control. Henry awakened to personal impotency at the very moment that he seemed most secure.

The loss of self-control and the return of the powerlessness of childhood were clear signs that his spiritual development was about to begin. In the terror of helplessness he had the chance to find saving grace. The devil was not his father or the horrors of the accident, it was his vain belief in himself. The vision was a sign that his rebellion had not gone far enough, for in truth all of his actions since the accident had reeked with self-justification. The psychic terror served symbolically to divide a life devoted to fulfillment of the self—at bottom a fleshly and materialistic goal—from a life devoted to the sacrifice of the self for spiritual selflessness.

The vision was not unlike hundreds of awakenings that occurred in New England and New York in the Jacksonian era. That Henry experienced the self-loathing and terror in England without the immediate stimulus of revival preaching did not negate its Americanism. To truly repent of the past was to understand that the source of corruption was ultimately in oneself, a deep felt conviction of many New Englanders who had conversion experiences and went on to become revivalists, reformers, and transcendentalists.[15]

Shortly after this seizure, or after what Henry called his "vastation," a London friend, Dr. James John Garth Wilkinson, introduced him to the writings of the Swedish religious mystic Emanuel Swedenborg. Henry found spiritual renewal in a creed that discounted the material reality of the flesh as well as the self. The vastation and discovery of Swedenborg seemed to divide self-indulgence and unease from selfless serenity. Henry became a philosopher-mystic who wrote religious and metaphysical treatises promoting the selfless way. He became a father who through educational experiment would lead his children toward the wisdom and spiritual development William of Albany had failed to provide.[16]

WITHIN TWO years after marrying, Henry had the opportunity to reconstruct the happy childhood he had missed. In January 1842 came his first-born son, William, probably named not after the despised William of Albany, but after Henry's brother William, who was also partly disinherited. A year later little Henry appeared, his father's namesake. In 1845 Mary delivered a third boy, christened Garth Wilkinson in gratitude to the Londoner who had introduced Henry to Swedenborg. In 1846 came the last boy, Robertson. And in 1848, as if to protest this male monopoly, Mary bore their only daughter, Alice.

By the time the family was complete, Henry Sr. had already begun what was to become an uncommon, even bizarre educational experiment. None of the children attended truly public schools. Before his experimentation ended—and it did not end because Henry wanted it to, but because the children grew up or events interceded—the elder James orchestrated a decidedly moving educational experience. Before he was ten, William had already attended several private schools near their comfortable Fourteenth Street home in New York City. Indeed, it was remarkable that the family remained for seven years in the Fourteenth Street residence, considering the frequent changes in schools and tutors. William lived longer there than in any single residence until at forty-seven he moved to Irving Street in Cambridge, Massachusetts. It was from the Fourteenth Street house that Henry Sr. resolved to keep his children from the public schools.

At first tutors came to the James home, but in 1850 eight-year-old William was enrolled in the Institute Vergnès not far from Washington Square. The Institute introduced both William and Henry, who attended with his slightly older brother, to French. Neither boy especially liked the place, as the Monsieur was stiff and easily irritated, while the other students were the culturally alien sons of wealthy Mexican and Cuban planters. Henry Sr. heeded his sons' discontent, and they were placed in Master Richard Pulling Jenks' school on Broadway near Fourth Street. William found the Jenks establishment more enjoyable, because an instructor taught art and allowed him ample time to sketch. An elderly Henry James recalled that his brother William was consumed with drawing: "As I catch W. J.'s image, from far back, at its most characteristic, he sits drawing and drawing, always drawing . . . and not as with plodding patience . . . but easily, freely, and . . . infallibly: always at the stage of finishing off, his head dropped from side to side and his tongue rubbing his lower lip."[17]

From about age ten, "Willy," as his family frequently called him, was more attracted to drawing than to any other activity—except, perhaps, roaming the streets of Manhattan in search of adventure, danger, or some boyish combination of both. William and Henry, who seemed to follow his older brother everywhere, were allowed to wander extensively. They ambled up Broadway where it was unpaved; they watched passengeres disembark from horse cars at Union Square; they went up Fourth Avenue to Twenty-third Street to observe dynamiting in preparation for the construction of the Hudson River Railroad, which would eventually take them to visit their paternal grandmother, Catherine Barber James, in Albany. And they stole away at every opportunity to Barnum's American Museum where they gawked and marveled at nature's oddities the biological anomalies that fascinated thousands in the decade before Darwin's *Origin of Species*. Perhaps William's earliest scientific interest was stirred by the curiosities that seemed to stare back at him in bottled, lifeless suspension.[18]

He also had opportunity to cultivate himself artistically. Henry Sr. accompanied his boys to local galleries, where they viewed American classics such as Emanuel Leutz's *Washington Crossing the Delaware,* and surveyed the early nineteenth-century painting of the Dusseldorf school. In 1853 father and sons marveled at the Danish painter Bertel Thorvaldsen's gigantic sculpture of Christ on display at the Crystal Palace Exhibition. The gallery-like interior of the Fourteenth Street residence complemented these "pictorial" evenings. Henry Jr. recalled that a Thomas Cole "covered half a side of our front parlor," while another huge canvas, Lefevre's *View of Tuscany,* loomed over a sofa. But Henry Sr. made his sons aware that in comparison with Europe's great collections, American art in the mid-nineteenth century was a distant second. Where were the works by El Greco, Rembrandt, Ruben? No matter how many New York galleries and exhibitions they attended, they could get at best only a taste of the great masters. "In Europe," as Henry Jr. later put it, "we knew there was Art."[19]

So in 1855 when their father suggested taking the family on an extended European trip, William and Henry excitedly concurred. Both boys were unhappy in their latest private school, a place called Forest and Quakenboss that specialized in Latin and arithmetic. Well-to-do Americans were now making regular, extended European crossings. Whenever the family visited Grandmother James in Albany, it seemed as if some affluent cousin had just arrived back from the continent. Mr. James was anxious to avoid the philistine coarseness of America. He wanted William and Henry in particular to have a "sensuous education," a thorough grounding in the great artistic and spiritual achievements of western civilization.[20]

But there may have been an even more compelling reason to cross the Atlantic. William was now thirteen, exactly the age when the elder James had suffered his accident and amputation; he was entering the period when his father had begun his great travail. Did the father note the parallel and decide it was time to monitor more carefully the eldest son's intellectual and moral development? Whatever the answer, it is clear that Henry had found spiritual direction in European trips. Twelve years had passed since the fright of his vision before the fire and his Swedenborgian renewal. Passage to the Old World had deep meaning for him, and he meant to share it with his children, especially with young William, who already showed unusual ability.

CHAPTER TWO

A Hungry Eye

He saw things . . . with what I think of as a hungry eye for the line or form or movement that expressed a meaning. One striking quality of his writing is that it often expresses a lively pictorial sense. Here is further evidence of a natural faculty of seeing that he sharpened by practice with pencil and brush.

Henry James III on his father's sketches

LATE IN June of 1855, the Collins liner SS *Atlantic* left New York harbor and after twelve days in rough sea docked at Liverpool. The crossing would become routine for young William James, who had already been whisked to Europe as a baby. The peripatetic James would travel to the Old World eight more times. One thing that made the 1855 trip noteworthy was its length: the Jameses would not return to America for almost three years. But what made it truly significant was the educational and cultural impact on a precocious, talented youngster who was just entering the turmoil and promise of adolescence. William was thirteen and a half when the James family left New York in June 1855, and he was nearly sixteen and a half when they returned in May 1858. He did not grow up in Europe; his prolonged adolescence continued through his late twenties. Yet he was clearly no longer a child. Europe made an impact on a forming, impressionable youngster which was just what Henry Sr. had wished—in the beginning.

Although they have dutifully recorded the 1855–1858 stay in Europe, William's biographers have downplayed the sojourn instead of seeing it as a truly catalytic period in William's development.[1] Other journeys, such as his Brazilian adventure in 1865–1866, and most of all, his 1867–1868 convalescence in Germany, have assumed much more interpretive importance. The fact that William did not compose notebooks and diaries during the 1855–1858 trip discourages making too much of it. Yet surely the early teens are one of life's notable passages, and it would be misleading to suggest that James somehow escaped the problems of self-definition in this stage of life. If James' life is to be interpreted as a continuum of fields of creative experience, then the first stirrings of creativity must be considered vital. How did William's intellectual gifts first make themselves known to him?

After arriving in Liverpool in early July, the Jameses went directly to London. Several days later they set out for Paris; after a short pause there, they traveled by rail to Lyon, and then, in a hired carriage, crossed the Jura mountains into Switzerland, arriving in Geneva in late July. The family encamped in a villa overlooking Geneva, with a view of Mont Blanc. By early August all of the children, save Henry, who was ill with malaria, and Alice, who was only seven, were enrolled in Achilles Heinrich Roediger's private school. Initially their father was much impressed by Roediger: "He is a man of great sense, and I should judge of enormous practical power, so that whatever he undertakes to do he is apt to succeed in. . . . I certainly never met a man in my life of more powerful personal magnetism."[2] But a month later Henry Sr. had cooled considerably and wrote to his mother that "the schools are greatly overrated. We do not find the advantages we had expected in them. . . . We have come to the conclusion that home tuition will be the best for all of them; that . . . it will also be greatly to the interests of the children both in morals and intellectual regards."[3]

At Roediger's William made rapid strides in French and began studying German. There was certainly no indication that William was dissatisfied. Perhaps Roediger was *too* magnetic and threatened the elder James' sense of authority over the children. William was particularly drawn to charismatic teachers, men who had a worldly success his father could not approach. Later young James would attach himself to the American artist William Morris Hunt and to the Harvard naturalist, Swiss émigré Louis Agassiz, both of whom had the professional stature Henry Sr. lacked.

By October 1855 the elder James had indeed soured on formal schooling in Geneva, and the family retraced their steps back through

Lyon to Paris. They stayed long enough to attend an art exhibition at the Palais de l'Industrie, an event that may have had considerable impact on William, as we shall presently see. Afterwards they recrossed the English Channel and returned to London, where Henry Sr. rented a house at Berkeley Square. As in Geneva, he was initially enthusiastic: "I find, " he wrote his brother Edward, "that the English schools have the reputation of being better than the Swiss, as to all the solid parts of education, and I hope that the boys will be brought around rapidly."[4] Mr. James employed a Scottish tutor, however, who gave the boys lessons in English history and guided tours of museums and monuments. In November the family moved to a furnished house in St. John's Wood, where the children romped in nearby Regent's Park and toured the Zoological Gardens. Their father seemed pleased: "The boys have a capital tutor, and were never so sweet and good. . . . I have no doubt we shall return home a well educated and polished family. . . . We have a home feeling in London that is very agreeable, and you do not feel yourself so constantly cheated as you are apt to in Paris."[5]

William and Henry also enjoyed London, particularly visits to the art museums such as the Pantheon and the National Gallery. They learned to appreciate the great allegorical themes of Rubens, Titian, and especially of the British painter Benjamin Robert Haydon, who was Henry Sr.'s favorite. According to Henry Jr.'s retrospection, he and William were enthralled with such Pre-Raphaelites as John Everett Millais and W. Holman Hunt, who were displayed at the National Gallery. The Pre-Raphaelites were the current rage, evoking romance with their exquisite lines and fairyland colors.[6]

Yet Henry Sr. soon tired of what he viewed as stuffy, pretentious English airs, and in May 1856 left London for Paris to rent a well-appointed house near the Champs-Elysées—a dwelling which a family friend recalled "had a courtyard with cobblestones and a pump over which hung . . . a never-to-be forgotten glamour!"[7] After less than a year in Europe, the Jameses had shuffled themselves from London to Paris to Geneva, back to Paris and London, and then back again to Paris. William continued to be tutored, this time by one M. Lerambert, whom he found overly refined and intellectual. He also drew with the French painter Leon Coignet, who was notable enough to have exhibited at the Luxembourg Museum; and he became deeply fascinated with the work of Eugène Delacroix. As William passed from school to school and tutor to tutor, an educational treadmill he later regretted, one constant emerged: his interest remained fixed on art and drawing.

In the summer of 1857 Mr. James decided to move the family once more, this time to the Normandy coastal resort of Boulogne-sur-Mer. William enrolled in the College Imperial, the equivalent of an American academy or private high school. For the first time he began formally to study science, an activity that Henry Sr. had been encouraging since the previous winter in Paris, when he had given his son a microscope inscribed, "William James from his father, Christmas, 1856."[8] The elder James observed, "Willy is very devoted to scientific pursuit, and I hope will turn out a most respectable scholar. He has been attending the College Imperial here all summer and one of his professors told me the other day 'that he was an admirable student, and that all the advantages of a first-rate scientific education . . . ought to be accorded him.' " Still, Henry was not overly impressed with William's scientific abilities, for he added, "He is however much dearer to my heart for his moral worth than for his intellectual."[9]

By winter 1857 the family was one again back in Paris at their stylish Champs-Elysées residence—but not for long. The financial panic of 1857 depressed the price of Henry Sr.'s American railroad stocks. Suddenly he became much concerned about extravagance, something the more practical Mary James had worried about from the beginning. Since their rent at Boulogne-sur-Mer was 200 francs as opposed to the 800 francs they were paying in Paris, the Jameses returned to the coast to take advantage of the off-season rates. William, Garth Wilkinson ("Wilky"), and Robertson ("Rob" or "Bob") were once again enrolled in the College Imperial, while Henry and Alice had tutors. Save for the brief return to Paris, the family was thus settled for the better part of a year at Boulogne, making possible the longest period of formal instruction in one place that William had yet received.

But Henry Sr. was rapidly tiring of Europe. As early as September 1856 he confided to a friend that "American disorder is sweet beside European order: It is so full of promise."[10] And in December 1857 Mary James wrote to her mother-in-law that it was "a daily battle" keeping Henry in Europe, "so strongly does he feel home pulling at his heart."[11] In May 1858 Mr. James booked passage for New York. Perhaps a major reason for the elder James' growing discontent was linguistic; unlike Henry and William, he could speak only English. While they were quickly adapting to European culture, he remained, despite his broad artistic and philosophical interests, fundamentally an American tourist abroad. After his vastation and introduction to Swedenborg, the Old World could not long sustain his interest. But how different was it for his two oldest sons?

THE PARIS that the James family first entered in the summer of 1855 was in the early stages of massive renovation. The renewal was the inspiration of Louis Napoleon Bonaparte, the president of the Second Republic, Napoleon III as he preferred, and of his collaborator, the prefect of the Seine, Georges (later Baron) Haussmann. By 1870 what had been a cramped, medieval-like city with its maze of narrow, crooked streets, inadequate sewage, and cholera-infested slums would be transformed into an expansive urban landscape crisscrossed by broad boulevards, graced by new museum facades and by improved parks like the huge Bois de Boulogne. But in 1855 the Jameses found themselves in a city of striking contrasts and confusions, with construction scaffolding on the Louvre, torn-up streets, and the Bois de Boulogne, which the family often visited, in the process of being transformed into the spacious park where Parisian society and commoners daily gathered.[12]

The family located in 1855–1856 in an apartment on rue d'Angoulime-St. Honoré, and in the spring of 1857 they moved to rue Montaigne. These addresses were in the western part of the city, on the right bank of the Seine, a fashionable area close to museums, art galleries, and the Champs-Elysées, still, however, adjoined to unexpected Parisian slums.[13] The city was radically altering its shape. And William, who was exchanging childhood for adolescence, found his own identity also in transit as he walked the streets of a Paris that changed daily before his eyes.

Upon their return to London from Geneva in October 1855, the Jameses had stopped briefly in Paris to attend the Universal Exhibition at the Palais de l'Industrie. The Exhibition highlighted two controversial contemporary French painters, Jean Auguste Dominique Ingres and Eugène Delacroix. Both had special rooms filled with paintings. Delacroix displayed forty-two canvases on loan from the Louvre, from the museums of Nancy, Lille, Rouen, and Versailles, and from private collectors. The Delacroix room contained a painting that seems to have served both in form and theme as a model for many of William's subsequent sketches. It was an oil painted in 1854 called *Lion Hunt,* a burst of color, an

ABOVE, Lion Hunt *by Eugène Delacroix. Courtesy of Syndicate de la Propriété, Paris.*

RIGHT, *sketch by William, ca. 1858–1860, showing the influence of Delacroix. By permission of the Houghton Library, Harvard University.*

indistinct yet unmistakable depiction of lions attacking horses and riders.[14]

James drew numerous animals attacking or devouring animals that are clearly influenced by Delacroix's fascination with this theme. *Lion Hunt* was the most vivid and controversial of Delacroix's ferocious animal pictures, and James almost certainly saw it at the Universal Exhibition in October 1855. Henry Jr. later maintained that the Delacroix that gripped William most was *The Bark of Dante,* which depicted Dante and Virgil afloat on the churning, hellish Marsh of Styx. It is more likely, however, that *The Bark of Dante* held more interest for Henry than for William; it was Henry who recalled the painting, and its historical allegory probably appealed more to his budding literary sensibilities.[15] Surviving drawings clearly show that William was attracted to natural violence uncluttered by classical personalities. There are no surviving sketches indicating that William was at all interested in representing Dante and Virgil, or anyone else, at sea.

Lion Hunt is not only closer to the themes of many of William's sketches, it also presented an avant-garde style that radically assailed contemporary aesthetic sensibility—much more so than did the photographic realism of the Pre-Raphaelite artists whom William admired but did not copy. Delacroix was not simply challenging the classicism of painters like Jacques-Louis David; others such as Ingres and Théodore Géricault had been doing so for years. *Lion Hunt* may have presented William with a special aesthetic perspective, one that anticipated a unique way of seeing, an original vision.

There is no direct evidence that James glimpsed his own creative style in *Lion Hunt,* for nowhere does he comment on the painting. Nonetheless, it is possible to imagine what would attract him—even at thirteen.[16] As depth psychologist Ira Progoff has remarked:

> When we have the personal experience of connecting inwardly with a work of art . . . we find that it activates a flow of emotions, thoughts, and sensitivities in the depths of us. That is one characteristic of a work of art: it has an evocative effect on human beings other than the person who created it. Sometimes, in fact, its power to evoke is so great that it stimulates recognitions of truth in the viewer that are greater and more profound than the artist originally conceived. But that only underscores the power of a

> work of art. It is able to go beyond itself. The process of creativity continues working in and through it after its own specific creation has been completed. The work of art has been completed, but its life is not finished. It retains a capacity of self-extending creativity that continues as an active factor in the world.[17]

What might William have experienced in 1855 when he wandered into the Delacroix room and gazed upon the Frenchman's most radical production? One contemporary critic, Maxime du Camp, wrote of *Lion Hunt*, "This picture defies criticism. It is a huge verbal puzzle in colors with no clue to the key word. . . . Color here has reached its ultimate degree of excess. . . . Delacroix is not the leader of a school, he is the leader of a riot." Delacroix's friend Charles Baudelaire wrote, "*Lion Hunt* is truly a color explosion. . . . Never have more beautiful, more intense colors penetrated the soul by way of the eyes." He suggested that "this painting, like sorcerers or hypnotists, projects its thought at a distance. The strange phenomenon is due to the power of a colorist, the perfect agreement of the tones, and harmony . . . between color and subject."[18] Critics found in the work a powerful freedom in which color was released from form, a freedom that tantalized the viewer to identify the figures—the lions, the horses, the hunters.

Lion Hunt evoked the mysterious relationship between color, form, and content—the great problem of how the eye and mind make the world. Of course this had always been a transcendent contribution of great painting: to bring color, form, and content into the eye and mind as a representation of the world, whether the representation be a landscape, a figure, or an abstraction. But Delacroix, like the innovative Fauves in the early twentieth century, saw the deception in our so-called true perspective. *Lion Hunt* showed that the world can be given to us in an almost unrecognizable way. It suggested that what is seen as the world might not be the world at all.[19]

Could this have been young James' first original insight: that visual forms were initially indistinct and unorganized, and only later developed into what was supposed to be the real world? Did he recognize in the swirl of line and color in Delacroix that "our space-experiences form a chaos, out of which we have no immediate faculty for extricating them"?[20]

We do know that James was learning to mix colors with Cogniet while infatuated with Delacroix. Hand, eye, and mind certainly were poised to discover new connections between the natural world and the artistic one. Eight years later when he began a notebook, "Index Rerum," which was "intended as a manual to aid the Professional Man in preparing himself for usefulness," under "Delacroix" James entered "See Art." The "Art" entry juxtaposed "Art vs. Nature," and explained the contrast in French, which translated reads: "To make a choice in Nature very cleverly we have made a law about it, because three fourths of the time nature goes without contrasts. It is therefore by inadequacy that we choose, because the means of art are limited, and it must always sacrifice one thing in order to set off another."[21] Delacroix stimulated his thinking about the relationship between art and nature, between the real and the unreal, between the objective world of nature and the subjective world of art.

One scholar has argued that William was obsessed with Delacroixian themes of natural violence because for him they were projections of murderous wishes toward Henry Sr.[22] But a less Freudian interpretation would be that William simply saw an exceptionally original expression in Delacroix, an expression that brought art, the mind, and nature into a new relationship. Delacroix emphasized a mysterious connection between line and color, form and content, and suggested that one did not have to simply view an art object as a true duplication of nature, immovable and wholly separate from the viewing subject, but might actually experience art as a moment when the art object and viewing subject were intimately related within a dynamic world. Many years later James would depart from conventional psychology and philosophy in a similar way. The world, he would conclude, is given through experience, not abstraction. Just as Delacroix related art and nature in the mind of the artist, so too were subject and object to be related for James in the stream of consciousness or in pure experience.[23]

In 1854, the year he painted *Lion Hunt,* Delacroix commented in his journal on the essence of the artist, interestingly in contrast to the scholar, seen as the artist's opposite: "When all is said and done scholars can do no more than find in nature what is already there. A pedant's personality plays no part in his work, but it is very different with the artist." For him, "It is the imprint which he sets upon his work that makes it the work of an artist—of an inventor." "The scholar," he continues, "discovers the ingredients, so to speak, but the artist takes ingredients that have no value in the place where they are, composes them and invents a unity, in short, he creates." He emphasizes what James' creative life would affirm: "[The artist] has the power of striking men's imagination by the sight of his creations and he does this in a manner characteristic of himself."[24] Delacroix may have inspired William by encouraging him to reconceive and recompose the visual presentation of nature.

Progoff again helps us to imagine how James may have received Delacroix's genius as an insight into his own: "In the flow of human experience, the point where two periods or units of life/time come together presents an open moment of possibility. At that point the events of the future are still in a formative stage. When events are perceived . . . they are not being seen in their completeness but in their process as they are moving out of their past into their future." As James considered Delacroix's art, he responded in the spirit of an open moment. He could do so more freely because of his life stage; he was an adolescent, caught in the open moment between childhood and adulthood, when gifted individuals are especially receptive to symbolic forms, to open moment images that become both unconscious and conscious signs of new powers and vocational options.[25]

THE ELDER James could not easily acquiesce to William's budding interest in new art forms precisely because such interest presented his son with the option of becoming a *professional* painter. In broad cultural terms, William was poised to participate in what has been called the professionalization of American culture, the penchant for career development and expertise that would increasingly characterize the nineteenth-century educational world.[26] This context offered "career" rather than his father's "morals and aesthetics" as the primary educational goal. Young James' fascination with art would not be simply intellectual, wholly a contemplation of nonmaterial ideals. And this troubled Henry Sr., who had rejected career even in the far less professionally conscious early nineteenth century. By the late 1850s, the son was far more comfortable than was his father with the upper middle class educational ambitions that began to characterize the new professionalism.

During the winter of 1856–1857, when William's infatuation with Delacroix had peaked, Mr. James made what Gay Wilson Allen has called "one of the most bizarre decisions in all his impractical educational experiments."[27] He enrolled William, as well as Henry and Wilky, in the Institution Fezandie. Headmaster Fezandie was a follower of the French utopian socialist Charles Fourier, a scientific-minded ally of Swedenborg. William found the place farcical because Fezandie and an assistant paced priggishly while demanding rote recitation. But there was a deeper dissatisfaction. His father was attempting to steer him away from art.

Several years earlier, Henry Sr. had expressed much admiration for the true artist. In 1850 he wrote that "the Artist is not good by comparison merely, or the antagonism of meaner men. He is positively good, good by absolute or original worth, good like God, good in himself, and therefore universally good."[28] But a painter could also be exposed to immorality, and Henry Sr. may have shared contemporary American prejudices toward actual—rather than ethereal—artists.[29] One biographer has argued that Henry Sr.'s shifting opinion of artists was related to an ongoing effort to manipulate William's career choices, in a futile attempt to resolve still-lingering oedipal conflict with his own long-deceased father, William of Albany.[30] But another explanation for the elder James' decision to place William in the Fezandie school virtually begs consideration. For all of Henry's fondness for great art as transcendent spirituality, he lacked the crucial element that would have enabled him to share William's enthusiasm for Delacroix, even if he had appreciated that painter's genius. He had never drawn; he had not spent countless hours sketching; he had no talent.

Art to Henry was more object than subject; it was something to be appreciated and discussed rather than to be created. True, Henry had earlier spoken of art as "the sphere of man's spontaneous productivity"; but that was an opinion, a didactive, essentially educative judgment.[31] William was experiencing art first hand, living it. Through discovery of Delacroix he was telling his father that he *was* an artist. Moreover, he was making the occupational commitment that had somehow always eluded Mr. James. William had discovered an original exemplar who had awakened a powerful inner sign pointing to his own intellectual-professional development. Henry Sr. could find no personal or cultural basis for sharing his son's experience.

By placing William at the height of his interest in Delacroix in a school devoted to Fourier, Henry Sr. moved his eldest son back along his own intellectual path. There was indeed a curious intergenerational echo here. William of Albany had never accepted Henry's religious flight from established, respectable Presbyterianism. Yet Henry saw in antinomian Protestantism, and eventually in Sandermanianism and Swedenborgianism, all of the emotional and intellectual excitement missing in his father's faith. He had first experienced this excitement in his own prolonged "open moment" during adolescence in the years following his accident. Just as William of Albany had ignored and opposed his son's discoveries, so too did Henry Sr. refuse to acknowledge and promote William's.[32]

Here was Mr. James with his repetitive, dense philosophy, his hit-or-miss tutors, his often ludicrous choice of schools, standing now not in favor of but in opposition to originality and the experience of art as a probable career. How could a father who had smothered his son with educational freedom from the earliest memories, a father who had always been so visible as to be underfoot, a father who continued to belabor ideas so familiar that they were now tedious, compete with the novelty of Paris and the genius of a great painter? He could not. Perhaps Paris reminded Henry Sr. of his own provincialism. Prolonged exposure to the Old World might alienate his children from him, poison them with charm, expose his intellectual limitations, and threaten his educational control.

For all of his knowledge of philosophy and religion, the elder James never learned to speak any language but English. By 1858 both William and Henry were fluent in French and could read German. Moreover, there was something corrupt in European manners, something decadent.[33] It was one thing to be cosmopolitan; it was quite another to be worldly. Mr. James certainly did not want his boys to become, as he once described Thomas Carlyle, "the same old sausage fizzing and sputtering in his own grease."[34] William and Henry were becoming too immersed; they swam in the Old World too naturally. William had already acquired a hungry eye, an artist's appetite not just for colors and form, but for a professional career as a painter as well. If William persisted in plans to be an artist, it would be far safer to pursue them in America.

CHAPTER THREE

Detached and Slightly Disenchanted

. . . a collection of the detached, the slightly disenchanted and casually disqualified.

Henry James on Newport society in the late 1850s

I have fully decided to try the career of a painter. In a year or two I shall know definitely whether I am suited to it or not. If not, it will be easy to withdraw.

William James to Charles Ritter
JULY 31, 1860

WHEN THE James family moved into a rented house on Kay Street in Newport, Rhode Island in the summer of 1858, it must have seemed almost a continuation of their European travels. Although they had recrossed the Atlantic and returned to the New World, they did not resume residence at their Fourteenth Street house in New York. That city could be terribly uncomfortable in the summer, and they had disembarked at the beginning of a particularly torrid one. By mid-July 1858 there were thirty-one reported cases of fatal sun stroke. Being used to the bracing sea air at Boulogne-sur-Mer, the Jameses were drawn to a locale where, according to the town newspaper, "night will always bring with it cool breezes from the sea to temper the heat, and make hours of sleep refreshing and invigorating."[1]

The family also had close friends who were permanent residents in Newport. Edmund and Mary Tweedy seemed like relatives. Henry Sr.'s sister Catherine and her husband, Robert Temple, had died suddenly in 1854. The Tweedys had agreed to raise their four orphaned daughters. William and Henry were particularly attracted to Mary, or "Minny," Temple, who would soon die of tuberculosis. William's artistic eye caught a special quality in the Temple girls. His only surviving oil canvas captured the oldest Tweedy girl, Katherine. Long after the family had left Newport, William continued to return for short visits, nearly always staying with the Tweedys.

Newport had other attractions besides climate and close friends. The town had about ten thousand permanent residents and another five thousand summer visiters who resided in four large hotels and assorted boarding houses. It was hardly a sleepy little village, but neither had it yet become the fabled Newport of the Gilded Age, the Newport of Ocean Drive and the Vanderbilt family "cottage," The Breakers, the summer playground of America's fabulously wealthy late 19th-century financiers and industrialists. For years well-to-do southerners had been coming to Newport to escape heat and malaria. It was the southerners who according to the local newspaper "first brought [Newport] into general notice, as a place of summer resort."[2] They mixed well with the predominately Yankee population and encouraged a relaxed, cosmopolitan ambience.

By the mid-nineteenth century Newport welcomed artists and intellectuals. The social register in the 1850s included Charles Norton, Henry Tucherman, John La Farge, George Bancroft, Charlotte Cushman, Louis Agassiz, William Greennough, George Calvert, and "the queen of any company," Mrs. Julia Ward Howe.[3]

Looking back, Henry James recalled a Newport that gathered in "a handful of mild . . . cosmopolites, united by common circumstances, that of their having . . . lived in Europe, that of their sacrificing openly to the ivory idol whose name is leisure, and that not least, of formed critical habit."[4] As one contemporary observer noticed, "It was not fashion which first brought people of luxurious tastes with means for indulging them to Newport. It was a satisfaction found in its air and scenery by people of a rather reserved, unobtruding, contemplative and healthy, sentimental turn little troubled by social ambitions."[5]

Here on an extremity of the North American Atlantic coast, a place that could not be reached by rail until 1864, a town with long-established trading ties to Europe, the Indies, and the American South, a town that depended on visitors for its commercial survival, came a family in transit. On eyeing Newport for the first time in 1860, James family friend Colonel Thomas Wentworth Higginson wrote: "How picturesque the old town looks as one glides down to it on a full tide; none of our seaside towns have roofs and gables that look so foreign."[6] Such a place was a suitable way station—an American town, but one that retained a European flavor that a family returning from three years abroad could readily appreciate. Newport was hardly Paris; yet to sixteen-year-old William James, poised to make art his vocation, it had a special charm.

AFTER SETTLING in Newport, William began to paint informally in William Morris Hunt's studio at 108 Church Street, a cozy apartment that stood behind his architect brother Richard's residence, Hilltop. Hunt was a native of Vermont with a Havard education and European experience. He had been a pupil of Thomas Couture in Paris but had become infatuated with Jean François Millet and the Barbizon painters, and their country themes of rustic peasant life. He was also a man of volatile temperament, which included periods of boisterious play as well as deep depression. He eventually took his own life.[7]

Perhaps the elder James felt that Hunt's Barbizon emphasis, in contrast to the artificial Parisian scene, would be a healthy influence on William. But his initial impression of Hunt hardly reflected it: "William Hunt the artist also is here established, and Willy begins to take lessons with him today. He is celebrated I am told, but his speech is broken and too excessively professional for me."[8] As a professional artist, Hunt presented William not only with an opportunity to continue to learn to paint; he also represented an alternative adult model, an accomplished, successful painter whose professional demeanor contrasted vividly with Henry Sr.'s eccentric, whimsical, and hopeless impracticality.

Moreover, shortly after beginning to study with Hunt, a worldly young artist with a professional air showed up at the Church Street studio. John La Farge was seven years William's senior, had spent considerable time in Europe, spoke fluent French, had studied briefly with Couture, and even ventured to criticize Hunt and the Barbizon school. Perhaps most significantly, La Farge thought that Delacroix was a great genius. William was much impressed: "There's a new fellow come to Hunt's class. He knows everything. He has seen everything—paints everything. He's a marvel."[9]

William's other Newport friends did little to counteract the vogue of Hunt and La Farge. James Steele MacKaye and Thomas Sergeant Perry came from families with new money and considerable influence. MacKaye was the son of Colonel James Morrison MacKaye, who founded the two remarkably successful companies, Wells Fargo and Western Union. Young Jim MacKaye also painted but eventually was drawn toward the theater. Tom Perry was the grandson of the War of 1812 naval hero Commodore Oliver Hazard Perry and the son of the recently deceased Christopher Grant Perry, who had practiced both medicine and the law. Tom tried painting, went on to Harvard as a tutor, became the editor of the *North American Review,* but did not settle into anything permanent. With inherited money he lived for years in Europe and returned to Boston in his mid-forties convinced he was a failure for not having developed a lasting profession. Both MacKaye and Perry, especially the former, were proponents of an active, career orientation in the arts. William's closest friends hardly saw their futures in terms of Henry Sr.'s idle reverie as a "student of life."[10]

Within a year the elder James found his ability to mold William's education seriously undermined. He wrote to a friend that "I have grown so discouraged about the education of my children here, and dread so those inevitable habits of extravagance and insubordination, which appear to be characteristic of American youth, that I have come to the conclusion to retrace my steps to Europe, and keep them there a few years longer."[11] Ironically, his sons, particularly seventeen-year-old William, seemed attracted to the same extravagance and insubordination that he had inflicted upon his own father while attending Union College. But this extravagance and insubordination meant more than a continuation of intergenerational rebellion. William was being drawn into a world of experiences that Henry Sr. had not shared, a world that he considered excessively professional, that mocked the life of an irrelevant eccentric. Despite considerable educational experimentation, Mr. James had failed to achieve the educational effect he had desired. His son seemed drawn to the self-indulgent, narrow professional life that he, for various reasons, had been unable to embrace.

SO AFTER a rough voyage on the SS *Vanderbilt,* the James family landed at Le Havre, went immediately to Paris, and within a few days crossed the now familiar Jura Mountains enroute to Geneva. There William enrolled in the local academy where he attended lectures in anatomy, studied physics, and became familiar with dissecting—presumably with the approval and encouragement of his father. But studying anatomy was as useful to the professional artist as to the scientist, and William gave no indication that he was being forced into something alien to his interests.

Young James composed his first notebook in Geneva in 1859–1860. It is a collage of names, book titles, drawings, aphorisms, and science notes. There is no sustained intellectual theme; no sign that he loathed science and longed to return to art. He did, however, inscribe aphorisms that prescribed formulas for success—an indication that whether he made a career of science or art, he did not intend to drift away into the fogs of Henry Sr.'s philosophical idleness. He quotes, for example, "Mr. Edgeworth," a "Lit. Char.," who "thinks that differences of intellect which appear in men depend more upon the early habit of *cultivating* the *attention* than upon the disparity between powers of individuals." He has copied the observation of the French naturalist Baron Georges Cuvier that "the man who has well studied the operations of nature in mind *as* matter, will acquire a certain moderation and equity in his claims upon Providence; he will never be disappointed in himself or others; he will act with precision and expect that effect and that alone from his *efforts* which they are naturally adapted to produce." And he notes Ruskin's observation that "it's no man's business whether he has genius or not; work he must whatever he is but quietly and steadily; and the natural and unforced results of such work will be the things he was meant to do and will be his best. No agonies or heart-renderings will enable him to do any better; if he be a great man they be great things; if a small man small things; but always, if thus peacefully done, good and right; always, if restlessly and ambitiously done, false, hollow and despicable."[12]

His father could hardly disagree with these sentiments, and would have praised William for entering them in a personal notebook—even though he scoffed at the value of professional success. Yet recording self-help and therapeutic thoughts in diaries or notebooks meant that William believed there was a road to future success, even greatness. The notebook reveals an unusually ambitious young man, one harboring hopes for truly exceptional accomplishment. Moreover, in the very act of writing a notebook William James took an important introspective step toward inquiring into the nature of reality. He had begun the literary mode that would eventually result in classic books and essays.

Although self-help philosophy predominates in this early notebook, there are a few jottings on science. William notes Cuvier's observation that "we must discriminate between affinity and analogy. Affinity is founded upon identity of plan. Analogy upon the grafting of the features of one plan upon the body of the other." There is a list of chemical compounds; drawings of polyps, lymph cells and blood circulation; a recommendation to use "silicate of Potassium for cementing china"; and a limewater and ammonia formula for curing chilblains. These scientific snippets are interspersed among book lists. William read or meant to read "Shakespeare's poems, Bacon's Essays, Taylor's holy living and dying 2 vols. $3 per vol. bound at Scribners, and Butler's Analogy in Aldine Classics." He also sent for "Gustave Planche Portraits d'Artistes 2 Etudes sur l'Ecole Francaise."[13]

If this thin notebook tells anything about James' creative life at the beginning of the 1860s, it shows a dabbling among several seemingly unrelated fields ranging from the natural history and biology of Cuvier to Faraday's electricity, from English literature to French art. James' mind seems unusually receptive, eclectic, and omnivorous. There was no sustained analysis, no fixation on one scientific question or problem. He clearly wanted to continue in art, but there was no repulsion from science. William's "open moment" continued, even as he pressured his father to allow him to resume studying art. And family affluence allowed him the luxury of prolonged experimentation; material circumstances encouraged James' open moment to become a chronic frame of mind.

Toward the end of the spring term William joined a social fraternity, the Société de Zofingue, in which he learned to drink beer and climbed in the Swiss Alps. In the early summer of 1860 Henry Sr. decided to move the family to Bonn, where the boys were boarded with separate German families to learn the language. William found mastering German "a mere process of soaking, requiring no mental effort," and thoroughly enjoyed the rich German fare and simple company.[14]

Then suddenly in July, Henry Sr. decided to return the family to Newport. He explained the decision to his friend Edmund Tweedy: "We had hardly reached here before Willy took an op-

William (at right) and friends in Geneva, ca. 1859–1860.
By permission of the Houghton Library, Harvard University.

portunity to say to me—what it seems he had been long wanting to say, but found it difficult to come to the scratch—that he felt the vocation of a painter so strongly that he did not think it worth my while to expend any more time or money on his scientific education! I confess I was greatly startled by the annunciation, and not a little grieved, for I had always counted upon a scientific career for Willy, and I hope the day may even yet come when my calculations may be realized."[15] Henry Jr. explained to Tom Perry that "we are going immediately to Newport, which is the place in America we all most care to live in. I'll tell you the reason. . . . Willie had decided to try and study painting seriously, and wished . . . to do so if possible with Mr. Hunt. . . . Besides that, I think that if we are to live in America it is about time we boys should take up our abode there; the more I see of this estrangement of American youngsters from the land of their birth, the less I believe in it."[16] But one wonders whether the estrangement belonged to the boys or to their father.

From Bonn William wrote Henry Sr. in Paris: "I wish you would as you promise, set down as you can on paper what your idea of the nature of art is, because I do not probably understand it fully, and should like to have it presented in a form that I might think over at my leisure."[17] Unfortunately his father's response has not survived. If William's reply accurately interprets the elder James' concern, however, it centered on the fear that the artist's aesthetic sense might harm his moral nature. "I do not see why a man's spiritual culture should not go on independently of his aesthetic activity," William countered; "why the power which an artist feels in himself should tempt him to forget what he is, any more than the power felt by a Cuvier or Fourier would tempt them to do the same."[18] He was sure that genius was fully capable of retaining a moral conscience. Nonetheless William treated painting as an experimental vocation, rather than as a moral one. "I am going to give it a fair trial, and if I find I have not the *souffle* give it up," he wrote to Tom Perry. "I don't hope to be anything *great,* but am pretty sure with good old Hunt as a guide, philosopher, and friend, to commit nothing bad. . . . Father has already fled this land where he can neither understand nor make himself understood."[19] At eighteen William James was both more vocationally flexible and literate than his father—particularly in Europe. And he possessed a generous quantity of something that had never far developed in the elder James—professional-intellectual ambition.

By the first of October 1860, the Jameses were once again settled in Newport, almost exactly a year since their departure. Henry Sr. seemed resigned and told Tweedy, "We look upon Willy's strong desire to return to Mr. Hunt as a Providential indication . . . and go where the spirit and the flesh alike draw us."[20] William might have settled comfortably into an apprenticeship with Hunt and gone on to do considerable if not great things in painting. Yet within less than a year, young James would be in transit again, this time to Cambridge and the beginning of a lifelong association with Harvard.

Part of the difficulty was Hunt. William Morris Hunt was certainly a competent, talented American painter, and during the family's first Newport stay William was quite taken with his personality and professionalism. Yet as was often the case with James, positive first impressions may have later turned to impatient criticism. What is lost sight of in acknowledging William's attraction is that Hunt, despite his talent and reputation, was not particularly *original.* William and Henry, who occasionally sketched with his brother, appreciated the relaxed atmosphere at the Church Street studio. And William often sought out isolated meadows and beaches as subjects for the landscape scenes that Hunt's Barbizon style encouraged. But learning to paint in pleasant surroundings was not the same as being in contact with genius, with a pioneer. There were probably half a dozen contemporary American painters who surpassed Hunt in originality and artistic impact.[21] In fact Hunt himself disparaged American painters and felt that even in Europe there had been "very few great painters."[22]

Buy why, if James knew that Hunt could not give him truly creative stimulus, did he agree to come back to Newport? Why did he not insist on studying painting in Europe? Even if Hunt had been a great genius, American painting in the mid-nineteenth century was just beginning to break away from the hold of portraiture. A young man who aspired to be an artist still faced a society that viewed artists as people of suspicious morality, still censored the use of live models, and did not encourage experimental techniques. There was nothing in Newport, New York, Boston, or Philadelphia that could begin to compare with the cultural and professional excitement that Paris could offer an up-and-coming painter.[23]

Desire to paint notwithstanding, William may have sensed that although he might be a talented artist, he could not be a great one. Newport, especially with companions like La Farge, MacKaye, and

Perry, would do nicely until he decided on an alternative profession. But if John La Farge's opinion of William's ability was accurate, there was little question about artistic potential. According to his biographer, La Farge recalled that "James drew 'beautifully' . . . repeating the word three of four times." Years after their Newport days the two men met again and reminisced, and La Farge was later to say that James "reminded me as we dined of our going out sketching together at the Glen, Newport, and of what I was painting then, and that I was not copying. On the contrary, I was merely using the facts to support my being in relation to nature."[24] That James had had insight into how creative painting emerged much impressed La Farge. William not only drew beautifully; he understood, as had Delacroix, that the artist brought his personality into the making of an art form. Personality and painting were intertwined in the creative act.

Why William was satisfied to return to Newport is linked to the question of why he left painting and decided to enroll in Harvard's Lawrence Scientific School in the fall of 1861. Biographers have posed various answers: that William was not really a painter, that he became ill, that he sacrificed an artistic career to satisfy Henry Sr.;[25] other writers on the era have suggested that the interruption of the Civil War made continuation in art untenable.[26] Yet if William saw increasingly that Hunt could not provide an original perspective—saw that Hunt, for all his professionalism, was not exceptionally creative—then James may simply have become intellectually impatient. Obviously the teacher may be mediocre and the student brilliant. La Farge, for example, surpassed Hunt, and William might have as well. But William was searching for a *medium* through which to express his genius; he was constitutionally restless and apparently did not see Newport as the appropriate context for developing his originality. This did not mean abandoning art for science so much as refocusing his creative sight. He began, for example, to examine the world of microorganisms while painting landscapes. Indeed, the family had barely returned to Newport before Henry Sr. asked Tweedy, who was in New York, to obtain "a Queckett's Dissecting Microscope: it will cost 37 shillings and 6 d: Willy needs it and will be much obliged."[27]

William also frequented Newport's Redwood Library, perhaps the best in New England outside Boston and Cambridge. Earlier he had shocked friends by reading pessimistic passages from Schopenhauer, and he was impressed with Ernest Renan's relativistic analysis of Christian theology.[28] But these were not signs that William was gravitating toward philosophy and religion, any more than purchasing a second microscope meant that he was becoming a scientist. His was an eclectic, omnivorous intelligence, an intelligence that explored various interpretations of natural, social, and metaphysical reality while simultaneously searching for a suitable medium of creative expression. As his Geneva notebook had first shown, James' vision easily shifted from one interest to another. In early 1861, James, rather than appearing a young man who reluctantly left painting to please his father, seems to have been experimentally inclined, interested in an astonishing variety of phenomena, poised to make his mark—and, incidently, if he could find one, to follow a career as well.

Even more significantly, William had already sensed that art and science were two sides of the same coin, or in our terms, two interconnected fields of experience. By moving from art to science he was not only broadening his intellectual scope, he was gaining insight into intellectual relationships. Rather than leaving one field for another, he was poised to connect field to field as a fundamental intellectual venture. A striking quality of James' intelligence would be its refusal to compartmentalize knowledge. Beyond growing anxiety about what career he would embrace were clear signs that he was following the natural bent of his genius.

THE CIVIL War, however, reordered the personal priorities of many American families. It seemed in the initial surge of Northern patriotism after the firing on Fort Sumter that William and Henry would join the tens of thousands of young men who were responding to President Lincoln's call for volunteers. "I have," the elder James wrote to a friend, "a firm grasp upon the coat tails of my Willy and Harry, who both vituperate me beyond measure because I won't let them go. The coats are very staunch material, or the tails must have been pulled off two days ago, the scamps pull so hard." Faced with such enthusiasm Henry Sr. mounted persuasive arguments against enlistment. "The way I excuse my paternal interference . . . is, to tell them, first, that no existing government . . . is worth an honest human life . . . especially if that government is likewise in danger of bringing back slavery again under our banner. . . . Secondly, I tell them that no young American should put himself in the way of death, until he has realized something of the good life: until he has found some charming conjugal Elizabeth to whisper his devotion to, and assume the task . . . of keeping his memory green."[29]

Three years later Mr. James reversed himself and allowed the younger brothers, Robertson and Garth Wilkinson, to join the fray. Wilky was seriously wounded. And when Bob, sickened by the carnage, threatened to desert, his father was unsympathetic: "I hope you will not be so insane. I conjure you to be a man and force yourself like a man to your whole duty. . . . It is the crisis of your character, and rely upon the Divine Providence to make it turn out prosperous and not adverse to you, by remaining manfully in your tracks."[30] The elder James found the military more suited to Wilky and Bob's capabilities. "They are none of them cut out for intellectual labour," he wrote Tweedy.[31] It is possible that Henry Sr. thought that the military experience would provide an appropriate mixture of discipline and ideals for his less intellectual younger sons, that the army offered a life-test that would both mature and enlighten them. Still, Mr. James' pacifism was clearly selective.

Henry Sr.'s about-face may have partly resulted from a growing conviction that the cause now demanded the fullest sacrifices. By the time Wilky and Bob had enlisted, Lincoln had issued the Emancipation Proclamation, thereby making the contest seem one of eliminating slavery as well as saving the Union. But did his initial refusal to let William and Henry enlist reflect only sincere pacifism and a desire to protect his favored sons? Can one safely assume that William and Henry really wanted to join the army?[32]

As war threatened, Newporters were somewhat undecided. Although the majority were to varying degrees opposed to slavery, the sizable influx of proslavery Southerners each summer kept the town from becoming charged with anti-Southern sentiment. Locally in the election of 1860, Lincoln won a scant plurality of 30 votes over his Democratic opponent, Stephen Douglas, who represented more moderate opinion. Lincoln's victory quickly polarized political feeling, however. As the Southern states seceded during the winter and early spring of 1861, Newport became decidedly more patriotic. By March 1861 there were regular evening demonstrations supporting the Lincoln government, with bonfires, torchlight parades, and cannon salutes. On April 13, 1861, citizens received the news that Fort Sumter had been fired upon. People swarmed into the streets and rapidly bought out all copies of both of the city's newspapers. By Monday, April 15, the governor of Massachusetts, William Sprague, called for Rhode Island to respond to President Lincoln's call for volunteers. The Newport Artillery met and quickly exceeded its quota of one hundred men. At Sayer's Wharf and along Thames Street, thousands waved farewell to troops who had boarded the steamer *Perry,* about to embark for Providence, the first stop on the way to the front.[33]

It would be hard to imagine William or even the gentle Henry not participating in the hoopla, or William not having thoughts about enlisting. As a voracious reader, a young man already inclined to gossip, and a keen observer of social life, James must have avidly followed the news. Would he have felt embittered, in lieu of his father's decision against enlistment, to read a late April 1861 editorial that exclaimed: "Soldiers of Rhode Island, your country has called for your service and you are ready"?[34] Perhaps not at all: for at the time of his country's great emergency, twenty-one-year-old William was making plans to enroll in Harvard's Lawrence Scientific School.

What was surprising, given Henry Sr.'s assertion that William and Henry were rabid to join Lincoln's volunteers, was that neither found a way to do so. True, Henry had suffered a mysterious injury while fighting a Newport fire six months earlier that he referred to as an "obscure hurt"; and back problems probably made enlistment unlikely.[35] William, however, had only vague neurasthenic complaints and would be well enough to enroll at Harvard in September. And despite continuing health problems, he traveled to the jungles of Brazil during the closing months of the war.

Even with less than perfect health, one suspects that had James really wanted to participate he would have found a way. It was not uncommon to "join up" even though parents forbade it. William had opposed his father on the issue of continuing in art. Was he so filled with guilt that he would not have dared oppose him once more, as his country cried out for volunteers? He would not have had to join the infantry in order to make a contribution. In early May the *Newport Mercury* announced that "in order to provide for the health and comfort of the Rhode Island volunteers now in the field, a corps of ten men will be formed to be called the Rhode Island Relief Corps to the Hospital Staff. . . . They are to assist the surgeons, and follow the regiment into the field to look after the wounded."[36] Here was an alternative to satisfy his father's pacifism, and one that would have kept him more out of harm's way. But William remained a civilian and was ever after silent about the whole matter; as his first biographer remarks, "If he could not act he preferred not to talk."[37]

But why did he not feel compelled to act? His education with the exception of the Hunt interludes had been mostly European. Of the six years between 1855 and 1861, he was abroad four. Further, he had been in Europe during the period of rapidly building sectional tension and had returned to a Newport largely unconcerned about polarized politics until immediately before the war. James had been absent during the crucial time when Northern minds and emotions were forging a militant sectional ideology.[38] William and Henry, as the latter observed, counted themselves among the detached Newporters, and by implication were not caught up in the hyperbole of mainstream American politics. William could not internalize the war in terms of an intellectual crisis; he did not face the quandary that many young New England intellectuals confronted when their romantic idealism met the shock and suffering of the Civil War.[39]

By contrast, Wilky and Bob were enrolled in the ideologically charged Sanborn School in Concord. Frank Sanborn was deeply involved in financing both the Massachusetts Kansas Aid Society and John Brown's raid on the federal arsenal at Harpers Ferry. His school provided the perfect indoctrination in militant anti-Southern, antislavery sentiment.[40] William could not view enlistment as a necessity because he had developed no ideology. Neither William, Henry, nor their father had been situated so as to catch the nationalistic fever that pushed thousands of thinking men unthinkingly into the war. Further, James almost certainly understood that whatever career he chose, his intellectual future could not be nurtured in the army. Later he would search for what he called the "moral equivalent of war"; now he sought, if not an intellectual battlefield, an appropriate challenge for his expanding powers.

Neither the war nor Hunt presented William with an irresistable opportunity. With the conflict now raging, and with the Northern defeat at Bull Run, another excursion to Europe seemed too risky. But there was another alternative. The Boston-Cambridge-Concord area had spawned more than its share of American geniuses. These environs had been the site of an American Renaissance.[41] Emerson and Hawthorne, along with lesser lights such as Henry Longfellow and James Russell Lowell, could still be heard lecturing or seen taking a walk. Harvard, with its new Lawrence Scientific School, moreover offered a young man a chance to learn natural science from the world-famous Swiss émigré naturalist, Louis Agassiz. Harvard was the one American place with an intellectual life that made Newport's fade into insignificance. It was a place where William could learn a profession, hence insuring that he would not share his father's floundering. Rather than simply another installment in the continuing story of James' vocational frustration, the move to Cambridge proved to be a momentous step toward original discovery and a true career.

William was following his own inclinations as a young man of genius. In Cambridge he would develop an understanding of science to complement his grounding in art; his fields of experience would interlock. He was still an artist; he would presently be a scientist. But in a crucial way he would be something else as well—a man whose career choices were strategies for exercising and developing a special genius. There was an underlying pattern in James' vocational shifts that empowered his originality. He did not so much abandon art for science as shift the center of his vision. This alteration of vision or intellectual focus from field to field was a necessary condition for his genius. Entering the Lawrence Scientific School meant that James had not so much changed careers as refocused his gaze.

CHAPTER FOUR

Who Made God?

To anyone [*into*] *whose hands this book may fall—if I am alive, do not read it. If I am dead, burn it and oblige Wm. James.*

To potential readers of the 1863 Notebook

BETWEEN 1861 when he left Newport and gave up making a career as a painter, until 1873 when he took an appointment as instructor of anatomy and physiology at Harvard, William James intensified his intellectual preparation. He began to participate in American higher education and entered the fermenting world of Victorian science. James built the foundation for an association with Harvard and nineteenth-century science that provided the context for future intellectual and social activity. Even as he struggled through a protracted vocational search, suffering despair and a variety of maladies, he also forged important intellectual associations and expressed thoughts that bespoke an extraordinary mind.

By enrolling in the Lawrence Scientific School in the fall of 1861, William became part of an early attempt to professionalize higher education at Harvard. In 1847 Abbott Lawrence, a wealthy Boston textile industrialist, had endowed the college with fifty thousand dollars to upgrade scientific studies. A year earlier, Harvard had hired the renowned Swiss naturalist Louis Agassiz, who became the leading light of the Scientific School. Agassiz, whose reputation and personality much attracted William, became engaged in intrauniversity conflict with a young man who had an ambitious, more structured vision of Harvard's future. But Charles William

Eliot would not become president of the university until 1869; in 1861 he was William's chemistry teacher.[1] Years later Eliot recognized that James came to Harvard at a pivotal moment: "Just before the Civil War," he observed, "the structure of Harvard University as a group of professional schools on top of the College became clearly visible. Harvard had already attained at that time something which neither Oxford nor Cambridge had then reached . . . a recognition of professional schools as an indispensable part of a university."[2]

Yet when William arrived in Cambridge in August 1861 and took a room near the Harvard Yard at the Corner of Linden and Harvard Streets, the school was not far removed from a simpler past. By the time he received an M.D. in 1869—the only degree William ever earned—there were just slightly over five hundred undergraduate students, forty-five teachers "of professorial grade," and five administrative officers. Courses were not yet numbered or arranged in the college catalog by department.[3] There were no entrance requirements for the Lawrence Scientific School. Students simply attached themselves to a professor in one field, and, if dissatisfied, switched to another professor in another field. James, for example, began with Eliot in chemistry and changed to Jeffries Wyman in physiology. It was a matter of simply following a mentor. There was no fixed period of study, although most students enrolled for a period of roughly two to three years, and arranged for an examination when they felt ready to "graduate." Entrance exams and a four-year curriculum did not come to Harvard until 1869.[4]

Cambridge was a town of some ten thousand in which the Harvard community formed what novelist WIlliam Dean Howells remembered as "a charming society, indifferent . . . to all questions but those of the higher education which comes so largely by nature." Howells recalled that "it was taken for granted that every one in Old Cambridge society must be of good family, or he could not be there; perhaps his mere residence tacitly ennobled him; certainly his acceptance was an informal patent of gentility."[5] This charming respectability was not lost on James. In 1863 William urged his family to join him, as he felt "the society here must be pleasanter than elsewhere, not so mercantile; the natural beauty of the place as soon as you recede a little from Harvard Square is great."[6] The place retained an exclusive yet liberal ambience that appealed to a young man nurtured on the special, even exotic educational experience.

William made a remarkably quick adjustment to Cambridge, to Harvard, and to living without his family. Shortly after arriving he reported experiencing several pangs of homesickness, but proudly added that "I am perfectly independent of everyone."[7] It helped that he met young Tom Ward, the sensitive son of Henry Sr.'s banker, Samuel Gray Ward. Until the end of the decade Tom was probably his best friend. By the middle of September William had fallen into a regular work schedule: "I get up at 6, breakfast and study till 9, when I go to School till One then dinner a short loaf and work again till 5 then Gymnasium or walk till tea and after that, visit, work, correspondence etc, etc, till ten, when I 'divest myself of my wardrobe' and lay my weary head upon my downy pillow."[8]

JAMES SPENT the first term studying chemistry with Charles Eliot, sitting in on Jeffries Wyman's class, "Comparative Anatomy of Vertebrates," and attending a series of lectures on natural history delivered by Louis Agassiz in Boston. Of these three, Agassiz quickly became the chief attraction, even though William was not officially Agassiz's student. Less than three weeks after arriving, William wrote Henry Sr. that "Agassiz gives now a course of lectures in Boston, to which I have been. He is evidently a great favorite with his audience and feels himself so. But he is an admirable, earnest lecturer, clear as day and his accent is most fascinating. I should like to study under him."[9] By Christmas William planned to study "with Agassiz 4 or five years." He had discovered a natural way of shifting interest in art to science:

> I had a long talk with one of [Agassiz's] students the other night and saw for the first time how a naturalist could feel about his trade in the same way an artist does about his. For instance, Agassiz would rather take wholly uninstructed people "for he has to unteach them all that they have learnt." He does not let them *look* into a book for a long while, what they learn they must learn for themselves and be masters of it all. The consequence is he makes *naturalists* of them, he does not merely cram them, and this student . . . said he felt ready to go anywhere in the world with nothing but his notebook and study out anything quite alone. He must be a great teacher.[10]

William thought he had found what he had ultimately failed to find in Hunt, a truly creative mentor. One wonders how Henry Sr. reacted to William's infatuation with a man who had the fame and professional skills to so capture his son's enthusiasm.[11]

Agassiz was in his mid-fifties and relatively late in his career when William came to Harvard. Born in Fribourg, Switzerland, he had received a Ph.D at the University of Zurich in 1829 and gained an M.D. in 1830 following study at Heidelberg and Munich. After briefly practicing medicine he turned to scientific research in Paris, befriended the Prussian naturalist Alexander von Humboldt, and studied fossilized fishes with Cuvier. In 1832 he became professor of natural history at the University of Neuchatel, where he wrote his master work, *Recherches sur les poissons fossiles* (Researches on Fishes' Fossils), whose five volumes were published between 1833 and 1844. *Poissons Fossiles* presented a new system of classification that at the time was widely acclaimed but was subsequently abandoned. He also did significant work in geology; his *Etude sur les glaciers* (Study of Glaciers), was published in 1840.

At the height of Agassiz's career, he was shattered by personal problems. A business venture—a publishing house he had established—went bankrupt, and his wife left him. He sought a new life in a new locale. In 1846, at the age of thirty-eight, he emigrated to the United States, gained the prestigious Lowell Institute Lectureship, and was appointed professor of zoology and geology at Harvard. Agassiz is remembered as a major opponent of Darwinian theory. He engaged in spirited dispute with two colleagues, Asa Gray and Jeffries Wyman, who were among Darwin's principal New World allies.[12]

Agassiz was clearly Harvard's star scientific attraction, an extraordinary teacher and an internationally recognized scholar, a grand protagonist for the theory of special creation. He assembled a vast Museum of Natural History and displayed thousands of specimens in zoology, paleontology, and geology. Indeed, he viewed science largely as the collecting, describing, and classifying of life forms, activities that underscored the incredible diversity and complexity of divine creation.[13]

Although enthralled by Agassiz, William also recognized the abilities of Jeffries Wyman. He told his father that "Prof. Wyman's Lectures on Comparative Anatomy of Vertebrates promise to be very good, prosy perhaps a little and monotonous, but plain and packed full and well arranged."[14] Wyman was seven years Agassiz's junior and a graduate of the Harvard Medical School, where he had lectured as a demonstrator of anatomy. An appointment as curator of the Lowell Institute enabled him to study comparative anatomy in Paris. After teaching briefly at Hampden-Sydney Medical College in Richmond, Virginia, he returned to Harvard as professor of anatomy in 1847. Wyman was also a collector and assembled a museum in Boylston Hall that emphasized his teaching areas, anatomy and physiology. Almost certainly James received his first sustained exposure to Darwinian theory in Wyman's class.[15] Edward Waldo Emerson, son of the great Concord sage and William's fellow student, remembered: "[Wyman] lived to have us come down and question him after the lecture—an unknown occurrence in any other classroom."[16] And Charles Eliot especially recalled Wyman's "type of scientific zeal, disinterestedness, and candor."[17]

William's most taxing scientific education in the first months at the Lawrence Scientific School came not from the lectures of Agassiz or Wyman, but in Eliot's chemistry course. In mid-September 1861 he wrote Henry Sr. that "this Chemical Analysis is so bewildering at first that I am entirely 'muddled' and have to employ most all my time reading up."[18] By late December he was still struggling: "Chemistry comes on tolerably, but not as fast as I expected," he told his parents. "I am pretty slow with my substances, have done but twelve since Thanksgiving and having thirty-eight more to do before the end of the term."[19] William reported that Eliot was "a fine fellow" and "a man who, if he resolves to do a thing will do it."[20] He did not, however, find him "a very accomplished chemist"—a judgment that implied correctly that Eliot's real talent lay not in science but in administration.[21]

William enjoyed chemistry more when given the opportunity to experiment, though sometimes with unexpected results. In the spring term he wrote his sister Alice that "I am now studying organic chemistry. It will probably shock mother to hear that I yesterday destroyed a handkerchief—but it was an old one, and I converted it into some sugar which though rather brown, is very good."[22] Over fifty years later Eliot recalled that "I enlisted [William] in the second year of his study of chemistry, in an inquiry into the effects on the kidneys of eating bread made with the Liebig-Horsford baking powder, whose chief constituent was acid phosphate. But James did not like the bread and found accurate determination of its effects three times a day tiresome and uncompromising; so that after three weeks he requested me to transfer that inquiry to some other person."[23] Besides, in testing the effects of baking powder on his own urine, he had a reaction that developed into an angry boil. Eliot recommended iodine painting, but William complained that it "seems only to prolong the boil, having dropped its use I now curse it aloud."[24]

James' less than successful stint in chemistry did not prevent Eliot from observing some extraordinary qualities:

> James was an interesting and agreeable pupil, but was not wholly devoted to the study of Chemistry. His excursions into other sciences and realms of thought were not infrequent; his mind was excursive, and he liked experimenting. I had received a distinct impression that he possessed unusual mental powers, remarkable spirituality, and great personal charm. This impression became later useful to Harvard University.

Eliot remembered that in "1863–64 James moved from [the] Department of Chemistry to Comparative Anatomy and Physiology in the Lawrence Scientific School and became a pupil of Prof. Jeffries Wyman," and that "his tendency to the subject of Physiology had appeared clearly during his two years [1861–1862] in the Department of Chemistry." Unfortunately, "his work was much interfered with by ill-health, or rather by something which I imagined to be a delicacy of nervous constitution."[25]

Bent toward physiology notwithstanding, James had not yet settled into any specific field. And to say that he was excursive and liked experimenting, while suggestive, tells us little about specific intellectual interest. What was he reading? What was he thinking about? Fortunately William continued to compose notebooks.

AS THE Civil War raged, while his younger brothers were fighting to save the Union, William indulged in the intellectual life, reading deeply and widely, safe from the turmoil and carnage that convulsed a nation. He began a lifelong habit of critically abstracting in his notebooks important contemporary scientific and philosophical works. It was in these notebooks, especially in 1863, that he first revealed the remarkable speculative abilities that characterized his mind thereafter.[26]

There is no direct notebook evidence that William read Darwin, although he was certainly introduced to "the developmental theory" when he studied with Wyman in 1863–1865. But clearly

William was much concerned with great late nineteenth-century speculations, with the nature of matter and force and the question of causation. In his 1862 notebook, Cicero's philosophical treatise, *The Tusculan Disputations* is juxtaposed against the theory of material causation proposed by the German physician and philosopher Ludwig Brüchner. James wanted "to see whether the ancients were as strongly subject to the idea of cause as we are. They did not require a causal explanation of the physical universe. See Cic. Tusc. . . . where a self mover is familiarly spoken of." There are scattered notes from Joseph Lovering's Introductory lecture on physics: "We shall treat of Acoustics, Electricity, Magnetism, Electromagnetics, Optics and heat from a mechanical point of view." There follow notes from a natural history orientation from Agassiz: "First requisite for the study = comprehension of the time required for geological changes. No preconceived ideas of the limitations of time. No going to work with *assumptions* of time before we know the facts" (1862 Notebook). The natures of force, matter, and time were speculative matters now explicable in terms of scientific fact.

There was one Cambridgean who stimulated William's theoretical thinking as much as any of his teachers. In the fall of 1861 James wrote his father that "in last year's class there is a son of Prof. [Benjamin] Peirce, who I suspect to be a very 'smart' fellow with a great deal of character, pretty independent and violent though."[27] Charles Peirce had received a B.A. from Harvard in 1859 and then entered the Lawrence Scientific School, where he concentrated in chemistry and graduated with the B.S. degree in 1863. Personal eccentricities probably prevented his being offered a professorship at Harvard, although he lectured there in the mid- and late 1860s. He was considered a genius, particularly in mathematics and logic. In the 1870s Peirce would publish several articles in *The Popular Science Monthly* that according to James made him the founder of pragmatism.[28]

In his 1862 notebook James quotes Peirce more approvingly than anyone. He agrees with "C. S. Peirce" that "the reductio ad absurdum can never be used in metaphysical discussion and rarely in scientific because it assumes that we know the sum of possibilities." Using Peirce's idea that finding metaphysical cause is impossible, James goes on to apply it to two great philosophers, one of whom represents the German idealism, the other the British empiricism that he would later challenge. "Kant works critically until he finds it unsatisfactory and then stops, at morality, freedom, God.

Hume . . . stops at cause. None succeed in leaving faith entirely out." Peirce underscored the persistence of the divine equation in all thought. "The *Thou* idea, as Peirce calls it, dominates an entire realm of mental phenomena, embracing poetry, all direct intuition of nature, scientific *instincts,* relations of man to man, morality, etc. *All* analysis must be into a triad; *Me* and *it* require the compliment *thou*" (1862 Notebook).

As William pondered Peirce's case for the inevitable subjectivity of supposedly objective thought, he composed his own speculations. Here is a very early example of theorizing, and one that draws a relationship between science and popular belief about reality:

> Those who oppose innate ideas, say they are generalizations from experience, one thinks necessary that which he has never seen violated. This applies to the *individual.* But the most general scientific ideas, as the laws of motion, etc., the unity of nature, seem almost necessary to the adept, yet are often in contradiction with vulgar prejudice and were slowly and laboriously evolved by successive generations. How long might it take for the law of inertia e.g. to petrify into an idea so familiar that to one born under its influence it would seem as necessary as that of cause? (to which it is related) In other words might not the op[eration] of in[ertia] id[ea], point to a historical growth of necessary truths as well as a formulation of individual experience. (1862 Notebook)

James realized that science is an intellectual construct, and that only time could make scientific laws into everyday assumptions about reality. He envisioned a dynamic historical relationship between scientific discovery and popular knowledge. Even as a twenty-one-year-old he could not keep theorizing isolated from practical effects. William did not want to simply discover ideas; he wanted to use them. But there is also evidence in his 1862 notebook that consideration of cause, exposure to Peirce, and sustained speculative thinking in general had altered his mental well-being.

After reading Balzac and Stendhal James refers to them as "immoral." There is a cryptic note, "Suicide morally/really considered." Reading Stendhal especially could have posed the question of suicide. This was the first indication that James might have considered taking his own life, a contemplation that by the late 1860s had become almost an obsession. But there was something else:

immediately before referring to suicide, James wrote, "The younger sons, not heads of great families are the true gentlemen. The latter have . . . responsibility to keep up the credit of the family." He had discussed father-son relationships with Peirce, for William notes that "He makes a test of any man's right to write upon freedom that he explain the authority of the father over his child. Unknown as yet—no one can even say what relation of father and child may be" (1862 Notebook).[29] Whatever the reason for William's growing seriousness, the mood did not quickly pass.

I HAVE FELT apathetic and indisposed for work. An illustration of the advantages of my plan of writing an abstract of what I read could not be sooner or more forcibly given me." William so describes his state of mind in early February 1863. He complains, "Have slept little lately and eyes felt weak all this day" (1863 Notebook). But writing book abstracts did not ease his discomfort. Here were early signs of a health problem that would periodically, now in one form, now in another, continue to plague him.

On February 4 he began taking notes on "Max Müller's book." Friedrich Maximilian Müller was a German philologist and renowned orientalist who later edited Sacred Books of the East in fifty-one volumes. But the book William read and abstracted was Müller's *Science of Language* (1861). He did not say that he was reading Müller for any particular class, although Wyman might have recommended a book that subscribed to an evolutionary view of language. William read voraciously and developed an independent reading program. He may have simply felt the book pertinent to what interested him. Science and philosophy were names for speculations about the nature of mind and matter. William's mind flowed freely over what he read, incorporating, discarding, and connecting. And his mind, rather than the subject matter, controlled his vision. Never, even in his deepest depression, did he ever allow another point of view to dominate or immobilize his selective, ever-active intelligence.

James begins by summing up the book: "The human mind is endowed with the power of forming rational conceptions, general ideas; and these by a law we do not understand it has the faculty of representing by articulate sounds. It has also a power we do not understand of *synthesis* or joining 2 or more ideas and conflating them in their natural relations as one." Men differed "from brutes by speech"; but "this [is] not a physical difference for many brutes can articulate. It is then caused by a mental difference. If it be true that we differ mentally from b[east]'s by the faculty of general conception, from this faculty must flow our power of speech." Words were simply "instinctive express[ions] of our general ideas" (1863 Notebook, February 4 entry).

Why would James be interested in a book that, among other seemingly esoteric philological ventures, traced the addition of the *d* of *loved* back to gothic usage in the fourth century? James spends more pages on *Science of Language* than on any other single book abstract in either his 1862 or 1863 notebook. But this was not strange for a man who would later consider the problem of whether thought was possible without language. Müller's book presented language as a physiological and psychological development. Speech was a product of physiological evolution, but the mind was the synthesizer; the mind gave language meaning. Müller was saying something important about the construction of a communicative world. James was fascinated and never forgot Müller, citing him in the famous chapter, "The Stream of Thought," in *The Principles of Psychology*.[30]

Although Müller emphasized the physiological similarities between the vocal chords of humans and of the higher apes, he argued that mentality differentiated men from animals. He quoted approvingly the confidences of Sydney Smith, an English clergyman, writer, and wit: "I feel myself so much at ease about the superiority of mankind—I have such a marked and decided contempt for the understanding of every baboon I have ever seen—I feel so sure that the blue ape without a tail [will] never rival us in poetry, painting and music, that I see no reason whatever that justice may not be done to the few fragments of soul and tatters of understanding that they really possess." The passage reminded James of how "science" could easily support the feelings of racial superiority that were evident in Americans; he notes, "Application to the anti-nigger feeling here." Drawing a parallel between dangerous applications of science and religion, he mentions a *Saturday Review* article that argued that the Scriptures supported the return of slaves to their masters (1863 Notebook).

In mid-February William turned to John William Draper's *Human Physiology* (1856), perhaps the period's leading physiology textbook and the first to include microphotographs. Draper was an American scientist with far-ranging interests in religion and history. He was an early convert to the evolutionary perspective and would later write *History of the Conflict Between Religion and*

Science (1874), which aroused considerable controversy.[31] James was particularly interested in Draper's contention that there were physical reasons for the development of everything in the natural world. He notes, "Argument in Draper . . . from similarity of development in individual and in the geological history of the animal series, attended with similarity of physical constitutions in each to prove that the career of development is guided by external physical causes" (1863 Notebook). William was clearly nurturing the developmental perspective from other writers besides Darwin.

As winter turned to spring James seems to have become more and more interested in the scientific problem of material causation. He has high praise for Englishman Thomas Henry Buckle, who sought scientific laws of history: "Buckle's noble enthusiasm for truth is inspiring." But even in the midst of considerable scientific reading he found time for more of Balzac, who delighted him with his extraordinary powers of observation: "Skipped through Balzac's 'Sys dans la vallee.' Wonderful! There never was such devotion of author to subject before. I will read all Balzac" (1863 Notebook).

IN MARCH 1863 James suddenly returned to Newport. His health had not improved, and he had decided to drop out of school until the fall. Increasing interest in science complemented declining health. He was also troubled about what bearing his studies would have on a career. Should he settle into the theoretical life, or find a mundane but more financially rewarding vocation? His mother, Mary James, pressed for something practical. William worried about being a financial drain on the family: "I am sorry to appear before you in the old character of a beggar. . . . Please send money immediately."[32] He considered the advantages and disadvantages of a career in medicine. Later he told a cousin, "Medicine would pay, and I should still be dealing with subjects which interest me—but how much drudgery and of what an unpleasant kind is there."[33]

The Newport convalescence did not mean a temporary end to intellectual pursuit. William undoubtedly discussed Henry Sr.'s latest book, *Substance and Shadow; or, Morality and Religion in Their Relation to Life*—he helped the elder James find a printer for the work. Mr. James dismissed material cause as secondary to spirituality in the true construction of reality. He may have suggested that William balance scientific with spiritual arguments. In any case, James' 1863 notebook contains the entry "Jonathan Edwards on Original Sin," dated April 1.

He resisted Edwards' Calvinism. Neither he nor his father could accept the idea that innocent children were condemned to hell simply because, like all humans, they had been conceived in original sin. "There is no *proof* that it is *just* we should be born to sin," William notes. Edwards was too dark; he left one with little to do but accept an unacceptable human predicament. By contrast, James warms to the pagan stoic philosopher Epictetus, who emphasized that true good was in oneself, that one's "first care should be the ease and quiet of his own breast" (1863 Notebook). Epictetus would not be the last stoic William would turn to for calm during periods of intellectual and emotional distress.[34]

Although James continued to make notebook entries, some thirteen pages covering the balance of the spring and most of the summer of 1863 are missing, either excised by himself or by a relative. By September he was back at the Lawrence Scientific School working with Wyman in the comparative anatomy laboratory. He wrote to Alice that "I work in a vast museum, at a table all alone, surrounded by skeletons of mastodons, crocodiles, and the like, with the walls hung about with monsters and horrors enough to freeze the blood."[35] By continuing with Wyman, who also taught in the Harvard Medical School, James came closer to choosing a medical career. Still, however, he wavered:

> I feel very much the importance of making soon a final choice of my business in life. I stand now at the place where the road forks. One branch leads to material comfort, the flesh-pots, but it seems a kind of selling of one's soul. The other to mental dignity and independence: combined, however, with physical penury. On the one side is *science*; upon the other *business* . . . with *medicine* which takes advantages of both. . . . I confess I hesitate.[36]

The pressures to choose a profession came not only from Mary James, but also from William's resolve to avoid his father's superfluous existence. James had moved from a relatively freewheeling desire to draw and paint toward a more controlled urge to theorize—controlled in the sense that fastening on problems in science and metaphysics seemed to generate an internal tension that was absent when drawing with Hunt. For the first time, career and intellectual anxiety became pressing. Intellectual labor and vocational indecision seemed to produce hitherto unexperienced anxiety—and later even the feeling that he was on the verge of madness. It was unlikely, however, that vocational indecision alone produced what would become known as James' psychic or spiritual crisis.[37]

In fact, James made several important decisions between 1861 and 1865. He left Hunt; he enrolled in the Lawrence Scientific School; he decided to study medicine. The early Harvard years could be interpreted as years of decision as much as of vacillation. Indeed, despite the appearance of vocational indecision, William nearly always avoided vocational inertia. But by 1863 a new seriousness, even urgency, was unmistakable.

HIS NEW intellectual intensity undoubtedly resulted also from his drawing further and further away from Henry Sr.'s perspective. For all of William's admiration for his father's philosophical acumen, he could not build his intellectual future on the transcendental spiritual reality Mr. James had depicted in *Substance and Shadow.* As young James gravitated toward the more substantial world of career and science, so too did he take more and more seriously the very intellectual tradition that Mr. James considered a chimera, a false God.

On September 10 James abstracted into his notebook German materialist Ludwig Brüchner's *Force and Matter* (1855). Following Brüchner, James wrote that "force and matter are inseparable. It is impossible to think of a pure force antecedent to matter creating matter; or of a force existing independently after the creation of matter." No one could conceive "of a force springing into existence creating matter and then merging itself in matter." Brüchner did not shrink from the final implication: "matter is imperishable" (1863 Notebook). Here was naked materialism: a system of thought that explained the world as entirely matter and motion. There was no need to look further into nature. Science was materialism.[38]

By interpreting Brüchner, James was attempting to clarify the consequences of positing a totally material world. He was much impressed with Brüchner's dynamic sense of nature, of the interaction of force and matter. It suggested a universe pervaded with natural energy—perhaps the same energized world as in Delacroix's *Lion Hunt.* James would later try to minimize the violence that intellectual abstraction did to the actual nature of things and forces in experience. His sense of reality necessitated an insistence on the seamless flow of force and matter.[39]

But Brüchner's materialism had a troubling implication, one that perplexed James throughout the sixties. Materialistic science was also mechanistic science. A dynamic, wordless interaction between force and matter meant that "the Universe is automatic." God was unnecessary, since "no external power is required to make things move or keep them in order." An automatic universe eliminated the need to seek metaphysical cause:

> The mind seeks a cause for everything, but in going back along the chain of causes we must finally stop at a first cause, which we must take for granted. And however remote we make this, we shall always have the same difficulty conceiving it. On the logical principle of not multiplying existences, we should not attempt to go further back than the physical Universe. If when forced to ask who made it, we are answered: "God," the very same reasons compel us to ask "who made God?" And so back *ad infinitum.* Referring to a first Cause satisfies us no more, logically, than assuming the Kosmos to be absolute. . . . The notion of a supernatural finger interposing in Nature and interrupting or correcting her course has all analogy against it, and not one positive reason in its favor. (1863 Notebook, September 10)

What place, then, did God or any nonphysical cause for nature have in the scheme of the universe? William's conclusion shows how far toward a thoroughgoing materialism he had traveled in the early sixties. The supernatural was "a kind of asylum for facts which we can not otherwise explain, but we see every year a little farther into the misty unknown, and every year natural, material laws are found to explain facts which were always thought inexplicable. . . . So that the old asylum is slowly and surely being emptied." There was no escaping "that Brüchner and Co. are in an inexpungable position if no other weapon than that of Cause is used against them" (1863 Notebook, September 10).

Yet James could not give Brüchner a complete victory. The reality of the material world and the fiction of everything else, from language to God, seemed self-evident; those who rejected materialism rejected modern science. William nonetheless posed a question that not only relegated him to the supernatural asylum, but anticipated the great place feeling would play in his intellectual world: "Whence then our feeling of the insufficiency of the Universe, whence our need of a God?" he asked. "Is it a logical need at all?" (1863 Notebook, September 10 entry). He seems to hint at a psychological answer; perhaps, considering his allusion to an asylum, a psychopathological one.

But as a whole the 1863 notebook does not attack the materialists. By and large William finds himself in agreement with "Brüchner and Co." He comes down hard, for example, on those who argued that God had created a link "between structure and function," that the deity had designed nature and then set it in motion. Men such as Agassiz and his teacher Cuvier maintained that structure in the natural world had existed before function: "The woodpecker's tongue is formed weeks before he touches the bark of a tree, so it is absurd to suppose that there is any *material* link between the two. . . . Since the link is not material, it must be mental,—it must have its source in the Mind of an Intelligence . . . and fitter of all its parts to each other." James thought this was silly: "In whatever way we look at this argument it seems to me to be a tissue of absurdities." And basically the trouble lay in referring to nature "our notions of the individual." The truth was that "nature only offers *Thing*." When men divided nature into parts, such as structural parts and functional parts, they engaged in fiction, not science: "The division is artificial" (1863 Notebook).

James had passed from a metaphysical discussion of cause to a consideration of biological nature. The notebook ends rather abruptly with an October 2 entry which surmises that "in every supposed case of sp[ontaneous] gen[eration] that has been carefully investigated the possibility of ordinary gen. has been ascertained." Therefore, "on the principle of not multiplying causes, we should assume that in every case where its *impossibility* was not proved, ordinary gen. took place" (1863 Notebook). By the end of 1863 William had sided with the Darwinists, not because he had intensively studied Darwin's *Origin of Species,* but from reading writers such as Müller and Brüchner, talking with young Charles Peirce, and studying with Wyman. Darwin may even have seemed rather intellectually timid in comparison to the now obscure Brüchner.

But whatever the strength of William's respect for Darwin—and respect him, along with the likes of Müller, Brüchner, and Charles Peirce, he did—James found himself more and more at odds with his father's assertion that the material world was illusion, the spiritual world reality. Visiting the James household must have required new intellectual diplomacy. The elder James must have seemed a relic of the transcendental past—a quaint, metaphysically obscure fellow, far out of step with the times, and not to be taken too seriously. Indeed, William was now clearly on a life course which would almost surely take him far from his father's expectations, both in terms of career and fundamental beliefs.

But if William could largely dismiss his father's point of view, he could not avoid the tension generated by taking intellectual problems seriously. There was both great excitement and unforeseen anxiety. He read avidly about nature and saw theory confirmed in Wyman's comparative anatomy class. Yet studying the materialists also forced him to discount what he read. If the written word was independent of nature, had nothing to do with the mechanical, inevitable, impersonal processes, what was its worth? How could anything that he read be in nature? His mind seemed detached from truth; yet he was compelled to read, analyze, and hypothesize in order to realize the truth. In reading the materialists James considered for the first time the awesome metaphysical problem of dualism, the subject-object split. Discovering a material, independent, and automatic reality divorced from any designs or thoughts, especially his own, put his individuality into question. James was too honest to exempt himself from the implications of materialism. His mind passed fully into intellectual matters, matters that would bring both misery and greatness.

CHAPTER FIVE

In the Original Seat of the Garden of Eden

No words, but only savage inarticulate cries, can express the gorgeous loveliness of the walk I have been taking. Houp lala! The bewildering profusion and confusion of the vegetation, the inexhaustible variety of its form and tints . . . are literally such as you have never dreamt of.

To his brother Henry from near Rio de Janeiro
JULY 1865

I never felt in better spirts, nor more satisfied than I do now with the way in which I am spending my time.

To his parents from the Amazon jungle
OCTOBER 1865

WHEN WILLIAM formally enrolled in the Harvard Medical School in February 1864, he did not really switch from study with Jeffries Wyman in natural science to preparation for the profession of medicine. Medicine was not yet a profession in the sense of entailing rigorous entrance requirements, an established curriculum, residency, or licensing.[1] Most American physicians had no formal medical education. According to one historian, "They had heard no lectures, performed no experiments, and more than likely, seen no dissections. They called themselves 'doctor' not because they were graduates of a university . . . but because they were in fact practicing what doctors of a university usually did."[2] Native physicians who wanted specialized medical education usually traveled to Edinburgh, London, or Paris. American medical schools were still informal and undemanding.

As with the Lawrence Scientific School, entrance requirements to the medical school were lax; a college degree was not a prerequisite. The two-year program included lectures, clinical experience, and visits to Massachusetts General Hospital. In their first year, students attended lectures on such subjects as anatomy, physiology, pathology, and surgery. In their second, they concentrated more on medical practice—disease etiology and clinical observation. The Medical School, which had been founded in 1782, was no longer located in Cambridge, but was situated on North Grove Street in Boston, "in order to secure those advantages for clinical instruction and for the study of practical anatomy which are found only in large cities," according to an 1878 book on Harvard.[3]

When William left his study of natural science at Lawrence Scientific School to study medicine in Boston, he did not really abandon Wyman. The latter taught in the Medical School, and his pupils were largely men who entered the medical profession, not a few of whom later became instructors at North Grove Street. Among them was Henry P. Bowditch, who in 1868 became America's first professor of physiology, a subject that would increasingly interest William. Bowditch, along with another medical student, James Jackson Putnam, engaged James in physiological discussions, discussions that tended to be about the brain and nervous system. Later their talks grew into friendship and regular summer sojourns in upstate New York that would greatly affect William's future.[4]

In medical school James complemented his growing theoretical knowledge—which seemed to interest him most—with microscopic examination of diseased tissue and the dissection of cadavers. His artistic talent served him well as he both sketched tissue cells and made anatomical drawings.[5] Medical study allowed James to focus on science, dabble with the art he still loved, and all the while acquire a profession. On the surface it seemed the best of all worlds.

William saw, however, that save for Wyman, French neurologist Brown-Séquard, and Oliver Holmes, Sr., who oversaw dissecting, the faculty consisted of unexceptional men. Moreover, he did not see much originality in the profession at large. He wrote to a friend, "My first impressions are that there is much humbug therein, and that with the exception of surgery, in which some-

Mixing art with science: LEFT, *notes from a lecture by Brown-Séquard, recorded in a Harvard Medical School notebook, ca. 1866. By permission of the Francis A. Countway Library of Medicine, Boston, Massachusetts.*

RIGHT, *sketch from a microscope slide of urine taken from a cadaver, drawn by William while in medical school, ca. 1866. By permission of the Houghton Library, Harvard University.*

thing positive is accomplished, a doctor does more by the moral effect of his presence on patient and family, than by anything else. He also extracts money from them."[6] As he studied medicine and began to think of himself as a doctor, the commonplace, mercenary aspects became more repugnant. Besides, William was not particularly adept clinically. Although he practiced self-diagnosis and frequently examined family members and friends, he often misread symptoms. Late in life James admitted to Charles Peirce that "my medical education has fitted me for only one thing, which is the recognition of my incapacity to have opinions either about diagnosis or treatment."[7]

Whatever the reasons for his dissatisfaction with medical school, William jumped at an opportunity to participate in an adventuresome alternative. The opportunity arose early in 1865 when Boston businessman Nathanial Thayer offered to finance a vacation for his overworked friend, Louis Agassiz. But Agassiz, eager to find new evidence to undermine the "developmental theory," turned the vacation into the Thayer Expedition, a full-fledged scientific expedition to Brazil. When William learned that Agassiz intended to include several student assistants if they could pay their own way, he was irresistibly tempted. The only problem was finding the money, since he was reluctant to ask Henry Sr. for funds. Luckily, William's Aunt Kate, his mother's live-in sister, offered to pay the lion's share, and his father contributed the balance.[8]

James was still much drawn toward natural history; in fact he had joined the Boston Society of Natural History the previous year.[9] Though he adopted the Darwinian viewpoint, he had never lost his admiration for Agassiz. Indeed, just before entering medical school, William had wanted to work for him: "If I can get into Agassiz's museum I think it not improbable I may receive a salary of $400 to $500 in a couple of years."[10] That salaried position never materialized. Now, however, William was to spend nearly a year with the man whose aura outshone that of all the others; one who had "more charlatanerie and humbug about him and solid worth too, then you often meet with." He was "[an] individual who wishes to be ominiscient, but his personal fascination is very remarkable."[11]

So in the spring of 1865 twenty-three-year-old William James forsook the tedium of medical school for the charisma of a famous scientist and an adventurous voyage. It was to be a familiar pattern: after the academic year always a journey—to Europe, or at least to upstate New York or New Hampshire. Travel broke the routine, provided new perspectives, new personalities, new creative opportunities. But the Brazilian Expedition did not immediately become the excitement-charged escape William expected. Perhaps he should have reconsidered when he discovered that the Expedition would depart on April 1—April Fool's Day.

JAMES' EXPECTATIONS were indeed high when the SS *Colorado* left New York harbor. The company, besides the incredible Agassiz, was interesting and varied. William's best friend, Tom Ward, was aboard, along with four other student assistants—Newton Dexter, Edward Copeland, Walter Hunnewell (the official photographer), and S. V. R. Thayer, Nathanial Thayer's son. Also aboard were the Expedition's official artist, James Burkhardt, and an Episcopal bishop, Alonzo Potter, who piously sanctified Agassiz's theory of special creation. William irreverently referred to Potter as "Bish," noting that "He and Prof. Agassiz furnish as good an illustration of the saying: 'You caw me and I'll caw you' as I ever saw." Potter urged one and all to "give up our pet theories of transmutation, spontaneous generation, etc. and seek in nature what God has put there rather than try to put there some system which your imagination has devised." William reported that "The good old Prof. was melted to tears and wept profusely."[12]

And there was a woman. Elizabeth Agassiz, Louis' American-born wife, took a special interest in William. But James' reactions to Elizabeth were mixed. He wrote to his mother that "Mrs. Agassiz is one of the best women I have ever met. Her good temper never changes and she is so curious and wide awake and interested in all that we see, and so very busy and spotless, that she is like an angel in the boat."[13] In his private diary, however, he wrote, "The excellent but infatuated woman *will* look at everything in such an unnaturalist romantic light that she don't seem to walk upon the solid earth. She seems to fancy that we are mere figures walking about on a stage with appropriate scenery." Later when making a return trip up an Amazon tributary to "the mosquitoes . . . of the cursed solimoes to whom I flattered myself I had bid an eternal adieu," he recorded that "Mrs. Agassiz said in the most enthusiastic manner: 'Well, James, you have a *very* nice time, won't you. I envy you.' Oh silly woman!"[14] Yet whatever her innocence, Elizabeth, as we shall see, probably understood the true importance of Brazil to William better than did any of the others.

As the expedition passed Virginia they sighted considerable

William—in front, far left, with cigar—with members of the Brazilian Expedition, 1865.
By permission of the Houghton Library, Harvard University.

smoke and speculated that it resulted from military activity around Richmond and Petersburg. They were passing the last great conflagration of the Civil War. The Confederates had set fire to Richmond to check the advancing Union army. For the first time, William caught a distant glimpse of the horror that for four years had convulsed the nation and nearly killed his brother Wilky. The expedition did not learn of Lee's surrender or of Lincoln's assassination until after they had disembarked at Rio de Janeiro. Even then, it was difficult to separate rumor from fact. The news of Lincoln's murder was interpreted by the party as a last-gasp rebel lie to undermine Union morale.[15]

William's festive mood evaporated as the *Colorado* churned toward Florida and the Caribbean. He became dreadfully seasick. "O the vile Sea! the damned Deep!" he exclaimed to his parents. "No one has a right to write about the 'nature of evil' . . . who has not been at sea. The awful slough of despond into which you are plunged furnishes too profound an experience not to be a fruitful one." He escaped excessive nausea, "but for twelve mortal days I was body and soul, in a more indescribably hopeless, homeless and friendless state than I ever want to be in again."[16] For the first time illness and despondency entered his correspondence. He began to ponder the relation of physical ills to spiritual well-being, to "states" of consciousness. Illness encouraged introspection.[17]

Although temporarily too sick to attend Agassiz's daily lectures in the ship's salon, William knew his adventurous mentor was bent on answering the Darwinists. Mrs. Agassiz recorded her husband's scientific objectives: "The origin of life is the great question of the day. How did the organic world come to be as it is? It must be our aim to throw some light on this subject by our present journey." Agassiz was quite specific about how to illuminate the origins question: "The first step . . . must be to ascertain the geographic distribution of present animals and plants." This entailed collecting large numbers of individuals, and being "careful that every specimen has a label recording locality and date, so secured that it shall reach Cambridge safely." Until scientists established "how far are species distinct all over the world, and what are their limits," nothing definitive about biological evolution could be stated. Without this knowledge, "all theories about their origin, their derivation from one another, their successive transformation, their migration from given centres, and so on are mere beating about the bush."[18] Agassiz's natural history had been upstaged by just one man. Now the *Colorado* and the Thayer Expedition would find and bring back the proof that the *Beagle* and Darwin had failed to find.[19]

Although William embraced the evolutionary perspective, he much admired Agassiz's meticulous observation of nature: "No one sees farther into a generalization than his own knowledge of details extends, and you have a greater feeling of weight and solidity about the movement of Agassiz's mind, owing to the continual presence of this great background of special facts."[20] The great scientist also offered knowledge and wisdom to an open-minded and grateful student. William reported to his mother that "I am getting pretty valuable training from the Professor, who pitches into me right and left and wakes me up to a great many of my imperfections. This morning he said I was 'totally uneducated.' He has done me much good already, and will evidently do me more before I have got through with him."[21] Agassiz had faults, but "his wonderful qualities throw them quite into the background."[22] He was the kind of great man James had imagined, a mentor on the high seas in search of scientific truth. James had attached himself to an exemplar who personified the drama of the fermenting Victorian scientific scene.[23]

Joy in Agassiz notwithstanding, after disembarking at Rio de Janeiro William had good reason to think that the decision to come to Brazil had been a colossal blunder. He found that his role would not be to speculate on the "developmental theory," or even to search for new life forms. He was to collect fish like a common laborer: "My whole work will be mechanical, finding objects and packing them, and working so hard at that and in traveling that no time at all will be found for studying their structure."[24] A few months of catching, sorting and packing specimens soured him on the entire expedition. "If there is anything I hate, it is collecting. I don't think it is suited to my genius at all," he complained to his parents.[25]

Irritation with collection was temporarily forgotten when he suddenly fell dangerously ill. The Agassizes had left him on the coast with Walter Hunnewell, the photographer, while they went to visit the summer palace of the Brazilian emperor, Dom Pedro. Hunnewell and James were to collect jellyfish in the shallow tides and off the beaches of Rio. He was first stricken with "unlimited itching of the skin, caused by flies and worst of all on both cheeks and one side of the neck by a virulent ringworm."[26] Then came a life-threatening fever. Hunnewell rushed him to a Rio hospital and for eighteen days acted as nurse, disregarding possible infection.

Brazilian physicians diagnosed smallpox; but Agassiz believed that William had contracted varioloid, a milder form of the dreaded disease. Whatever the correct diagnosis, his eyesight deteriorated and he feared blindness. Yet within several weeks the symptoms had all but vanished. In early June he wrote to his parents, "I was released yesterday. The disease is over, and granting the necessity of having it I have reason to think myself most lucky." Mr. and Mrs. James must have been relieved when they read, "My face will not be marked at all, although at present it presents the appearance of an immense ripe raspberry."[27]

Return to health, however, did not change his mind about the folly of having shipped to Brazil. He admitted that "coming was a mistake. My forte is not to go on exploring expeditions. I have no inward spur goading me forwards on that line as I have on several speculative lines. I am convinced now, for good, that I am cut out for a speculative rather than an active life,—I speak now only of my *quality;* as for my *quantity,* I became convinced some time ago and reconciled to the notion, that I was one of the lightest of featherweights."[28]

William had turned an important corner. There would be a radical change in the tone of his letters home. It was as if he had crossed an important personal Rubicon, and found on the other side a positive premonition of the future. Indeed, his Brazilian recovery from varioloid and blindness was the first of James' "twice born" experiences, to be followed by the far better-known one of the early 1870s in Cambridge. The illness had given him time to focus his vocational plans. He could now reflect on a variety of educational experiences: drawing with Hunt; the lectures, reading, and laboratory work at Harvard; and lately, collecting in Brazil. Now the creative intellectual life seemed a matter of destiny. Yet the phrase "several speculative lines" was ambiguous. What speculative lines? And what kind of a career did speculation presage?[29]

IN MID-JULY, in the middle of the mild Brazilian winter, William wrote to his brother Henry from a pristine jungle glade some twenty miles from "damnable Rio," an idyllic spot he called the "Original Seat of Garden of Eden." Reclining on a flat stone, he experienced an unexpected, wondrous feeling: "I almost thought my enjoyment of nature had entirely departed, but here she strikes such massive and stunning blows as to overwhelm the coarsest apprehension." Back in Newport James would have sketched the scene. Now he drew with words: "The bewildering profusion and confusion of vegetation, the inexhaustible variety of its forms and tints . . . are *literally* such as you have never dreamt of. The brilliancy of the sky and clouds, the effect of the atmosphere, which gives their proportional distance to the diverse planes of the landscape, make you admire the old gal Nature."[30]

To almost lose his sight had been to almost lose touch with nature. Now nature consumed his attention and he wanted to share her return with a brother who appreciated descriptive nuance.[31] Brazilian doldrums had been transformed into a startling sensitivity to tropical landscape.

Toward the end of July William joined the Agassizes on a trip up the Atlantic coast to Pará—today Belém. There the party turned inland at the mouth of the Amazon and began what would become, for William, a four-month-long exploration of "the Amazons"—the immense river system coursing for hundreds of miles to the Peruvian Andes. He not only decided not to board a ship from Pará bound for New York, thus ending the tedium of collecting, he had changed his mind about the expedition's value: "Now that the real enjoyment of the expedition is beginning and I am tasting the sweets of these lovely forests . . . I find it impossible to tear myself away, and this morning I told Prof. I would see the Amazon trip through at any rate."[32] William seemed to *see* in nature something that had hitherto surfaced only when he viewed Delacroix or drew a beach scene or landscape in Newport. He still sketched an occasional Indian or monkey, but the Brazilian scene presented James with a mysterious, elemental reality that begged to be described in words.

The party moved up the river on the steamer *Icamiaba* to the remote village of Santarém, some three hundred and fifty miles from the coast. At Santarém, William, Newton Dexter, and their Indian guide separated from the Agassizes and went by canoe up the Rio Tapajós to collect fish. A week later they rejoined the expedition on the upper Amazon at Manaos. From early September until William departed for the United States in December, he traveled up various tributaries collecting fish, now and then reconnoitering with the Agassizes on the *Icamiaba*. This was an unexcelled wilderness experience: nothing, including subsequent back-packing trips in the Adirondacks, could compare with the remoteness, the discomforts. The Garden of Eden could also be considerably less than a tropical paradise: "Although I covered my feet and face, I could not doze on account of the myriads of mosquitoes whose screaming almost drowned the noise of the paddles." And leaving the canoe for campsite did not always bring relief: "Slept little owing to a sudden invasion of mosquitoes who sung but did not bite, and to the noise [of] the cats mad in the yard." (The cats he refers to were jaguars.) On these occasions James yearned for the comforts of civilized Cambridge: "Heaven grant it may come to pass!"[33]

But discomfort did not bring illness or depression. From Tefé on the Amazon, William wrote to his parents that "my health at present is probably better than it ever was in my life. . . . I never felt in better spirits, nor more satisfied than I do now with the way in which I am spending my time." The drudgery of collection, his illness and the prospect of his losing eyesight had temporarily darkened his mood, had obscured the expedition's true benefit. "The fact was that my blindness made me feel very blue and desponding for some time. I only rejoice that I was saved from acting on my feeling; for every day for the last two months I have thanked heaven that I kept on here, and put the thing through instead of going prematurely home."[34]

Just before departing from Brazil, in a remarkable letter to his mother William tried to explain why the trip had been a success. Something in the wilderness experience had altered his sensibility:

> You have no idea, my dearest Mother, how strange . . . home life seems to me from the depths of this world, buried as it is in mere vegetation and physical needs and enjoyments. I hardly think you will be able to understand me, but the idea of people swarming about as they do at home, killing themselves with thinking about things that have no connection with their merely external circumstances, studying themselves into fevers, going mad about religion, philosophy, love and sich [such], breathing perpetual heated gas and excitement, turning night into day, seems almost incredible and imaginary.

Cambridge and Harvard seemed unreal; indeed civilization itself appeared fictional, lost in some dimly remembered past which in fact was only eight months distant. Brazil had evoked a mood "so monotonous, in life and nature, that you are rocked into a kind of sleep," so that "the old existence . . . has already begun to seem to me like a dream."[35]

What James experienced as he paddled hour after hour down endless rivers, observing luxuriant jungle, suffering incessant mosquitoes, and listening to exotic night sounds, was an overpowering correspondence with natural surroundings. Never before had he sustained such a direct dialogue between nature and his extraordinary observational powers. Earlier he had been fascinated with the Delacroixian image of nature, with violence, brilliant coloring, unexpected, mysterious form. Now he found himself actually in the midst of a Delacroixian setting, inside the natural mystery, within "the wonderful, inextricable, impenetrable forest" that "plunges down into a carpet of vegetation which reaches to the hills beyond, which rise further back into the mountains."[36] For several years at Harvard he had been among those who, as he had told his mother, studied themselves into fevers and went mad over religion and philosophy; those who, encased in civilized forms, had mistaken an artificial life for a real one. His creative field had been fixed on book and laboratory science, his person caught in a derivative rather than an original environment. Now William beheld direct, unmediated reality. His eyes and mind were poised to translate the immediacy of nature. Later he would spend enormous theoretical energy describing pure experience.[37] James' descriptive sensibility, his proclivity to juxtapose the theoretical to the natural, was fine-honed in Brazil. As he experienced the wilderness, his mind opened to descriptive possibilities; landscape and state of mind became newly related.

Biographers have interpreted James' Brazilian journey as one more chapter in a succession of vocational failures; since William did not pursue collecting and became ill, the Agassiz trip becomes another false start along the scientific path that would eventually disappear altogether. This interpretation is based on the assumption that William's vocational struggle was a kind of Russian roulette, a hit-or-miss affair: eventually William would discover that he was a psychologist; eventually he would discover that he was a philosopher.[38] But there was a vital link between the Brazilian trip and James' emerging world picture, his interconnected creative fields, the center of his vision. The distinct change in the tone of his journal and letters, from the negativism of the period before his varioloid to his optimism afterward, marks an new awareness of his special destiny.

TWENTY YEARS later Elizabeth Agassiz, whose enthusiasm and naiveté had both amused and irritated James, asked, "Dear William, Do you remember the afternoon when you and I passed each other in separate boats; as I floated out . . . into the sunset glow . . . and you floated into the hidden water way in the forest: As you went by you said to me 'Is it real or a dream?' " William recalled the scene. "Indeed I do remember the meeting of those two canoes. I remember your freshness of interest, and readiness to take hold of everything, and what a blessing to me it was to have one cultivated lady in sight, to keep the meaning of cultivated conversation from growing extinct. I remember my own folly in wishing to return [to Cambridge] after the hospital in Rio; my general greenness and incapacity as a naturalist afterwards, with my eyes gone to pieces. It was all because my destiny was to be a 'philosopher'—and the fact is which then I didn't know . . . that if a man's good for nothing else, he can at least teach philosophy."[39]

James failed to take up the romantic clue Mrs. Agassiz had offered, and preferred retrospectively to view the trip as a clear sign that he was headed toward philosophy rather than natural science. But at the time when they recalled the trip, in 1885, William had just been appointed professor of philosophy at Harvard, which appeared to seal his vocational fate. His remark to Mrs. Agassiz was more a statement about his vocational status in the eighties than a transcription of his actual perspective in the mid-sixties. Elizabeth had remembered his state of mind, recalled his inquiry into reality, the questioning of what was real and what dream. James was at a crossroads in his education, seemingly ready to take up "speculative lines"; and what was real would become a fundamental and persistent speculation.

There remains a passage in William's Brazilian diary that deserves noting for the unusual symbolic power it projects. It comes closer perhaps than anything in his correspondence to his family to capturing what Brazil meant to his creative and vocational future. He had been canoeing up the Paraná, a sizable Amazon tributary, had made camp, and after dinner decided to explore. He came upon an unusual sight:

> I suppose the beach had recently connected with the left bank, but the river rising had already made it an island with a broad channel between it and the shore. I took a long walk over it and found neither gulls nor eggs, but two enormous silvery trees which in their passage down had got their trunks across each other and then their branches catching the bottom had come to anchor in this spot. The

> river going down had left them high and dry. No, not dry exactly for a deep pool had been excavated beneath them by the current which was now filled with green stagnant water and covered by minute flies. O to be a big painter for here was a big subject.[40]

The action of water would indeed become a big subject in James' psychology. A steam became the most memorable metaphor in *The Principles of Psychology,* in "The Stream of Thought": "Consciousness, then, does not appear to itself chopped up in bits. Such words as 'chain' or 'train' do not describe it fitly as it presents itself in the first instance. It is nothing jointed; it flows. A 'river' or a 'stream' are the metaphors by which it is most naturally described. In talking of it hereafter, let us call it the stream of thought, of consciousness, or subjective life."[41] When did James begin to think of consciousness as sensibly continuous? Did he sense that thoughts flowed and shaped the mind as the Paraná had shaped the channel and islands?

There is also a rough correspondence between the Paraná description and a passage in an essay in which James expresses his final thoughts about the value of psychic research. Again, as on the Paraná, James juxtaposed waterlogged trees with the mind:

> Out of my experience [with psychic research] one fixed conclusion dogmatically emerges, and that is this, that we with our lives are like islands in the sea, or like trees in the forest. . . . The trees . . . commingle their roots in the darkness underground, and the islands also hang together through the ocean's bottom. Just so there is a continuum of cosmic consciousness against which our individuality builds but accidental fences, and into which our several minds plunge as into a mother-sea reservoir.[42]

These analogies aside, there is an even more telling symbolism in the Paraná passage. William's creative life in 1865 seemed as entangled as the "silvery trees . . . which had got their trunks across each other . . . and . . . had come to anchor in this spot." The speculative lines that he forecast as his vocational future were as intertwined as the trunks and branches of the silvery trees left high and dry on the Paraná. Art, natural science, medicine, and philosophy had become interwoven as interrelated strands of his creative life. His mind was beginning to exhibit its most fundamental quality: James could not help relating and connecting the separate strands of his educational experience; or rather, his mind interrelated them naturally. The natural scene he described on the Paraná, the one he yearned to paint, suggested not just the interrelatedness of nature, but of his mind as well. Did he glimpse a vision of his special future in those trees? Did he see that in *not choosing* between art, natural science, and speculative lines he had "anchored" his genius in an interdisciplinary mode?

William found natural settings indispensable to his creative life. On one occasion he attributed creative insight to the natural scene and mood that catalyzed his thinking in an isolated wilderness camp high in the Adirondacks.[43] He did not mold his frame of mind, his psychology, or his metaphysics solely from laboratory experience, conversation, or book knowledge. Fields of experience were inclusive and expansive, and the fields that stimulated his thinking were often the natural ones, the wild variety. New England gentlemen-intellectuals were in general becoming increasingly enthusiastic about wild, unspoiled wilderness, and the older transcendentalist perspective had not been completely obliterated by a more empirical understanding of the natural environment. Rather than a familiar foe, nature was becoming an ally, a healer of the spirit in the face of ever more alarming, onrushing urban and industrial growth.[44]

In one sense James never left the Brazilian jungle. A wild, unfixed nature continued to figuratively anchor his world picture. Near the end of his life he told a more disciplined German colleague: "I am satisfied with a free wild Nature. You seem to cherish and pursue an Italian Garden, where all things are kept in separate compartments, and one must follow straight-ruled Walks. Of course Nature gives material for those hard distinctions . . . but they are only centers of emphasis in a flux for me; and as you treat them reality seems to me all stiffened."[45] The Amazon scene provided an unforgettable wild garden, one with lasting symbolic import, the kind of memory that encouraged him to equate the artificial with the unreal, whether in civilizations or in metaphysical theories. Indeed, without powerful images of the natural world, whether of streams, fields, or mother-seas, James could not have created his special psychological and metaphysical world.

By February 1866 he was back in the civilized excitements of Cambridge, surer of but not yet fixed on a vocational or creative course. In fact James was entering a most difficult period, one that combined great insight with what he would soon call "the worst kind of melancholia."

William at the time of his spiritual crisis, ca. 1869.
By permission of the Houghton Library, Harvard University.

CHAPTER SIX

The Antinomian Root

My old trouble and the root of antinomianism in general seems to be a dissatisfaction with anything less than grace.

Diary entry
APRIL 1868

JAMES CONTINUED in medical school after returning from Brazil in December. It was too late, however, to enroll in lecture courses, so he began observing patients at Boston's Massachusetts General Hospital in February 1866. He remained there throughout the spring and early summer, taking a room on Bowdoin Street in Boston. From mid-July to September he joined the family at Swampscott, a coastal resort just north of the city. By early November the Jameses had moved to 20 Quincy Street in Cambridge. The Quincy Street residence, just a few houses from the Agassiz manse, would be William's American home for the next decade. He had hardly settled in, however, when he embarked on his European sojourn of 1867–1868. The late sixties and early seventies were among the most emotionally and intellectually volatile years of his life. But though caught in the grip of great spiritual depression, James was also wonderfully creative; he strove closer toward his special perspective.

At first he seemed content. "I am so settled at home that I don't feel as if I'd ever been away," he wrote to brother Wilky, who was now living in Florida. "Even a white female face can no longer charm me as it did in the first days."[1] Hospital work seemed to agree with him: "I have been at work for a week now. I take to it quite naturally." To Wilky he made a bizarre but quite scientific request: "If you can get a young alligator's head boil it so as to clean the skull and send it on to me. . . . Agassiz confiscated all we got on the Amazon."[2] Twenty-four-year-old William, now a veteran of a genuine scientific expedition and doing practical scientific work in both hospitals and laboratory, seemed headed for a career in medical and/or natural science.

But William's contentment was momentary. Nearly always in the first rush of enthusiasm for a new environment he exaggerated the pleasurable and made himself vulnerable to a depressive reaction. In truth he had resolved nothing about career choice and felt pressured to find a suitable profession. "I am conscious of a desire," he wrote to Tom Ward, "I never had before so strongly or so permanently, of narrowing and deepening the channel of my intellectual activity, of economising my feeble energies and consequently treating with more *respect* the few things I shall devote them too. This temper may be a transient one . . . but something tells me that practically, my salvation depends . . . on following such a plan." He strove to find his "centre of oscillation"; he needed to "know some *one* thing as thoroughly as it can be known, no matter how insignificant it may be."[3] James understood that finding a profession required him to do what his father and his own education had thus far prevented: to specialize.

The trouble was, of course, that James had oscillated to many centers over the past ten years—to art, to natural history, to laboratory research, to medicine, to metaphysical concerns, to communing with nature. To borrow a Jamesian concept, he had in fact begun to build the base for a pluralistic specialization. Moreover, "career" was particularly problematic for American intellectuals in the mid- and late nineteenth century. They could not simply and automatically embrace the increasingly narrow, specialized training that the new professionalism demanded. James understood, however, where success would lie in his immediate future, and he feared he was too broad, too thin, too transparent to enter a profession. But he could not put aside the "general subjects." In his letter to Ward he confided that "the only fellow here I care anything about is Holmes, who is . . . perhaps too exclusively intellectual but sees things so easily and clearly and talks so admirable that its a treat to be with him."[4]

Oliver Wendell Holmes, Jr. was the son of Oliver Wendell Holmes, the American author and physician who was among William's professors at the Harvard Medical School. The younger

Holmes was a Civil War veteran who took a Harvard Law degree in 1866 and would go on to teach jurisprudence at Harvard in the early 1870s while editing the *American Law Review*. He eventually went on to the Massachusetts Supreme Court, becoming chief justice in 1899. President Theodore Roosevelt appointed him associate justice of the United States Supreme Court in 1902, an appointment he held into the Franklin Roosevelt administration in the 1930s. Of the exceptionally bright young men James befriended at Harvard, Holmes would achieve the highest office and broadest public recognition.[5]

Given their separate vocational paths, one at first wonders what drew William and Wendell Jr. together, especially in light of the fact that they both had eyes on the same young woman, Fanny Dixon, whom Holmes later married.[6] The elder Holmes reported favorably on William; he was impressed with James' ability to take intellectual matters "far beyond the facts of anatomy."[7] Then, too, Wendell Jr. had been deeply interested in art during the period when William painted with Hunt. In 1860 he had written a short piece for the *Harvard Magazine* on the Rennaissance painter Albrecht Dürer.[8] Holmes had sympathized with Dürer's strength in the face of melancholia, a condition he himself was experiencing at the time. James came to a greater appreciation of stoicism through Holmes, even though he could not accept the latter's fatalistic metaphysics.

Wendell and William sat in Holmes' room until the wee hours smoking cigars, sipping whiskey, and engaging in speculative discussion. Holmes challenged James to rethink theoretical questions, encouraged him to reconsider the intellectual problems that he had broached in the notebooks of 1862 and 1863. Their conversations began in the winter of 1866 and were interrupted by William's hiatus in Europe in 1867 and 1868. William found an intellectual strength in Holmes that buttressed him in the face of increasing intellectual insecurity and emotional depression, once telling him "I don't know whether you take it as a compliment that I should only write to you when in the dismalest of dumps—perhaps you ought to—you, the one emergent peak, to which I cling when all the rest of the world has sunk beneath the wave."[9]

William's admiration originated partly from deference to Wendell's more profound experience rather than from fundamental philosophical compatibility. Holmes was a Union Army veteran; he had been wounded; he had come to believe that life and death were matters of chance and that the universe was ultimately indecipherable. While William had been collecting fish in Brazil, Holmes had been dodging Confederate bullets—except for the one that nearly killed him. The Civil War schooled Holmes to accept a capricious reality and taught him the self-delusion of putting humanitarian sensibility at the center of philosophy. If war meant that innocent young men had to die, so be it. The glory was in the mystery, even in the suffering. "The faith is true and adorable which leads a soldier to throw away his life in obedience to a blindly accepted duty, in a cause which he little understands, in a plan of campaign of which he has no notion, under tactics of which he does not see the use."[10] Like a good Calvinist, Holmes thought that suffering was inevitable and that final meanings, whether authorized by an inscrutable God or by an indifferent cosmos, were beyond human understanding; it was folly to ignore such a universe or to invent another.

James disagreed. He could not abide a universe which appeared both fixed and irrelevant. Such a world denied that shifting feelings and underlying temperaments were essential ingredients in everyone's notion of reality. There was a fundamental emotional component in metaphysical interpretation, whether the philosopher recognized it or not. He confided to Henry Jr. that Holmes was often filled with "cold-blooded egotism and conceit."[11] Moreover, as Holmes steeped himself in the law, his metaphysical interests declined at the very moment when William's became more intense. He became, in one essential sense, the kind of narrow professional that William knew was most likely to achieve worldly success. Holmes explained to James that "if a man chooses a profession, he cannot forever content himself in picking out the plums with fastidious dilettantism . . . but must eat his way manfully through crust and crumb—soft, unpleasant, inner parts which, within one, swell, causing discomfort in the bowels. Such has been my cowardice that I have been almost glad that you weren't here lest you should be disgusted to find me inaccessible to ideas . . . alien to my studies."[12] Wendell's narrowed, professional resolve would, by 1875, end their philosophical conversations.

Yet while they speculated, there was also important intellectual agreement. Both considered themselves foes of all philosophies that denied natural, physical reality. In the spring of 1868 when he was engaged in an intense examination of the relationship between mind and art, William wrote to Wendell: "Your image of the ideas being vanishing points which give a kind of perspective to the chaos of events, tickleth that organ within me whose function it is to dally with the ineffable. I shall not fail to remember it, and if I stay long enough in Germany to make the acquaintaince of ary a philosopher, I shall get it off as my own, you bet!"[13] He admitted that "I am tending strongly to an empiristic view of life" and asked, "How long are we to indulge the 'people' in their theological and other vagaries so long as such vagaries seem to us more beneficial on the whole than otherwise?"[14]

Holmes' response is lost, but the probability that all ideas including his own had nothing to do with an empirical reality would have little bothered him. William, however, feared for himself as well as for society. That *all* ideas had nothing to do with empirical reality posed an overwhelming threat to intellectual life, made the life of the mind irrelevant—a monstrously depressing prospect.[15] Holmes countered James' romantic Brazillian revery, suggesting a tough-minded, logical empiricism that demanded a fatalistic metaphysics. William had been tending toward a materialistic empiricsm since the early 1860s. But he now began to consider not only the metaphysical, but the personal, social, and creative consequences of embracing empirical science.

After the Civil War the Jameses were never able to recapture the remarkable family life of the prewar period. Robertson and especially Wilkinson were transformed by the Sanborn school and wartime experience. In 1863 Wilky was seriously wounded in Captain Robert Gould Shaw's gallant but ill-fated attempt to take Fort Wagner in South Carolina.[16] After the war Wilky and later Bob traveled again into the Deep South to continue the idealistic struggle and make their fortunes as well. In March 1866 Wilky attempted to establish a "yankee plantation" in northern Florida. He was one of dozens of New England veterans, usually officers, who migrated south to continue the war's work. Now that slavery was abolished they hoped the region could be reorganized as a haven for independent freeholders. Ex-slaves deserved the opportunity to farm. Yankee planters would encourage local whites to cooperate with their former charges and establish economic independence for all. They also expected that inexpensive land and profits from cotton would bring them quick wealth.[17]

Henry Sr. had advanced Wilky what at the time was a fortune—$15,000—and the latter bought land in Alachua County near present day Gainsville. By late 1866 Bob had arrived. "Our work progresses slowly but surely," Wilky wrote his father, "and I think we are going to make a crop. Every indication so far goes to prove the Northern men with assiduity and good common sense and caution will double the amount of cotton that the Southern . . . [planters harvested]."[18] While Wilky informed Henry Sr. about financial matters, he confided to William his condescending if well-intentioned impression of the newly freed slaves: "The blacks jog on for the most part in their same peaceable shambling way. Some of them are wide awake to their condition, but the masses of them bow and scrape in their humility before the Almighty Anglo-Saxon. . . . I consider it a great privilege to be able now to lead these people right."[19]

William lauded the idealism; but what he really admired, indeed envied, was his brothers' commitment. They, like Wendell Holmes, had found a life task, a "centre of oscillation." Their struggle seemed to point toward success and underscored his own failure to settle into an occupation. "Thus you may see how our life glides away in levity and mirth while you are mingling the sweat of your brow and tears with the cotton-yielding mother-earth. But it will not always be so. Your day will come and then where shall *we* be?"[20] Within a year, however, sanguine hopes for Florida success had turned to complete failure. Insects destroyed the first harvest, and the initial good will of local whites turned into implacable hatred upon their discovery that Wilky had served as an officer for a black regiment. The Florida fiasco proved the first of several disasters that jolted the younger James brothers.[21]

Family fortunes were shifting, and not for the better. Alice James suffered from severe symptoms of a chronic nervous disease then diagnosed as hysteria. If overly excited in conversation, she fainted, or worse, lost control of her physical coordination. Alice was whisked away under a local doctor's care to a New York City hospital where she remained for several months, at times visited by worried parents. For much of her relatively short life she was in and out of spas and sanitaria or nursed by her faithful friend, Katharine Peabody Loring.[22]

In the wake of the Florida failure and Alice's collapse, William wrote Robertson that "I confess sometimes the prospect of our scattered and in various ways decapitated family of old is a little disheartening . . . but hang it, the world is as young now as it was when the gospel was first preached, and our lives, if we will make them so, are as real as the lives of any one who ever lived."[23]

But an open, unfinished world was more temperamentally fitted to William than to Wilky, Bob, and Alice.

By 1867 William was exceedingly restive. What had he accomplished in life? He was now twenty-five. His younger brothers had risked much—their lives in war, and their energies, hopes, and father's money in Florida. Even the delicate Henry had managed to write book reviews and had just published a story in the *Atlantic Monthly*; he was rapidly finding *his* center of oscillation. Of all of his siblings, only Alice was less productive than he; but her sex, youth, and ill health meant that her failure could not console him. Daily he soured on medicine as a satisfactory career. He mocked intern manners and himself as well: "Their toadying the physicians, asking to run errands for them, etc. this week reaches its climax; . . . I have little fears, for my talent for flattery and fawning, of making a failure."[24]

His health began to deteriorate. For months a pain in the small of his back had festered. Later he believed that family members suffered from "dorsal insanity."[25] Well-to-do New Englanders were flocking to European spas to cure maladies ranging from the common cold to back pain, gout, rheumatism, and tuberculosis. The James family worked out a European round-robin of convalescence, as one sibling after another took turns drawing from family income to restore failing health.[26] William convinced his parents that Europe would not only ease the back problem but would further his education as well. He planned to convalesce in Germany, a country which had in addition to superior spas the foremost physiological scientists in the world. If health permitted he would attend lectures at the universities of Berlin and Heidelberg. By mid-April William was aboard *The Great Eastern,* the world's largest passenger steamship.

Yet beyond dissatisfaction with medicine, a bad back, and a dull Cambridge, William's flight to Europe was prompted by growing intellectual discontent. The speculative lines he intended to explore had reached a dangerous impasse. Did ideas still matter? Could he adopt the new science and yet continue to speculate, or did realism dictate that he narrow his creative field to a specific body of knowledge? In truth William's intellectual future was in passage too as *The Great Eastern* churned toward Liverpool.

UPON DISEMBARKING, James traveled straight to Paris. "I was very much stunned with the immensity and magnificence of

Paris," he wrote to his father shortly afterward. "They have made great alterations in our old haunts and opened two superb boulevards . . . which entirely upset my topographic associations." He hurried on to Dresden, on the Elbe in eastern Germany, where he hoped to be near both "the baths" and the universities. Dresden also had an opera house, a picture gallery, and a reputation for inexpensive living. But despite these advantages, James found little to recommend the place. "I have been just a little over two weeks settled in Dresden," his letter to Henry Sr. continues; "there is not the slightest touch of the romantic, picturesque, or even *foreign* about living here."[27] He complained to his mother from Frau Spangenberg's rooming house that "life is so monotonous in this place that unless I make some philosophical discoveries or unless *something* happens my letters will have to be both few and short."[28]

The daily routine was relaxed but dull: "I get up and have breakfast, which means a big cup of cocoa and some bread and butter at 8. I read till 1/2 past 1 when dinner which is generally quite a decent meal, after dinner a nap, more *Germanorum* and more read[ing] till . . . I go, generally to the Grosser Garten a lovely park outside the town. . . . Often I go and sit on a terrace which overlooks the Elbe. . . . My great resource when time hangs heavily on my hands is to sit in the window and examine my neighbors."[29] His doldrums continued into midsummer. "I have read nothing since my last, but medicine, Lerve's Aristotle (in Germ. trans.) and some essays full of talent by Herman Grimm and have no 'ideas.' "[30] Yet as was usually the case, James' opinion that he had read nothing meant that he had failed to read anything that stimulated his imagination.

William's failure to make "philosophical discoveries" or find stimulating "ideas" coupled with continuing back problems made him generally unhappy, if not miserable. He sought out a Dr. Carus who advised that he make the short forty-mile trip to Teplitz and undergo the regimen of mineral water and thermal baths. Unfortunately the treatment did not ease his back and even occasioned a new discomfort—he complained to Henry Sr. of "a chronic gastritis of frightful virulence and obstinacy." Physical affliction and intellectual inertia encouraged thoughts of self-destruction. "Although I can not exactly say that I got low-spirited, yet thoughts of the pistol and dagger and bowl began to usurp an unduly large part of my attention," he admitted to his father.[31] Indeed, thoughts of taking his own life had surfaced well before leaving Cambridge. "All last winter [1866–1867], for instance, when I was on the continual verge of suicide it used to amuse me to hear you chaff my animal contentment," he wrote to Tom Ward. He lamented that "sickness and solitude make a man into a mere lump of egotism without eyes or ears for anything external, and I think, notwithstanding the stimulus of the new language etc. that I have rarely passed such an empty four months as the last."[32] If only he had Wendell Holmes' ability to abstract himself from personal sensibilities!

In September he made plans to leave Dresden and Teplitz for Berlin to study "the Nervous system, and psychology."[33] Perhaps Berlin would break what he told Holmes was "a tedious egotism."[34] Once immersed in the exciting discoveries of contemporary science, he would perhaps gain that illusive "centre of oscillation" that refused to focus amid the crosscurrents of changing interests, ill health, and shifting geographical setting.

At first the change seemed to work wonders: "I am a new man since I have been here, both from the ruddy hues of health which mantle on my back, and from the influence of this live city on my spirits. Dresden was a place in which it always seemed afternoon," he wrote to his brother Henry.[35] His old Newport pal, Thomas Sergeant Perry, had shown up, and even though William was not yet well enough to attempt laboratory work, he had "blocked out some reading in physiology and psychology." He wrote to Ward that "it seems to me that perhaps the time has come for Psychology to begin to be a science . . . some measurements have all ready been made in the region lying between the physical changes in the nerves and the appearance of consciousness. . . . I am going to study what is already known, and perhaps may be able to work at it. Helmholtz and a man named Wundt at Heidelberg are working at it and I hope if I live through this winter to go to them in the summer."[36]

By September, 1867, James was clearly drawn to this new experimental field, a professional-creative scene that would at last provide both intellectual excitement and a future career. He wrote enthusiastically to fellow medical school student Henry P. Bowditch: "The opportunities for study here are superb. . . . The physiology lab. with its endless array of machinery, frogs, dogs etc., etc. almost 'bursts my gizzard' when I go by."[37] He advised that "one cannot expose himself too early to the contagion of the thorough, serious, German way of working."[38]

Continuing scientific studies seemed particularly attractive in an atmosphere alive with new physiological thinking—thinking that disavowed the vitalism of the mid-nineteenth century that ascribed the functions of living organisms to forces distinct from strictly chemical and physical properties. From men like Du Bois-Reymond James absorbed pathbreaking science first hand.[39] James understood that any scientific role he might play would be within an up-and-coming field, rather than within an established tradition.

Yet his enthusiasm lasted only a few months. In early 1868 he left Berlin, failing to finish the term, and returned to Teplitz and the soothing waters that had earlier refused to heal his back and had brought on gastritis. James was in the throes of a personal crisis that could not be resolved through the study of German science any more than it could be cured in the baths. It was a time of troubles that deepened as the European convalescence lengthened. To the bright, urbane, multilingual young man who in January 1868 had turned twenty-six, Berlin had suddenly become "a bleak and unfriendly place."[40]

TO HIS father William explained the sudden flight from Berlin as an escape necessitated by both health and educational considerations. He wanted to be well enough by summer to study physiology at Heidelberg with the greatest living German scientist, Herman Helmholtz, a pioneer in optics, physics, and biology.[41] To Robertson James, he was more pointed, confessing that "this accursed thing in back has now lasted for 13 months, it scatters all my plans for the practice of medicine." William still thought, however, that with a regimen of "hot baths, douches and hot mud-poultices" he would "recover a sufficiently strong back to be able to do laboratory work . . . and perhaps be able to get a professorship somewhere."[42]

But hope for a strong back proved only that; by late March his parents learned that the baths, if anything, had made things worse. His mother wrote to his brother Robertson, "We got a letter from Willy a few days ago saying that the Teplitz treatment this time had been an entire failure." Incredibly, William intended "to try another cure in April milder than the last, and after this experiment, if he is not better he will come at once home."[43]

What was troubling William was more complicated, however, than a persistent backache. For all of his genuine interest in the new physiology and his rising professional hopes, nothing particularly original had emerged within his own mind. James was far too intellectually ambitious to simply follow a scientific career. He needed to make scientific discoveries; he needed to approach physiology and psychology originally. Yet what was clear to William in 1867? He had tended, as he told Holmes, toward the empirical view—yet he could not become a thoroughgoing materialist. The life of the mind, consciousness, feeling *had* to have a place in reality. The speculative life needed expression to find focus.

Furthermore, James' crisis was not strictly intellectual, any more than it was wholly vocational. Although his letters and diary give no direct evidence that his troubles were in part related to sexual frustration, there are good reasons for supposing they were. William, after all, was in his mid-twenties, interested in the opposite sex, and therefore almost certainly concerned with finding sexual release. Unless he led to a secret life with ladies of ill repute, or practiced the self-denial of a cloistered monk, he must have occasionally indulged the forbidden Victorian vice of masturbation—a recourse which, given contemporary attitudes, probably further intensified his dissatisfaction, self-loathing, and depression. Further, frustrated sexuality would have done nothing to unblock his creative inertia. Interestingly, after marrying he had few back complaints, and he became a prolific scholarly writer.[44]

Shortly after arriving in Berlin in late September 1867, William wrote to Henry Sr. asking for the latter's article "Faith and Science," which had been published in the *North American Review*. He wanted his father "to feel how thorough is my personal sympathy with you, and how great is my delight in much that I do understand of what you think, and my admiration for it." Yet their perspectives were greatly at odds: "You live in such a mental isolation that I cannot help often feeling bitterly at the thought that you must see in even your own children strangers to what you consider the best part of yourself."[45] There followed an exchange between father and son clearly showing that William craved independent creative power.

The elder James believed that William's inability fully to comprehend "Faith and Science" had a simple cause: "It arises *mainly* from the purely *scientific* cast of your thought just at present and the temporary blight exerted thence upon your metaphysical wit." William was temporarily bogged down "in this scientific or peurile stage of progress." The deficiency had a typical characteristic: "You believe in some Universal quantity called Nature, able not merely to *mother* all the specific objects of sense, or give them the subjective identity they crave to our understanding, but also to *father* them or give them the objective individuality or character they claim in themselves." This resulted in making "the idea of creation . . . idle and superfluous to you, for if nature is there to give absolute being to existence, what need have we to hunt up a creator?" Indeed, Henry implored that "the creature" had an obligation to "bring him[self] up to the level of the creator."[46]

William could not accept a doctrinaire assertion "that such and such *must* be the way in Creation, as if there were an a priori logical necessity binding on the mind. This I cannot see at all in the way you seem to." He objected to the insistence that there must be a God when it was simply an "a posteriori hypothesis." Even more vital, however, was something that touched deeply on the use of his intellectual powers: "It refers to the whole conception of Creation, from which you would exclude all arbitrariness or magic." William asserted that "the creator must be the all, and the act by which the creature is set over against him have its motive within the creative circumference. The [creative] act must therefore contain an arbitrary and magical element."[47] Creation was not simply God-given; the creature, man, was necessarily and contingently creative. Man had the opportunity to create as a flesh and blood creature in the creative scheme of things. He was not on earth simply to prove the existence of God, or to deny material nature. The son leaned toward naturalism, the father toward eschatology.

James could accept his father's condemnation of natural man as selfish, and his view of spiritual man as transcendent, as "a history" of creation, but not as creativity itself. For William, the natural experience of actually creating was the essential reality; he was more the participant than, like his father, the awed spectator. Creative man was "emanation"; he was issuing, coming out, emerging—he was in a permanent chrysalis. "To me he can be no true creature outside of that condition; inside of it he may be anything."[48] To have been left metaphysically between flesh and spirit was intolerable to Henry Sr.; his personal history of rising from the selfish flesh to the unselfish spirit was for him the great theological-metaphysical point. To William, reality could never be so unambiguous. He was trying to tell his father that choosing between science and God, or nature and spirituality, would not clear his way, but would be a creative dead end.

In March William settled in again at Frau Spangenberg's boarding house in Dresden, where he embarked upon a broad reading program. Among the authors on his reading list for 1868 were Wundt and the psychopathologist Liebault. Yet literary writers were by far more numerous and included Schiller, Goethe, Sand, Dickens, Shakespeare, Balzac, Dumas, and Maury.[49] Although he did not give up scientific reading, his attention was clearly directed toward aesthetics and art; in considering "several speculative lines," he returned to his first interest.

Among all of James' diaries and notebooks, scholars have been overwhelmingly interested in the diary which covers the period 1868–1873. Interpretation has centered on the problem of why James was depressed, why he considered suicide, and why after April 1870 he seemed to emerge from despair.[50] But William's diary is more than a record of melancholia and recovery; it is also a creative document that contains early theoretical statements that would ever after mark his perspective. One crucial point needs emphasis: despite illness, lingering oedipal conflict, and vocational indecision, James took a fundamental step toward an original vision in the spring of 1868; he took this step years before he defeated melancholia.

His mother knew. Mary James sensed that despite the continued litany of health complaints, her son was beginning productive, creative work. "Willy is now at Teplitz taking another cure," she wrote Robertson. "He writes very long letters to Harry about books and pictures, remarkable letters full of originality and power of thought." Mary emphasized—in striking contrast to William's biographers—that "although he is . . . far from well . . . he is full of mental activity and vigor—He had also written 1/2 dozen papers for publication[,] principally reviews of scientific works."[51]

Within a week after beginning his diary, William confirmed to Henry his mother's suspicion that something significant was being generated in Dresden, something in his original field, art. "I have been more absorbed this time in the general mental conception of different school[s] and pictures than in their purely artistic qualities," he related. "I have come to no conclusion of any sort, and see that to do so would require a study of all the galleries and antiquity cabinets of Europe. . . . Still the problems are fascinating."[52] No longer content simply to study physiology, his mind now focused on interpreting the qualities of art objects. James slipped in the spring 1868 from creating art as intrinsic expression to using art as a catalyst for speculation. Visual and theoretical powers united and pressed, haltingly but persistently, toward original thinking in a new creative field.

Still filled with "apathy and restlessness," James began studying art at the Dresden gallery. He was particularly interested in distinctions between classical and modern art—"modern" meaning anything from the Renaissance into the nineteenth century:

> It struck me the other day that among works of plastic art a division (with respect to the 'Weltanschaung' involved) might be made between such as Raphael's and M. Angelo's on the one hand and those of the Greeks and Venetians on the other. The former points expressively and with consciousness on the part of the author to the existence of something ineffable beyond the picture, which it is the best function of the picture to make us feel; the latter doing something similar of course for the philosopher who looks at them from without, but executed by the artist with no such thought, but complete and rounded in themselves. Perhaps a scheme of criticism of works [in] the first class might be formed on an analysis of the manner in which this indication of the ineffable is consciously to the author executed. This is worth thinking about.[53]

As he sat in the gallery, James' speculative powers revived. He was both perplexed and fascinated: "I went yesterday again to the collection of casts. All I have written or may write about art is nonsense. Perhaps the attempt to translate it into language is absurd—for if that could be done what would be the use of art itself." (Diary, April 11, 1868). He began to explore a theoretical problem that would occupy him until the end of his life: the relationship between consciousness and object, between subjective and objective reality.

"I feel myself forced to inquire while standing before these Greek things what the X is that makes the difference between them and all modern things, and I clutch at straws of suggestions that the next day destroys," he noted. Then he discovered something significant:

> I was much struck by . . . looking at Rietschel's group of the dead Christ and his mother—a remarkable and successful thing. How I longed for old Hunt to be there to hit off the thing in a few of his smiting sentences, and put me on the track of its failure perfectly to succeed. But as I glanced around it at the Greek things I saw instantly that one *effect* of the difference was that if the Madonna's nose were knocked off or her face gnawed away by the weather and if the Christ were mutilated the *essence* of the thing

> would be gone. Whereas it makes hardly any difference in the Greek things. The cause of their existence (I mean the idea of the artist) lies all through them and can bear any amount of loss of small details and continue to smile as freely as ever.

There appeared to be something worthwhile in Dresden after all. "I am getting demoralized here but I think I am kicking up something too" (Diary, April 11 and 12, 1868).

What he was kicking up was original insight. Something in modern art created a state of mind, a dissonance, a subjective distance that was absent in Greek art. James was making sophisticated psychological distinctions as he gazed at works of art. And as he cast those distinctions into language he juxtaposed two fields of creative experience—art and psychology.

The evening of the day he recorded these thoughts he attended a performance of *Hamlet* and was made even more acutely aware of creative struggle. "Here again is the problem which I have had before me for the last few days," he observed in his next entry. "Is the mode of looking in life [of] which Hamlet is the expression a final one or only a mid stage on the way to a new and fuller classical one[?]" The creative suspense was almost unbearable. "The fullness of emotion becomes so superior to any possible words, that the *attempt* to express it adequately is abandoned, and its vastness is indicated by slipping aside into some fancy, or counter sense—So does action of any sort seem to be to Hamlet inadequate and irrelevant to his feeling" (Diary, April 13, 1868). And William, too, desperately sought creative fulfillment; hence, the lapse into "fancy, or counter sense"—restlessness and depression.

James tried again to express the state of mind that separated himself, Hamlet, and modern artists in general from the Greeks. He returned to the casts: "Rietschel's thing good again but the thought of it tends to one point, requires the spectator to put himself in a particular mood to be sympathetic, and so would as a constant companion be very 'aggravating.'" With classical art, however, no mood was necessary. "The Greek things never have any point—the eye and the mind slip over and over them, and they only smile within the boundry of their form. They may stand for anything in the scale of human being" (Diary, April 14, 1868). This was more than art criticism. Without fully realizing it William was lapsing into introspection, into his unique mode of psychological description.

On April 21 he read Schiller's "Essay on Grace and Dignity," after which he commented: "My old trouble and the root of antinomianism in general seems to be a dissatisfaction with anything less than grace." And the next day after reading Schiller's "On the Pathetic," he observed: "Our moral reason demands in every given case that such and such a definite act should be performed . . . ; the aesthetic sense, or imagination on the other hand merely *craves* and what it craves is 'to maintain itself in play free from laws.'" He pondered the relationship between moral law and creative freedom. "Into the former I think an element of personal fear will always be found to enter in a more or less subtle matter; while the latter is without this. How to explain the latter on utilitarian or Darwinian principles is a problem worth investigating" (Diary, April 21 and 22, 1868).

William recognized a tension that persisted the rest of his life. The moral law demanded that one follow social convention, find a vocation, develop a profession. But grace or aesthetic sense or imagination demanded free play. The antinomian root was not for William so much a desire to achieve theological grace as to find an imaginative rejuvenation, the creative state of mind that the Greek artists had possessed as an inward, almost instinctive quality. Following the moral law, which in the broad social sense meant adopting a conventional profession, did not necessarily mean pursuing the creative life. Schiller sharpened a growing realization that the true creative life, like grace itself, was freely bestowed, not earned. As Emerson before him, James felt himself to be in the "optative mood"—receptive, wishful, and expectant for the flowering of his genius.[54]

In late May William wrote a long letter to Tom Ward explaining that he now had "a better insight into what makes the Greek things so peculiar to us" as well as a "real acquaintance" with Goethe and Schiller. He also revealed he had gained "the friendship of a young American lady here in the house, who has stirred chords in this dessicated heart which I long thought had turned to dust."[55]

The "American lady" was Kate Havens, an accomplished pianist who was also boarding with Frau Spangenberg. She was William's first European romantic interest, and apparently Miss Havens had feelings for him, though her health probably prevented a continuation of their relationship.[56] William reported that she "is a prey to her nerves and is in a sort of hysterical hypochondriac state, but her mind is perfectly free from sentimentality and disorder of any sort—and she has really genius for music."[57] Indeed, Kate and her music heightened the effect of William's aesthetic revival, producing emotional overflow, self-loathing, and hopes for his own creative salvation. He wrote, "To night while listening to Miss H's magic playing . . . my feelings come to a sort of crisis[;] the intuition of something here in a measure absolute gave me such an unspeakable disgust for the dead drifting of my own life for some time past. I can revive the feeling perhaps hereafter by thinking of men of Genius. It ought to have a practical effect on my own will" (Diary, May 22, 1868).

He resolved to take limited practical action—to try a pragmatic therapy to contain an enormous imaginative appetite. "Every good experience ought to be interpreted in practice. Perhaps actually we can not always trace the effect, but we won't lose if we *try* to drop all in which this [is] not possible." He urged himself to "Keep sinewy all the while—and work at present with a mystical belief in the reality interpreted somehow of humanity" (Diary, May 22, 1868).

The important point is not that James adopted a practical, active approach, so much as what *caused* him to move toward a pragmatic strategy of coping with inner problems. Clearly in the spring of 1868 he faced a creative crisis which arose from powerful desires to express the world naturally, as the Greeks had, without artifice or pretension. He wanted imagination and the world to be in proportion; he wanted ideas and their objects to be naturally related.[58] Yet he did not know how to accomplish what seemed a vain, hopelessly romantic ideal. Intellectual desire and what were surely sexual feelings for Kate Havens had transformed the Dresden-Teplitz scene from the dull, relaxed place it had seemed the previous fall. Through all the anxious excitement his back pain persisted.

In June William decided to go to Heidelberg, where he hoped to attend Helmholtz's lectures. But after less than a week, severe dorsal pain forced a return to Dresden. Then he moved on to try the baths at Divonne, France, in Savoy. There are no diary entries from late July 1868 until February 1869, and he confined himself mostly to letter writing.[59] Despite the baths, from Divonne he reported to Bowditch that "I have been growing worse in every respect . . . since being here."[60] And to Henry he announced an ultimatum: "I am determined to get well by next Spring. If I die for it."[61]

James neither got well nor died; he went home. "I have a better chance of getting well in the quiet of home, than tossing about like a drowned pup about a pond in a storm," he explained to Bowditch.[62] But his return to Cambridge did not end his time of troubles.

ALTHOUGH RELIEVED to be back on Quincy Street with family after a year abroad alone, James faced immediate pressure. He resolved to finish medical school, for which he needed only to write a thesis and take the oral examination. "I find I only have 6 weeks to do my thesis in," he told Bowditch. "Said thesis on cold will not contain . . . original suggestions. Time is too short and I only aim at squeezing through."[63] But the practical problem of finishing his thesis did not prevent him from reading medicine. Indeed, he told Thomas Ward, "It grows so interesting that I find myself regretting there is no chance for me to stick to it."[64] In March 1869 Oliver Wendell Holmes, Sr. passed William on the final oral examination; James could now claim to be a bona fide medical doctor—the M.D. was the only degree James ever earned.

His study was not limited to medicine. "I am swamped in an empirical philosophy," he wrote Tom Ward. "I feel that we are Nature through and through, that we are *wholly* condition, that not a wiggle of our will happens save as a result of physical laws, and yet notwithstanding we are in rapport with reason.—how to conceive it? Who knows? . . . We shall see damn it. We shall see."[65] James once again took up the theoretical quest; but now he emphasized philosophy.

Sometime in 1869 or 1870 he joined a group of young Cambridge intellectuals who dubbed themselves the "Metaphysical Club." Among the regular members were Charles Peirce, Wendell Holmes, Jr., Chauncy Wright, Francis Abbot, and John Fiske.[66] Peirce, Holmes, and Wright articulated the case for a strictly empirical science. Wright was the most extreme and espoused a materialistic nihilism. James later remembered that Wright argued that "behind the bare phenomenal facts . . . there is *nothing*."[67] In 1873 William wrote "Against Nihilism," an essay he never saw published, in which he critiqued Wright's thoroughgoing, uncompromising positivism.[68] By then he was convinced that any philosophy that broke the world down into unrelated, unconnected phenomena denied common sense.

He also read two British authors who in one way or another

challenged a hard-line empiricism—David Masson and Samuel Bailey. William was impressed with David Masson's *Recent British Philosophy* (1865), which espoused a "philosophical centrism" that classified and divided idealists from empiricists. He studied and recorded his response to Samuel Bailey's *Letters on the Philosophy of the Human Mind* (1855), which argued for a natural realism. To Bailey all experience was immediate, and consciousness in effect disappeared as a mediator for experience. In a series of notes on Idealism, William wrote, "Bailey acts as if to define a fact were to repeat it, as if to answer the question: what is a zebra? one must become forthwith a zebra, or at least refuse to say 'striped' but cry continually 'zebra, zebra.' "[69]

During the same period William delved into Alexander Bain, who, in *The Emotions and the Will* (1855) had criticized the narrow associationism of John Stuart Mill. Bain's theory of belief was particularly important to him, since it maintained that belief was inseparable from action, a position he was to adopt. Yet William did not abide by Bain's assumption that consciousness was reducible to physical processes. After the remarks on Bailey, William noted, "Bain and Wright seem to think that we may give to representation primitively a merely physical existence (i.e., not let it refer out of itself and then gradually by 'association' show how this outward reference grew up" (Notes on Idealism).

Sometime in 1870 or 1871 James also considered the philosopher of evolution, Herbert Spencer. Although William does not mention what book or books he read, he may have obtained the first volume of *Principles of Psychology*, published in 1870–1872. He remarked that "Spencer at one time seems bent on proving [the mind] to be a mere product, is down on space etc. as involving an intrinsic . . . quality, and yet on the other hand in his doctrine of the unknowable . . . he of course implies that all the qualities of thought are due to an original *NATURE* in the subject" (Notes on Idealism). Here was an early example of what by the late seventies would become a full-fledged assault on Spencerian psychology.

William was drawn to British empiricism, but had real doubts about its application to consciousness. Empiricism did not satisfactorily explain the way the mind made the world. Could he explain the relationship between consciousness and their objects more convincingly?[70] As in Dresden, creative possibilities seemed at hand but he could not realize their fulfillment.

EARLY IN 1870 James lapsed again into deep depression. On February 1, 1870, he wrote in his diary, "a great dorsal collapse about the 10th or 12th last month has lasted with slight interruption till now, carry[ing] with it a moral one."

> Today I about touched bottom, and perceive plainly that I must face the choice with open eyes: shall I frankly throw the moral business overboard, as one unsuited to my innate aptitudes, or shall I follow it, and it alone, making everything else merely stuff for it? I will give the latter alternative a fair trial. Who knows but the moral interest may become developed. Hitherto I have given it no real trial, and have deceived myself about my relations to it, using reality only to patch out the gaps which fate left to my other kinds of activity, and confusing everything else together. I tried to associate the feeling of moral degradation with failure, and add it to that of the loss of the wished for sensible good end—and the reverse of success. But in all this I was cultivating the moral interest only as a means and more or less humbugging myself. (Diary, February 1, 1870)

William felt deeply that to seek "morality"—and I believe he meant not simply ethics but an ontology centered in consciousness—had meant putting morality and materialism in opposition. Hence, when he leaned toward materialism he felt degradation and failure. Now, however, he wished to pursue morality for itself. This was more, however, than opting for the moral life; it was also a statement of intellectual ambition and creative design. By opting for morality he was really deciding to originate a new, more metaphysically convincing notion of consciousness.

The death of his cousin Minny Temple in March confirmed James' growing resolve and sense of direction. "By that big part of me in the tomb with you may I realize and believe in the immediacy of death!" he wrote, deeply touched by her death. "May I feel that every torment suffered here passes and is as a breath of wind—every pleasure too. Acts and examples stay. Time is long. One human life is an instant." He asked, "Is our patience so short-winded, our curiosity so dead or our grit so loose, that that one instant snatched out of the endless age should not be cheerfully sat out[?] Minny, your death makes me feel the nothingness of all our egotistic fury" (Diary, March 22, 1870). Had William known that Minny recently had written to a friend that William "is one of the very few people in this world that I love,"[71] his feelings could not have been more heartfelt. Total commitment was now his duty: "Since tragedy is at the heart of us, go meet it, work it into our ends, instead of dodging it all our days, and being run down by it at last. Use your death (or your life, its all one meaning)" (Diary, March 22, 1870).

On the last day of April he made the practical intellectual-emotional commitment that biographers have made so much of:

> I think that yesterday was a crisis in my life. I finished the first part of [Charles] Renouvier's 2nd Essay, and saw no reason why his definition of free will—the sustaining of a thought *because I choose to* when I might have other thoughts—need be the definition of an illusion. My first act of free will shall be to believe in free will. For the remainder of the year I will abstain from mere speculation . . . in which my nature takes most delight and voluntarily cultivate the feelings of moral freedom, by reading books favorable to it as well as by acting. (Diary, April 30, 1870)

He did not really decide to stop speculating, however, a point often ignored: "After the first of January [1871], my callow skin being somewhat fledged, I may perhaps return to metaphysic study and skepticism without danger to my powers of action" (Diary, April 30, 1870). Embracing action did not mean giving up theorizing.

Renouvier's will to believe was surely important to James' optimism for the future. But far more critical to understanding his center of vision was the fact that will was but a strategy in achieving original speculations. James' diary shows clearly that he wanted to create something so badly that life itself was but a medium to achieve that goal; it was not really worth living except in the experience of creation, the acting-out of his special genius. Far from being grounded in pragmatism, his psychology and metaphysics were grounded in an insatiable craving to bring some fresh speculative perspective into the world. Pragmatism was merely a means to that end. The compass of James' desire dictated the depth of his depressive reaction. No strictly pragmatic intellectual would have been plagued with what James called "the worst kind of melancholia."[72]

To speculate, to philosophize without a center of vision, taxed James' sanity as much as descending into the slough of inactivity. "To be responsible for a complete conception of things is beyond my strength. To make *form* of all possible thought the prevailing *matter* of one's thought breeds hypochondria." He admitted, however, the considerable scope of his plans. "Of course my deepest interest will as ever lie with the most general problems." By February 1873 James had hedged his vocational bets: "I decide to stick to biology for a profession in case I am not called to a Chair in Philosophy, rather than to try to make the same amount of money by literary work, while carrying on more general philosophic study." But prophetically he added, "Philosophy I will nevertheless regard as my vocation and never let slip [a] chance to do a stroke at it" (Diary, February 10, 1873).

In April Charles Eliot offered William an instructorship in anatomy and physiology. James accepted; he had already rationalized taking a post that would give only partial reign to speculative goals. "A 'philosopher' has publically renounced the priviledge of trusting *blindly,* which every simple man owns as a right—and my sight is not always clear enough for such a duty." Still the appointment presented him with a chance to work in an up-and-coming field. "You can't divorce psychology from introspection, and immense as is the work demanded by its purely objective physiologic part, yet it is the other part rather for which a professor is expected to make himself publically responsible," he had noted in his diary on February 10.

James' crisis in the sixties and early seventies was fueled by a powerful imaginative ambition, the desire to have grace bestowed in the creative act. He wanted the moral life, and he opted for an active, purposeful will. As important as belief in the will and the moral life were, however, they were not in themselves enough. They remained subordinate to the predominate imaginative ambition. Unlike his father, William would not settle for the simple, selfless life. He was not only of another age; he wanted more than conversion to faith, more than the contemplative life. Within his anxiety James desired a special destiny. He wished to be chosen, to be offered grace as the saints of old had been, but in a new secular setting. In the rapidly developing science of physiological psychology he had found a fertile field in which to sow. In August 1873, for a salary of six hundred dollars, William began a thirty-four-year teaching career. But the storms of personal crisis were still blowing.

CHAPTER SEVEN

Tragical Marriage

To wander from the outward order that keeps the world so sweet is in a word to throw yours upon the tragical.

To Alice Howe Gibbens
SUMMER 1876

I did mean, God knows, to make my life serve his. To stand between him and all harmful things.

Alice Gibbens James to Francis R. Morse, on William's death

IN MID-JULY 1878, Mr. and Mrs. William James boarded a Boston train and began their honeymoon, or more properly in those days, their wedding journey. They were bound for Saratoga Springs, New York, a fashionable resort town just south of Lake George and the upstate Adirondack wilderness. Saratoga Springs certainly had attractions for the recently wed—several outstanding hotels, Congress Park for concerts and strolling, casino gambling, and horse racing. The groom had reserved a suite at the Grand Union, the largest and indeed the grandest of the resort hotels. "If you're in for all the life you can get, and all the distinguished visitors, and the big politicians, and style and jewelry, and a full band all the while, you want to go to the Grand Union," novelist William Dean Howells had one of his characters remark. "Thats where *I'd* go if I was in Saratoga for a good time."[1]

In 1878 the Grand Union had been recently renovated, and its

Alice Howe Gibbens at the time of her engagement to William, ca. 1878.
By permission of the Houghton Library, Harvard University.

owner, the New York City department store magnate Alexander T. Stewart, had indulged in Victorian extravagance. The mammoth dining room required the attention of thirty-five cooks, two hundred waiters, and twelve carvers. Daily rations included fifteen hundred pounds of beef, twelve hundred quarts of milk, eighty chickens, and two hundred and fifty quarts of strawberries. The hotel employed hunters to make sure an ample supply of fresh game and fish was in readiness. The ballroom with a twenty-seven-foot ceiling was illuminated by three huge electrically lighted crystal chandeliers that cast their glow on elegantly attired guests who waltzed on the polished eighty by sixty-five-foot dance floor. If, as the *New York Times* quipped, "At Saratoga, Americans are taught to understand the essentials of a good hotel," then at the Grand Union the essentials were, even for the European-traveled Jameses, surely a bit overwhelming.[2]

From Saratoga William and Alice continued to Lake George on their way northward to the cool mountain air of the Adirondacks. Their destination was Keene Valley, nestled amid looming peaks, wild glens, rushing brooks, about fifteen miles from Lake Placid. During the 1870s increasing numbers of affluent easterners journeyed into the Adirondacks to escape the frenzy of urban-industrial culture. A contemporary account maintained that the Adirondacks offered "the perfect relaxation that all jaded minds" sought.[3]

William may not have been jaded when he led his bride from a lush resort into the wilderness, but he certainly suffered from the common Victorian complaint, neurasthenia. Two years before meeting his bride-to-be, William, accompanied by Cambridge friends James and Charles Putnam and Henry P. Bowditch, had trekked the Adirondacks and discovered Keene Valley. The next summer the four hikers had purchased a plot of land from two Keene Valley farmers and built a rough story-and-a-half cabin which they christened "the Shanty." In 1876 the farmers moved several miles up the valley, and the Cambridge chums bought their house. It was to this rude setting that the newlyweds came by carriage after disembarking from the train at Westport, about a hour distant. And it was here that James would come, again and again throughout his life, to ease the academic burden, the nervous fatigue that invariably increased at the end of the winter term. Ironically, this place which came to be synonymous with relaxation was also the place where in 1898 he irreparably strained his heart.[4]

WILLIAM WAS introduced to Miss Alice Howe Gibbens, an attractive, brown-eyed, brown-haired young lady in her late twenties, by the unconventional amateur philosopher Thomas Davidson. Davidson and Henry James, Sr. were members of the Radical Club, a Boston Unitarian gathering that discussed spiritualism and psychic phenomena. Both Henry Sr. and Davidson thought William would be interested in Alice, but William credited the latter with the introduction.[5] Davidson, an idealist strongly drawn to religion, later became enamoured with an Italian monastic order. He was also familiar with Keene Valley and later established Glenmore, a philosophical summer school, several miles from the Shanty. James had special feelings for Davidson, not only because of the introduction to Alice, but because the latter's metaphysical skill challenged his own. They discussed the cosmos summer after summer at Keene Valley and Glenmore.[6]

The courtship was troubled by more than the usual ups and downs of romance. Alice's family had suffered a terrible domestic tragedy. Her father, Dr. Daniel Gibbens, had tried unsuccessfully to settle his Bostonian family in California in the mid-1850s. He had purchased land near Santa Barbara, but was pushed out by an unscrupulous local grower.[7] During the Civil War he had been an army physician and afterwards had taken an administrative position with the Union occupation army at New Orleans. His family expected him back in Boston for Christmas 1865. Instead they received a telegram that Mr. Gibbens' body was being returned from Mobile, Alabama. He had apparently shot himself. Elizabeth Webb Gibbens, Alice's mother, was devastated, and Alice, the eldest of three daughters, shouldered many family responsibilities. Daniel Gibbens left a small inheritance and the family, after hearing accounts of cheap room and board in Germany, left Boston for the continent in 1868 (Alice could not know that her path nearly crossed William's as he took the baths at Teplitz and pondered sculpture at Dresden). By 1873, their money depleted, the Gibbenses returned to Boston, where Alice and a sister were employed as teachers in Miss Sanborn's School for Girls.[8]

Even though eleven years had passed since her father's suicide, Alice still bore the scars of disrupted family life. And although she was attracted to the thirty-four-year-old Harvard professor who came from a socially prominent family, there was no rush to the altar. Although none of the surviving letters indicate that William told Alice that he had himself contemplated suicide, their heavy tone makes it probable that at some point he did. Moreover, despite the James family's social standing and financial resources—which must have seemed considerable—Alice could not have helped seeing that his family suffered substantial maladies. Henry Sr. could be passed off as merely eccentric, but what of Henry Jr. with his chronic back problem, Robertson's alcoholism, and the reclusive, sickly Alice? Was this a family she wanted to join?

William had thought himself unfit for matrimony. In 1869 he told Robertson, who had recently become engaged, that "I account it as a true crime against humanity for anyone to run the risk of generating unhealthy offspring." In his own case he concluded that "I myself have long since fully determined never to marry with anyone, were she healthy as Venus of Milo, for this dorsal trouble is evidently something in the blood. I confess that the flesh is weak and passion will overthrow strong reasons, and I may fail in keeping such a resolve; but I *mean* not to fail." He felt that he had a moral obligation not to marry, one strengthened by the prevailing Victorian fear of hereditary degeneracy: "I want to feel on my death bed when I look back that whatever evil I was born with I kept to myself, and did so much toward extinguishing it from the world."[9]

Picking a marriage partner was no casual matter in the 1870s. Young men learned that bachelor sexual habits influenced the health of future children. Manuals warned that self-abuse led to impotency, lost manhood, and degenerative offspring. Phrenology was a common "scientific" guide for choosing a partner on the basis of the brain's "soul chambers." Physicians urged young men to avoid women with nervous temperaments, or decidedly intellectual interests. There was great cultural pressure to make the appropriate choice.[10]

William had been attracted to sensitive, intellectually inclined, even nervous young ladies. Minny Temple's untimely death had touched him deeply; he had developed a friendship with Kate Havens, who suffered from neurasthenia; and he had seen his own sister slip deeper and deeper into nervous collapse. These were precisely the kind of women a wise husband-to-be should avoid. Most of all he understood that lapses into melancholia, nervous fatigue, and his chronic back ailment had compromised his own fitness, and probably relegated him to the life of a bachelor. Although somewhat improved by the mid-seventies, William still doubted that there would be a wife and family in his future.

But that was before he had met Alice, an event that quickly replaced reason with passion, as William had predicted might happen. In the summer of 1876 he wrote to her from Keene Valley, displaying his heart on his sleeve: "To state abruptly the whole matter! I am in love *und zwar* (forgive me) with yourself. My duty in my own mind is clear. It is to win your hand, if I can." And he asked Alice if there was any reason for not pursuing the courtship. "What I beg of you now is that you should let me know categorically whether any absolute irrevocable obstacle already exists to that consummation." Perhaps anticipating rejection, he explained, "You may ask what I gain by writing now, since I propose no prompt settlement. I seem to myself to gain everything, time, clearness, and what to me under my present unlucky cerebral circumstances, is all important, a definite plan of action."[11] From the beginning James let Alice know that should she agree to marriage, it would be to a man suffering from protracted mental distress.

For Alice these sudden intentions scrawled on a postcard from a man she barely knew, penned in a remote mountain camp in upstate New York, must have been unsettling. James, however, anticipated her unease. "Sooner or later you should have to know the state of the case between us, and if it shock you now, you will feel in a week or two that it is better open than concealed. If the important thing is that you should begin to know more of me, surely *this* advances that end!"[12]

It was portentious that James should announce both his love and ill health by letter. Over the thirty-two years of their married life he wrote Alice well over fourteen hundred letters and postcards from places as distant from Cambridge as California and Germany.[13] He courted Alice by mail throughout their married life. Of course in the nineteenth century the educated classes were accomplished and frequent correspondents. Nonetheless, unlike his own father, Henry Sr., William was away from his wife a good deal, and at such seemingly awkward times as during the infancy of their first child, Henry. Through writing to her James seemed to renew his romantic enthusiasm for Alice, a mood he may have had difficulty sustaining amid everyday domestic cares and confusions.[14] In these missives, often pervaded with domestic small talk, he included important commentary on mood, health, and intellectual interests. And even as he courted Alice he launched into a serious discussion of the institution he so summarily suggested they might enter.

James wanted marriage placed within a larger structure of belief. Defining the grounds of marriage was not simply an attempt to arrange a satisfactory relationship. Union with Alice would be the ideal center of a broader understanding of social obligations. Their liaison was at the heart of William's conception of the moral life. His creative future embraced a moral burden, one that Alice would have to share.

IN THE summer of 1876 James explained in a long letter to Alice the kind of marriage he envisioned. "There is a natural 'plane' of life whose prosperous results justify themselves and therefore refer to no end outside of themselves," he began. "It is the open, healthy, powerful, normal life of the world, salubrious, unconscious—the only one that can be in any way *officially* recognized by society." The healthy world was universal and "all departures from the standard of wholesomeness it establishes, are crimes against its law. . . . I for one have immense piety towards it." To enter into matrimony without being a part of this great state of healthfulness imposed special conditions. "The marriage of unhealthy persons, can only be forgiven by an appeal to some metaphysical world 'behind the veil' whose life such events may be supposed to feed." Yet if one or both partners were unhealthy and justified their union on high metaphysical ground, they must realize there would be no conventional approval. "But to the healthy natural sense such a world is problematic [and] whoever falls back on it for justification of what is outwardly a social crime, does so at his own peril, and should expect no countenance, perhaps even no tolerance from public opinion." There was but one inevitable result that must be accepted by both partners. "To wander from the outward order that keeps the world so sweet is in a word to throw yourself upon the *Tragical*."[15]

If James' announcement of amorous intentions had shocked Alice, his thoughts on the implication of their potential union must have made her deeply uneasy. For William was saying that should they join, they could never, because of his ill health, have a normal, happy, healthy marriage. He divided society into two—a large conventional sector that unreflectively went its normal, healthy way; and a small self-conscious sector, bent on expressing higher principles and resigned to tragedy. This distinction between what might be called the worlds of the healthy and the sick found an important place in James' published writings. Over twenty years later, in *The Varieties of Religious Experience,* James would bifurcate society into the "healthy-minded" and "sick-minded," taking special pains to illustrate the specialness and suffering of the latter.

And in *Pragmatism* (1907), he identified two basic approaches to reality, "tough-minded" and "tender-minded"; those who embraced the former were mundane and better equipped to handle everyday life, while those who embraced the latter were more sensitive, but less able to withstand life's hardships.[16]

William was only being honest. For over a decade he had suffered ailments ranging from backache to eyestrain, fagged nerves, and melancholia. Alice needed to understand that marriage to a thirty-four-year-old neurasthenic who was prone to depression would not, as the years progressed, produce normalcy. Yet he also attempted to structure the emotional basis for their marriage, to chart their future relationship, to rationalize certain domestic behavior. It would be risky to contend that James was trying to manipulate the bride-to-be into confusing the neurotic with the tragic, thereby freeing himself from normal, "healthy" responsibilities, while persuading Alice to resign herself to accepting aberration. William also embraced the tragic view of life; he never maintained that ill health excused one from moral responsibility. Nonetheless, James was delineating marriage roles, the compass of matrimonial expectation.

He explained that the "spiritual ends" of a tragic marriage "may run *across* the ends of nature, that its laws may defy those of the outward order—in other words that the deepest meaning of things may be revealed only in the *Tragical.*" He would sacrifice the natural and normal for nonmaterial ideals: "To abandon my private spiritual advancement for the sake of obeying a natural average would be worse than a *crime* against nature—it is a *sin* against the holy ghost—metaphorically speaking. I say, I stand by to matrimony, and I feel that it is . . . well and good for me."[17]

There would be no double standard. The healthy, normal Alice would be burdened as well. "But if the other person thru heedlessness, superficiality, or insensibility, slide into the natural crime, without the imperative spiritual need which alone can make it be forgiven, what thanks has she? She falls prey to the 'Tragical,' this time without atonement." James did not want a marriage in which only he sacrificed for lofty pursuits while his wife attended to domesticity. She as well must bear the idealistic cross. "Thus she must not take me unless she find it spiritually laid upon her as a tragic duty to do so." And if Alice refused the matrimonial burden, he had the perfect rationalization: "If she does not find it so and spurns me, I have the refuge of thinking that I suffer for the sake of the law of nature, which is sacred too."[18]

James added in a postscript that there was "one thing I fear may be misunderstood." The plea for a tragic marriage was not a spur-of-the-moment strategy, something concocted to influence her decision. It was not the first time he had "admitted . . . the righteousness of stepping out of what I've called natural law. I have always admitted it in principle, while firmly believing that none but very exceptional circumstances would permit it in practice." The "inward conflict" between the natural and spiritual "related merely to the decision whether *this* was for me such a critical case. In deciding as I did I cast my lot with the 'tragic' instead of the 'normal' side of life." These were his terms, this was his stand: "And now, most indulgent Miss! I have unpacked myself, and will unless interrogated forever after hold my peace."[19]

In the absence of her letters to William, how Alice reacted to all this can only be guessed. It is clear that she did not immediately agree to his proposal for a "tragical" marriage. Rather than refusing, Alice probably postponed, thus encouraging William to take a stoic stance. But the suitor was still miserable. "Love may be a disease, it may mean nothing or worse than nothing, but it shall not poison me. I shall save my soul. I will feed on death and the negation of me in one place shall be the affirmation of me in a better," he wrote to her that fall.[20] This sounds much like a paraphrase of his climactic diary passage of 1870—"My first act of free will shall be to believe in free will."[21] If Alice did not return his love, then the loss would turn him toward a higher cause; out of suffering would come a better self. "This is your undivided work, and you ought to feel happy for it. No other experience of my life has ever shadowed forth an idea of it."[22] Unrequited love made James squirm; the situation had to be quickly resolved—one way or the other.

By spring 1877 Alice still had not made a decision. William's high-minded talk about marriage as tragedy, in truth a condescending approach to a less seasoned mind, diminished. He became at once more demanding and more dependent: "To have you recognize me, to have your truth acquiesce in my better self from hence forward [is] the only possible goal of my conscious life." He pleaded, "You *will* do it, you will value me, care for me, so long as I know my place and keep it, for then what there is of me is *real* and it is all yours."[23] On the one hand his fate was in her hands, but on the other hand he continued to indulge suffering and express martyrdom. "What is the use of religion unless it enables us to *postpone* solutions to Eternity? Until then that you neither doubt nor judge me, is the sole hope of W. J."[24] He loved Alice and was miserable in this state of indecision. But he steadfastly refused any union not based on mutual sacrifice to higher spiritual and metaphysical principles.

At the end of April an understanding was reached. William was "simply shocked at the thought of a being like you countenancing one like me. You *ought* not to be willing to do it."[25] She had finally recognized that the "real" William was the unhappy, long suffering, unhealthy man who out of love and deference to a higher destiny was willing to commit the "social crime" of matrimony. She came to realize what a fateful choice she must make. If she accepted James' terms she could never hope for normal domesticity; if she rejected them she doomed him to lonely self-abnegation.

Alice had realized that she could save him from himself, from morbidity, depression, and the darkness that had threatened to destroy his will to live. He told her that a great weight had been lifted from him and, he hoped, from Alice as well. "Dearest friend, now that you *recognize* me, the horrible lonesome pang is removed from my existence and I feel as if I could happily look upon you, whatever you decided to do. So do respect *yourself* now. You can find no profounder way of showing respect to your Wm. J."[26]

Strangely, however, after Alice agreed to marry him, William began to act as if he feared their union. When the 1877 winter term ended, he decided to take a Newport vacation—alone. "I may find it best to go away this week without even seeing you," he wrote, explaining that "for your sake as well as mine now it behooves me to be *hygenic* before everything. I am still used up with wakefulness, and I must at any cost get gradually back to good habits in that respect by fleeing every disturbance."[27]

Paradoxically, the great happiness that Alice's presence provided also produced overstimulation and insomnia: "Strange that my peaceful rides with you should disturb me, but they do so really through an excess of the peaceful feeling they leave behind them." Alice was thereby introduced to what would become habitual throughout their marriage. At the end of the academic year, he usually went alone to Keene Valley, Europe, or some place distant from Cambridge to restore his nervous health. But how bewildering it must have been when he wrote that spring, "Don't think my own Alice, that it is doubts like those I told you last Sunday, which trouble me now. Never have I felt so intimately near to you as during the past couple of wakeful hours."[28] And yet he had left her. She began to learn, even before their wedding, that distance and exotic places were essential for James, for his health, and for the vitality of their relationship.

Alice would also learn that his neurasthenic complaints and depression were far more than typical Victorian dissatisfactions, for they were intertwined with his metaphysical preoccupations—with his very intellect.

In early June 1877, James sent Alice an extract from an 1873 metaphysical notebook. It was written, he explained, "many years ago when I was going thru the pessimistic crisis."[29] He sent the notebook extract partly to explain why it was that he needed to both possess and flee from her. But in the midst of attempting to sort out the rationale for his seemingly strange behavior, James also outlined the crux of his protracted intellectual dilemma. Moreover, he anticipated fundamental Jamesian concepts—the sanctity of choice, the predominance of temperament in philosophical creeds, the necessity of chance in a worthwhile universe. His relationship to Alice, its depth, pathos, and passion, encouraged creative expression. Significantly, it was not until their courtship that William began to prepare substantial psychological and metaphysical essays for scholarly publication.[30]

Sometime in 1873 William had gone to a circus in Boston. It is fitting that the notebook extract begins with images from the natural world, with the "sight of elephants and tigers at Barnum's menagerie." He had just begun teaching physiology at Harvard and was especially attuned then to thinking about the heirarchy of nature. Perhaps he even remembered visiting Barnum's as a child in New York. Moreover, when he sent Alice the extract, he was at work on an article published a year later as "Brute and Human Intellect," an essay which explored the reasoning distinction between humans and dogs.[31] The natural world as represented by circus beasts became the occasion to introduce a persistent metaphysical problem, one that occupied James for decades. The elephants and tigers were "so individual and peculiar, yet stand there so intensely and vividly real, as much so as one's own, so that one feels again poignantly the unfathomableness of ontology, supposing ontology to be at all." Though he and the animals shared a natural reality, there was a chasm between them. "I can never hope to *sympathize* in a genuine sense of the word with their being."[32]

Subjective feeling for the beasts and utter separation from their reality troubled him. "Their foreignness confounds one's pretension to comprehend the world—while their admirableness undermines the . . . moral frame of mind in which one says the real meaning of life is *my* action. This great world of life in no relation with my action is so real!" How to cope with this great ontological gulf? For William the question was not simply an interesting abstraction; it became an obsessive concern—this world with no *relation* to his being. Most people, he had concluded, "consider a part of the not me as kindred to the me in quality [and] a part as foreign or hostile." And when the "not me" became "sufficiently great or powerful to threaten them or the part in which their interest lies, their attitude becomes one of true dignity." Many "become militants, cleaving to their good they defy the contradictory element in its behalf, refuse to compromise, truckle or connive, and when the situation becomes strained enough, are ready to be martyrs" (1873 Notebook excerpt). Objective reality—the "not me"—could be interpreted as an affront to personal values, and in opposition one could embrace various religious and political creeds. Indeed, there were important variations in the way people responded to this incredible and profoundly frightening separation between self and the world. By defining how others approached the chasm he had sought his own solution.

"For the devout believer," the extract he sent Alice explains, "the details of things are . . . hostile, the sum total he believes to be akin to his inner nature"; nature, the cosmos, would have an overriding unity regardless of the separateness of individual facts. On the other hand, "the modern empiricist and the pessimistic idealist"—by the mid-seventies they had for James become mostly Darwinists—were "like leopards, all the kinship is in details, the total is foreign." But neither approach to duality satisfied William. The devout believer could only find "a lofty and noble sort of conception of the world" by "neglecting the shabbiness of details, and constancy of failures and accidents, etc." All the devout believer needed to do to construct unity was to pick the materials for "such a one [and] string them properly." All the rest was "accident." And the empiricist and pessimist, what James called "the other kind of religious man," simply emphasized "the perishability of all things, etc., etc, and says they are the key to the essence, I will string philosophy through them—and he acts as legitimately as the former man (the devout believer)" (1873 Notebook excerpt).

The large truth for James about both the religious and empirical solution to dualism, to the subjective-objective chasm, was that both were contrived. "Only those elements enter into our experience which we agree to attend to—the rest are ignored. Scientific discovery is often nothing but the recognition of a fact that long stood knocking at the door of our senses, but was ignored. Once attended to, it is seen to be an indispensible part of experience, of the world" (1873 Notebook excerpt). These translations of reality, thus exposed, were more choice than necessity, more concoction than truth, and hence unconvincing.

He told Alice in 1877 what he had told himself in 1873; namely, that the two great intellectual traditions in the western world, one religious and idealistic, the other scientific and empirical, did not explain the great gulf between self and world—did not, in short, explain reality. Regardless of whether one colored the world optimistically or pessimistically, idealistically or empirically, these world pictures were never disassociated from the private interests and sympathies of the individual. That meant that there could be a third view; indeed, theoretically there were as many views as there were persons. James was free to construct a philosophy that satisfied his own sympathies. Personal temperament undergirded and directed how one perceived reality. In 1879 he would give published expression to what he called "the sentiment of rationality."[33]

In revealing to Alice what his philosophical alternative had been in 1873, James also showed that the plea for a tragical marriage was firmly based in a moral principle. He had not solved the great problem of dualism, the unbridgeable distance between the "me" and the "not me." And he had appealed to the stoic tradition for a precedent for coping with the troubling chasm. "But if as in Homer, a divided Universe be a conception possible for his intellect to rest in, and at the same time he have vigor of will enough to look the Universal Death in the face without blinking, he can lead the life of moralism," he had written. From the notebook Alice learned that for at least four years William had lived with a heroic, tragic world picture, one that confronted dualism and "Universal Death" with an active, living moral commitment—"a militant existence, in which the ego is posited as a monad with the *good* as its end, and the final consolation only that of irreconcilable hatred—tho Evil slay me, She can't subdue me, or make me worship her. The brute force is all at her command, but the final protest of my soul as she squeezes me out of existence gives me still in a certain sense the superiority" (1873 Notebook excerpt).

Whatever vicissitudes lay ahead in married life, Alice now knew that "such people can laugh at fate, are flexible, sympathize with the free flow of things, believe ever in the Good but are willing that it should shift its form." In the light of William's recent flight to Newport, his plea for conflating moral tenacity with flexible freedom had special relevance. Individuals with these heroic qualities, his letter urged, "do not close their hand on possessions. When they profess a willingness that certain persons should be free they mean it not as most of us do, with a mental reservation, as that the freedom should be well employed and other similar humbug, but in all sincerity, and calling for no guarantee against abuse, which when it happens they accept without complaint or embitterment as part of the chances of the game. They let their bird fly with no string tied to its leg."[34]

And the years ahead proved that William would certainly fly, often leaving Alice to attend to domestic chores. For all of his metaphysical and moralistic sincerity, William was giving his wife-to-be a clear signal that an uncompromised priority of their "tragical marriage" would be his whim, a prerogative he exercised frequently.

But he evidently felt that Alice needed a still more vivid statement, a description of his deepest springs, the bedrock of his temperament. This same letter to her includes probably the best surviving expression of the core of his emotional life, a life that would drive him ever onward in ceaseless, restless mental activity:

> I have often thought that the best way to define a man's character, would be to seek out the particular mental or moral attitude, in which when it came upon him, he felt himself most deeply and intensely active and alive. At such moments there is a voice inside which speaks and says "this is the real me!" And afterwards considering the circumstances in which the man is placed, and noting how some of them are fitted to evoke this attitude whilst others do not call for it, an outside observer may be able to prophecy where the man may fail, where succeed, where be happy, and where miserable. Now as well as I can describe it, this characteristic attitude in me always involves an element of active tension, of holding my own as it were, and trusting outward things to perform their part so as to make it a full harmony, but without any guarantee that they will. Make it a guarantee—and the attitude immediately becomes to my consciousness stagnant and stingless. Take

> away the guarantee, and I feel . . . a sort of deep enthusiastic bliss, of bitter willingness to do and suffer anything which translates itself physically by a kind of stinging pain inside of my breast bone. (Don't smile at this—it is to me an essential element of the whole thing!) and which although it is a mere mood or emotion to which I can give no form in words, authenticates itself to me as the deepest principle of all active and theoretic determination, which I possess.[35]

Seven years later James would elaborate on the physical basis of emotion in his controversial essay "What is an Emotion?"[36] He now gave Alice prophetic insight into the remarkable, willful, and at times exasperating man with whom she would live and for whom she would sacrifice for over the next thirty-three years. She would witness frequent goodbyes that were often strained and tearful. The physical freedom was obviously more his than hers. For all of his the high-minded talk of tragical marriage, the tragedy seems to have been more Alice's than William's. And although it is true that William occasionally shared his intellectual life with her, she was always more a helpmate than an intellectual equal. Their matrimony owed as much to a quite conventional Victorian double standard as it did to James' special union of high moral and metaphysical ideals.

At great personal cost, Alice did indeed "recognize" William; she recognized that without tragical marriage, in all probability he would not be able to fulfill a great intellectual destiny. She took on the burden of his melancholia, a malady based in temperament and in his refusal to accept either a mystical or empirical resolution of the tragic split between the self and the world. William later acknowledged that marriage to Alice was the really effective therapy in pulling him out of his philosophical depression. After rereading letters to Alice penned during their courtship, James wrote that "a stranger, reading those letters . . . would think of a man morally utterly diseased. That you, having read them, should still have been willing to risk things with me, seems to me now most surprising." There was no philosophical turning point, no great book, no Renouvier that saved him from himself. "I have got out of that whole frame of mind (moral disease) not by any acute change or act of discovery, but simply gradually and by living with you." Rereading the letters was "a strange real solid experience, that enables me to see what a newly made over man I am. They bring back all the *real suffering* of that time." And he lamented what posterity can only underscore: "How vile a thing it was that I destroyed the early lot of yours! I would give anything to have them now."[37]

CHAPTER EIGHT

Creating the Realm of Consciousness

I confess that in the past few years . . . I have felt most acutely the difficulties of understanding either the brain without the mind or the mind without the brain. I have almost concluded in my moments of depression that we know hardly anything.

From Notes for the Baltimore and Lowell Lectures
1878

BETWEEN 1868 and 1878 James turned failure into success, despair into optimism, and illness into health. Here emerges the central James legend: the story of before and after, of faith overcoming doubt, of action eclipsing inertia. Nor was his travail, for all its intellectual power, far removed from the great American self-help ethic that trumpeted the adage, "God helps those who help themselves." Despite his genuine humanity and remarkable intelligence, James was nonetheless a splendid New World example of self-centeredness cloaked in moralism. The legend, however, is largely the creation of James' interpreters, who have told over and over his ascent from depression to optimism, from determinism to free will, from inertia to action. There is, as we have seen, cause for such an emphasis. But within James' own mind something perhaps even more essential was happening.

In terms of the broader prevailing intellectual context, James is generally interpreted in relation to the ideas to which he responded—Darwinism, empiricism, idealism, pessimism.[1] There is no question, for instance, that he appreciated the revolutionary importance of Darwin to science, to philosophy, and even to popular values.[2] But by the late 1870s William had embarked upon an intellectual campaign to establish a unique psychological and philosophical point of view, one which stood on its own foundation regardless of the other standpoints in vogue. In this sense, Darwinism was instrumental rather than essential to James. It did not dictate the parameters of his thinking any more than did British empiricism or German idealism, except as an intellectual medium his originality chose to mold. Genius must manipulate; and James began to manipulate ideas with remarkable success.

In 1878 his "Remarks on Spencer's Definition of Mind as Correspondence" and "Brute and Human Intellect" were published in the *Journal of Speculative Philosophy.* These were his first major scholarly essays, and they marked his public emergence as a major thinker. Both articles challenged Herbert Spencer's notion of mind as an evolutionary product consisting of associative mental habits, to mind as a faculty that responded to environmental stimuli in predetermined patterns. James posited an alternative: an interested, disassociative mind willfully selecting rather than simply reacting to environmental stimuli. Consciousness attended to its own interests; an active, choice-seeking consciousness was as much a product of evolution as determined mental habits. He restated his case for an interested consciousness the next year in *Mind,* in "Are We Automata?"—portions of which were eventually incorporated, along with portions of "Brute and Human Intellect," into *The Principles of Psychology.*[3] These essays established James' reputation as Spencer's major American opponent. They cast him in the role of a psychologist-philosopher who defended the mind's creative potential against a determined destiny. They outlined foundational positions on mental functioning that James spent the next thirty years elaborating.[4]

James had been inquiring into the nature of the mind throughout the early and mid-1870s. Even before taking a teaching position in January 1873 he had been exploring the scientific terrain in which his psychology would eventually emerge. In 1871 he took a university course of lectures titled "Optical Phenomena and the Eye." By 1872 he was using Henry P. Bowditch's laboratory on Grove Street in Boston for physiological experiments—although we do not know exactly what kind. William also attended Bowditch's lectures at the Medical School and engaged Bowditch and another former Harvard Medical School student, James Jackson Putnam, in discussion on the brain and nervous system. He wrote to Henry Jr., who had gone to Europe with Alice and Aunt Kate in the fall of 1872, that he liked "paddl[ling] round in [Bowditch's] laboratory. It is a noble thing for one's spirits to have some responsible work to do. I enjoy my revived physiological reading greatly."[5]

In January 1873, James began teaching physiology to some forty-five Harvard undergraduates. By the fall of 1874, he was offering "Natural History 3," a course on the comparative anatomy and physiology of vertebrates. After Jeffries Wyman's death in September 1874, James was put in charge of Wyman's Museum of Comparative Anatomy. One year later he offered a graduate course that dealt with the relations between physiology and psychology and that included experiments earlier carried out in Bowditch's laboratory. In 1876 he set up the first American laboratory to explore relations between physiology and psychology. As his first doctoral student in psychology, G. Stanley Hall recalled, it was a modest beginning: "In a tiny room under the stairway of the Agassiz Museum [James] had a metronome, a device for whirling a frog, a horopter chart and one or two bits of apparatus."[6]

Between 1870 and 1876 William concentrated—with time off for a European trip to Italy between August 1873 and April 1874—on the study of physiology. He inquired into the structure and functions of the lower and higher brain. And he was especially interested in the interconnections between the brain and consciousness in animals and humans. His creative field at this time was brain physiology, an area he related to psychology. He considered Darwin and Spencer's ideas important and relevant, but did not graft his system onto theirs; indeed he considered Darwinism mistaken in its approach to psychology.

Of course Darwin had provided the great scientific thunderbolt of the late nineteenth century. But others had also been at work undermining the long-standing belief that the mind and body were independent. Physiological psychology was developing rapidly in Germany, where through laboratory measurement men such as Herman Helmholtz, Gustav Fechner, Ernst Weber, and Wilhelm Wundt were engaged in distinguishing simple sensation from thinking. "Reaction-time psychology" seemed to close the imagined gap between mind and body, just as evolutionary theory had narrowed the distinction between men and animals.[7] Into this fermenting science that placed great pressure on prevailing belief in the sanctity of the human soul plunged William James. Like Darwin, he advocated a perspective whose implications carried well beyond the properties indigenous to his science. During the late 1870s his physiological psychology spoke with a brilliant authenticity that began to set him apart, just as he had hoped it would.

THERE IS a striking example of James' originality in science before 1880 that has never been published: the well over one hundred pages of his hand-written notes for the 1878 Baltimore and Lowell Lectures.[8] William delivered the lectures in February and November 1878, first at Johns Hopkins in Baltimore and then at the Lowell Institute in Boston. There were six Lowell Lectures on "The Brain and the Mind," but the list of individual presentations was never published, nor were the lectures themselves. The Baltimore occasion came five months before his marriage to Alice Howe Gibbens. Anticipating extra expenses, James wanted to augment his Harvard salary while building a professional reputation. Johns Hopkins' president, Daniel Gilman, may have proposed the lectures as a lure to bring James to the Baltimore school as professor of psychology.[9]

The Lowell Lectures of 1878 were the first of three Lowell presentations James would eventually deliver.[10] The Institute had been serving as a public forum since 1836 under the auspices of the Lowells, one of Boston's oldest and wealthest families. James' former mentor, Louis Agassiz, had lectured there, as had the elder and the younger Holmes. So would philosophical colleagues Charles Peirce and Josiah Royce. Such was the Institute's prestige that President Theodore Roosevelt held forth there as well. The Lowell audiences represented the educated, upper, and middle classes of Cambridge and Boston, the area's intelligentsia. If James were well received, his up-and-coming reputation would be given a boost.[11]

James' notes suggest that he approached his audience as a concerned citizen who wanted to separate truth from fiction, science from pseudoscience, fact from hearsay. "One is obliged," they begin, "to form *responsible* opinions concerning a variety of matters about which irresponsible opinions are very frequently and loudly expressed." Contemporary science was alarmingly and untraditionally ambitious: "Now physics and chemistry are running together, whilst geology, zoology, astronomy and human history all seem to be coalescing into a vast system called the theory of evolution, whose aims whatsoever they may be are not modest, and to whose results no laboratory verification but only the large vague circumstantial evidence can be applied." And he challenges any so-called science that posited "a vast and fundamental theory—the theory that all the feelings of which we are directly conscious are built up of unconscious units. The theory in fact that mind as well as matter has a molecular composition."

The materialist-thinking evolutionists had exposed the mind to dangerous uncertainty. "Hitherto it has been generally supposed that however phenomenal outer things might be, in our feelings, at any rate, we know the thing in itself." For example, "When we were joyful the joy *was* just what we felt it to [be]." Or if "a man tells you he is cold, cold he is." James focuses on what he must have imagined was his audience's most basic assumption about reality—the truth of personal emotional or physical feelings: "To say now that pain is not really what it seems but is really composed of other smaller feelings that we know nothing of, seems to reduce all things to chaos again."

How had it happened that feelings were no longer considered what they seemed to be, but rather were seen as minuscule bits and pieces of something akin to impersonal matter? "Certain philosophers analyzing our finished thought about *things* come to the conclusion that it is wholly made up of sensations. They are the bricks whilst space and time are the mortar that join them together into unity." Philosophers of this school spoke of sensations as "perfectly definite bricks, absolute and distinct packages. This comes of starting from the study of objects." Although James does not mention names, the empiricism of Locke, Hume, and Mill is implicated; the British treated the mind as an object and had hence misconceived its true nature.

Nineteenth-century physiology, however, had rejected this view of sensation as perfectly definite bricks—that is, as defined, discrete bits of sensation. James notes that "the great lesson of H[elmholz] is that a sensation is the most fluctuating and indefinite of mental occurrences." He explains that a sensation "is whole plastic in the hands of our thoughts of things." Sensations were apt to be "ambiguous" and "enter into many objects, . . . according to the thought of the object well after its felt quality." The mind could make a sensation a mere sign, which it in turn transformed into a meaning: "The sensation merely wakens the mind from its slumber, and when the mind puts forth its power it may do so in the way of overriding, altering or ignoring altogether the sensation."

The new scientific physiology was not a reductive materialism. James did not want his audience to confuse Helmholtz, Fechner, Donders, and Wundt with Spencer and Darwin. He was using German science to countervail the reductive materialism of British empiricism. He did not want to simply build his psychology on the base of British empiricism; it was a foundation that he would, with German assistance, radically reconstruct. Moreover, in his lecture notes James links the new physiology to Kantian idealism:

> I merely wish to call the attention of persons who are fearful of the reduction by physiology of higher and lower powers of the mind [,] that H[elmholtz] and W[undt], put forth a doctrine . . . which has been hailed by the followers of Kant as the most striking experimental verification ever made of a doctrine originally deduced a priori, the doctrine of innate potentialities in the mind which sensations merely awaken into exercise.

To those that still revered the transcendentalism of Emerson, the allusion to Kant would be reassuring.[12] But to directly connect what was viewed by many as materialistic physiology to the loftier philosophic tradition was to risk much. Either he would have to "prove" the connection, or risk being as irresponsible as the materialist evolutionists he attacked.

James turns next to Helmholtz and optics to specifically illustrate the relationship between sensation and the mind. He subscribed to the latter's proofs that double vision is less a matter of sensation than of perception.[13] The mind, for example, had more to do with the production of double vision than had the retinal image. There were limits to "the power of thought to make a double image single"; but "these limits seemed traced by the habitual experience of the race." The critical point was that the "third dimension [perspective] in general is a retinal quale [feeling], like the sky—a particular determination of position therein can only be called a *thought* of the objective reality, which here as everywhere has the *power* of altering the sensation." "We must" he notes, "give mentality the proper perspective" to insure that "images are seen single." The mind initiated the field of vision, not sensation. In this sense the mind controlled sight. Retinal sensation in the eye simply provided information. James would elaborate on "the spatial quale" in a major article the following year, as well as in the longest chapter in *The Principles of Psychology,* "The Perception of Space."[14]

Not only was sight not sensation, the mind was not the sum of sensations. To reduce mind to sensation led to the philosophical skepticism that James and, he trusted, his audience wanted to avoid. But complex question of the origin of sight leads James further along the speculative path. Vision was a matter of perception, and perception belonged to the domain not only of psychology but of philosophy as well.[15]

James' critique of the sensationalist understanding of sight was an important preliminary for the main event, a discussion of the brain and its relationship to consciousness. But to be objective was difficult. In his notes he vividly depicts an ideologically charged, emotionally volatile scientific environment, one of which the public not only took notice, but took sides:

> In these recent days we hear a great deal of the marvellous achievements of science, how Darwinism has made us understand so much about animal and vegetable forms, and how in particular the physiologists by the deep insight they have been acquiring into the nervous system and brain, have to a great extent banished the mystery which used to hang about the action of the mind and constituted a new psychology which explodes and renders obsolete the old views of metal action all based on a priori speculation. Whilst this is triumphantly repeated by the sectaries of physical science, it is as indignantly denied by another class of persons. The latter fancy that they see the most brutal materialism lurking behind what the former call enlightenment and scientific progress, like some hideous heathen idol whose form is dimly seen through the glare of fireworks and golden dust and dazzling vapours of incense with which its followers continually fill the air before it.

Darwinism had generated a great new division of opinion, in which "both sides are confident and often bitter." "Each one of us," James observes, "probably is to some extent a partizan in the matter." And "the worst of it is that in this matter of the brain and the mind people are ready to become very eager partizans on a very slender basis of study." Both camps were superficial: "Those whose highest flights are articles in the *Popular Science Monthly* will talk of the exploded superstitions of introspective psychology, and those who have hardly opened a treatise in physiology will declaim against the degrading sophistries of medical materialists."

To check these unschooled reactions, James proposes to lecture six hours on "a single subject—the brain, and . . . see exactly how much recent investigations have explained its action and in particular how much they may be said to have cleared up or made less mysterious the action of consciousness in each one of us." He hopefully proposes that "a sober review of this kind ought to do good to the over hasty partizans of both sides" since this approach "may perhaps make the scientific ones feel themselves prematurely sanguine, and the others feel over jealous and timid."

This was an admirable objective. But James could barely claim established scientific credentials, let alone a renowned reputation. Only one scholarly article of his had yet appeared, and he was still an assistant professor. Moreover, he had not really specialized in physiology or in any other branch of science, or, in truth, in anything at all. Masterfully, he turns his lack of reputation in any one science into an opportunity to speak authoritatively for several. "I have," he admits, "for some years past . . . been inclined to deplore the rather wide surface over which my instruction had to be spread. I have been obliged to teach a little anatomy, a little physiology, a little psychology; and I have felt that where one's wisdom tried to cover so much ground it must needs be thin at any given spot."

Spreading oneself thin academically, however, avoided a great peril in specialization: "A physiologist may form the most careless opinions in psychology and keep a good conscience. The only thoroughness obligatory on him is physiological." The same was true with other disciplines: "The philosopher . . . who follows the subjective method may in like manner [ignore] the erudities of the physiologist. He does not feel the dignity and the difficulty of their problems any more than they do his."

A transdisciplinary education made James especially aware of the intellectual isolation inherent in specialization. He illustrates the point with an analogy bourgeoise Bostonians could well appreciate. "Each [discipline] . . . wishes to shove back the fence and reduce the size of his neighbors lot for the benefit of his own and the disputed boundary leads to the jealousies we have seen." But James' universality, his experiences in art, natural science, medicine, physiology, philosophy—indeed his "fields" or "lots" of creative experience—allowed him to ignore boundaries. "I think we ought tonight to aspire to the attitude of one who should own both lots. He does not care where the fence stands and being master of all the land tries to cultivate every square foot of it impartially."

In lesser hands this tactic might have earned him the label of an intellectual monopolist bent on appropriating several specific sciences for himself. He quickly shows, however, that everyone has an intimate stake in breaking down false scientific boundaries: "We are each alike proprietors of a body and a mind. We are as much interested in having a sound science of one as of the other. As proprietors of a body we ought to feel the insufficiency of every theory of the mind which leaves the body out. As owners of a mind we ought to feel the worthlessness of all explanations of our own feelings which leave out that which is most essential to be explained." The body-mind field was a self-evident connection. Who could deny it?

His notes suggest that in his lectures James took his audience into his confidence and made them *feel* the seriousness of contemporary intellectual issues. Science was not a private matter. "A teacher must form responsible opinions; a reader need not. But the *difficulties of deciding a question* are not often felt by one who is not responsible for his decisions." Since 1873 he had taught science at Harvard. A teacher-lecturer had an additional burden. Rather than simply deciding, he needed to decide *rightly.* James did not simply want science, he wanted responsible science.

HE BEGAN with an anatomy lesson, a discussion of the physiological properties of nerves, spinal cord, and brain in lower and higher animals—in frogs, eels, dogs, and humans. Indeed, the opening chapters of *The Principles of Psychology* duplicate the general outline of the Lowell Lectures: first, a statement on the scope and context of his discussion; second, the anatomy of the brain; third, the mental role of habit formation.[16]

The mind functioned as a rational machine because "appropriate acts result from previous acts of intelligence so often repeated that lower centers have learned the trick." These acts impacted on the brain's lower centers and with repetition created a habit pattern. He gives examples: "Cases in human beings—music, stand, walk, ride, skate, read aloud, etc. Conscious first—then lower center stuff! Grow to modes." Even the simplest acts in animals and humans seemed learned: "Dogs, children, [and] deaf don't wink." Inherited instincts were important. But the question of instincts aside, "whether parents' acts leave [a] trace or not, man's *own* acts do. Every act modifies *and* makes easier."

The key to the smooth running mental machinery was well-established, unbroken habits. Exceptions, lapses, and delays impeded mental industry. "We think when tempted [we can] start afresh tomorrow as well as today." That, however, was "self-deception," for after the slip, one was "not the same man." Efficient mental habits insured proper behavior, a cultural necessity to his Victorian audience.

Though he emphasizes the importance of keeping habits unbroken, James does not reduce the mind to habit formation. The upper brain, or cerebrum—what William called "the hemispheres"—was free "for higher flights" once lower brain habits were in place. In fact, he remarks, "no habit [is] wholly performed by lower centers even in dogs." Always the brain's higher centers are in "hemispheric superintendence." But the hemispheres could free themselves from having to think when performing the mundane but necessary acts of walking, eating, or talking. Once routine mental habits were established, the hemispheres could indulge in creative activities, and the mind lost its "machine character."

The brain functioned in separate yet interrelated worlds—the world of automatic action carried out by the lower centers, and the world of spontaneous action taken over by the higher centers. These separate yet connected functions were derived from the distinctive anatomical structure of the nerves, spinal cord, medulla oblongata, and the cerebral hemispheres. In primitive organisms and in the lower centers in animals and humans, there was a "simple scheme of action: [a] direct loop line." That is, a stimulus always resulted in a reflex loop which left a trace, which scooped out a path. There were few sensations and few paths, and the lower centers were "slaves of present sensation"; they could not "deliberate, pause, postpone, weigh—prudence [is] impossible." In some life forms, being the slave of present sensations had utility. A polyp, for example, did not need prudence.

While a lack of high intelligence was most appropriate for polyps, such a deprivation would destroy the human species. The hemispheres provided spontaneous action, which implied choice and judgment—activities with obvious survivial benefits for mankind. Scholars have rightly emphasized James' defense of spontaneous intelligence.[17] But he never defended spontaneity in order to *eliminate* the concept of mental automation. Indeed, lower center habits were critical if the higher centers were to be freed for creative work.

James' understanding of how the brain functioned complemented the emerging American urban-industrial context. Factories and offices wanted workers with the automatic habits whose development he advocated. Yet the new routines—the exaggerated emphasis on professional standards, office efficiency, factory production, and the proper use of time—made spontaneity, choice, and freedom seem dearer.[18] The Darwinian view of consciousness did not balance habit and adventure. Indeed, as James had correctly seen, for Darwinists the mind was already built up, with nothing to do for itself. James' psychology was more culturally accommodating.

But important problems about the mind remained. It was now necessary to make "a psychological digression" so as to "make details of [the] sketch complete." How could the brain's higher center store something as fleeting and insubstantial as an idea? How did mental images revive themselves? James reminds himself to "remark on something analogous." There were reoccurring images "even in [the] organic world" such as "bruises and scars." And there were others, like the "violin, razor [presumably in reference to the beard and shaving], phosphorescence, mirror, clock at night, string round finger, and muscles at night." While these "quickly perish *here*, [they] last for years in hemispheres." Mental images became instantly permanent. Once an image was given over to the higher centers it was "imperishable." He concludes, "In hemispheres there must be separate pigeon-holes into which ideas, or several elements which compose them are switched off. If each acquisition is laid up in its own cells and fibres, no need to wipe slate." In the higher brain, "Impressions [are] not superposed—[but] lie side by side."

The brain was a sophisticated storage center that called up information as demanded. Experiments showed that the brain retrieved and retained complex images which drugs could obliterate. Administering an anesthetic, for example, "resulted in paralysis of volitional movements." And blocking the auditory and optic centers resulted in the loss of hearing and sight. Physiological psychology maintained that "each motor and sensory element has its home. When that [is] destroyed representation [is] wiped out; remaining representations [are] not enough to constitute thought." Localized brain functions accounted for various mental activities.

The brain as a storehouse of mental images, however, was not necessarily what it appeared to be. Physiologists, though seemingly believable, were not to be believed. James refuses to explain mental activity strictly though brain physiology. "Psychology the Torch" he notes. But he does not mean the conventional psychology, for there was "no faculty process resultant." And he quickly dismisses phrenology, although he recognizes its value as an art. Cryptically, he anticipates his own theory of the mind by observing that the "entire brain [is the] outlet of a funnel." What flowed out of the funnel was *not* necessarily dependent on the funnel itself.

James rejects a physiologically determined mind not as a partisan, but as a man of science, one who had sifted all available evidence. "Were there no conflicting evidence physiologists might fairly claim to have explained the succession of our thoughts by the changing stream of nervous action, as it shoots through the cortex awakening now one set of localities now another." Unfortunately, such an explanation was skewed, "I *regret* however to *dampen* the enthusiasm which in some of you the evidence brought forward may have aroused, by saying now that that evidence was picked." There were, for instance, "in *human* pathology innumerable cases . . . which contradict the notion of immutability of fixed localities for motor and sensory processes"—cases reported by his old medical school instructor Brown-Séquard and the English physiologist Hughlings Jackson. But if the seats for higher mental powers were not localized, "Are we then to say that we know nothing of the brain and everything is chaos!" "By no means!" he answers. Indeed, "This hopeless conclusion would be very wrong. It is not chaos. The matter is only more complicated than it first appeared."

Regrettably the next portion of the notes to the Lectures—some twenty-three pages—is missing. There remains only a tantalizing generalization: Immediately before the excised section, "There is a tendency for [mental] functions to *localize* in particular tracts, but it is like that of *water* to flow." This is particularly suggestive of James' later description of the "stream of thought," in which he would maintain that consciousness "is nothing jointed; it flows." Twelve years prior to the publication of *The Principles of Psychology,* James may have made one of his greatest discoveries: the notion that "within each personal consciousness, thought is sensibly continuous."[19]

WE NEXT find William reviving the earlier attack on the Darwinists, particularly John Tyndall and Thomas Henry Huxley, who argued that "the bodily event is the condition, the mental event the consequence" and believed that "what we esteem as highest is at the mercy of what we esteem as lowest and must ask its permission to exist." In the end what did these "illustrious" men say? "Nothing but this: that since frog's legs twitch and automatic habitual acts may be performed by men and yet outwardly appear rational, then when rational acts appear with consciousness, this consciousness is unconcerned with their production." To read the power of independent consciousness out of mental life was irresponsible; unfortunately, these were the "wild and whirling utterances" not of madmen but of respected men of science. As a result, "science will *lose* all credit which has distinguished her from metaphysics—[for] converting 'maybes' into 'mustbes.'"

By attacking the credibility of these illustrious men of science, James was building his own scientific authority. Moreover, he was implying that his system was culturally as well as evolutionarily more useful. He wanted to show that consciousness economized "nervous machinery." He is concerned in the lectures with how the conscious mind made the lower centers more efficient, and he utilizes the prevailing Helmholtzian conservation of energy theory, which assumed that there was a limited quantity of nervous energy. Efficient nerve function benefited from parsimony. In broad social terms, James' suggestion that the conscious mind could control nerve behavior touched the Victorian concern with nervous disease—with neurasthenia. There was a therapeutic message intertwined in James' science. His audience would appreciate a psychology that promised to preserve nervous energy, especially in an America ever more inclined to value efficiency not only in the work place, but for itself.[20]

When the nervous system was "made simple by deprivation of Hemis[pheres]," what remained "formed a machine remarkable for regularity and accuracy," one "responding to few stimuli but to them strongly and well." With the Hemispheres added, however, the animal was "capable of reacting on slight variations in environment," to "faint sounds, motions of points of field of vision." In the face of claims for an automated physical chain of causation, James notes that "these [slight variations] were responded to, not directly but *indirectly* through processes whose subjective side we call "considerations," and which are composed of complex and shifting vibrations in the hemis[pheres]." These shifting vibrations in the hemispheres corresponded to the physiological consistency of the brain's matter, which was like jelly. Rather than being predictably associative, the hemispheres were the "most instable of bodies."

Here was an indispensable point. Jamesian psychology depended on the instability of the brain's higher centers; "their very essential *function* necessitates instability. They are there to adapt the animal to minute variations in [a] complex environment." Know the hemispheric development and one could predict mental behavior, and vice-versa—almost as a natural law: "Where [the] animal responds to few circumstances [hemispheres] small. Where he reacts to many, to distant and future as well as present, like man, [hemispheres] large and all controlling." Such a large-hemisphered brain was potentially dangerous because "caprice" was its law; it was "happy go lucky, hit or miss." Is James here trading a fully determined brain for an erratic, uncontrollable one?

James quickly reemphasizes scientific responsibility. The brain's volatility could be checked, paradoxically, by its instable product—consciousness. "If consciousness can load the dice, can exert a constant pressure in the right direction, can feel what nerve processes are leading to the goal, can reinforce and strengthen these and at the same time inhibit those which threaten to lead astray, why consciousness will be of invaluable service." The ability to attend to its own best interests was in the *nature* of consciousness. "It feels the goal, it knows the interests, it is the sole stand of use. Apart from some C[onsciousness] staking, positing, creating some particular end as good, we could not talk about the brain being useful or efficient, its actions being appropriate at all." Consciousness made goodness possible.

If consciousness could pick and choose, could control its own interests, than *any* meaning derived from consciousness was quite arbitrary and open to doubt—even Darwinian meanings. "Mr. Darwin's C[onsciousness] looks on and says survival is the *summum bonum* and then measures the utility of every creatures organs by this end. Such a reaction of the N[ervous] S[ystem] is efficient, such an instinct appropriate, such a habit fit, etc. In short he decides which of the animal acts is worthy of approbation and encouragement as measured by the interest in survival." Yet this was only Darwin's consciousness *choosing* a particular explanation. One could say "survival or pleasure or what not is my end, and no action shall be called useful, no nervous tendency encouraged which fails to minister to this end." The great truth was not Darwin's decision to interpret nature as evolutionary survival; the greater revelation was that consciousness itself decided what was to be considered important and what was to be ignored; "I think if we survey the field of consciousness from lowest to highest, we shall find that it always seems to be *comparing* and *selecting*."

William's concluding and perhaps most forceful point is that Darwin's belief that men, like other animals, survived through adaptability was not the crucial discovery. Deciding that nature's objective purpose was survival was but a single example of the selecting power of consciousness. In a revealing passage, James notes that "Impartial C[onsciousness is a] nonentity. For consciousness . . . is busy *picking out* from data offered it by [a] lower stage, some one item, noticing, emphasizing attending to, pursuing it—ignoring all the rest." Jamesian consciousness created itself. What emerged was an active, aggressive, selective, essentially restless consciousness—the mind as busybody rather than spectator, a mind that selected its interests while attending to its own experience.

The view of the mind that James first presented in the Baltimore and Lowell Lectures of 1878 was clearly a prelude to several of his most important ideas. By criticizing Darwinian psychology he advanced an early criticism of late nineteenth-century deterministic thinking, one which he never tired of making, whether in science or philosophy. He articulated the vital relationship between brain physiology and mental activity in a way that implied that the latter was an evolutionarily acquired, functional ability to make choices that benefited the human species. Nature did not predetermine the mind to work in settled, predictable patterns, but had provided the wherewithall for self-interested choice-making.

Thus in 1878 James gave notice that his professional, scientific interests would be centered on the preservation and extension of the mind's freedom to act in its own interests. But the Baltimore and Lowell Lectures also provide a large hint toward a fuller understanding of William's intellectual biography. His emphasis on self-direction implies that the development of his own ideas had not been a matter of responding to this or that scientific theory, so much as of his individual consciousness busily selecting now this, now that. His mind *selected* his science. Not only did the Baltimore and Lowell audiences hear James attack Darwinian psychology; they were privy to an extraordinary personal statement. It was as if he were saying, "this is the way my consciousness works and therefore, creatively, this is what I am." Originality was the mind; and the mind had the responsibility of creating a true mental science rather than indulging in opinions or partisanship. In this sense James fashioned a psychology not through reaction to Darwin or Spencer, but through describing the mind working—*his* mind working.

James avoided developing a psychology that would shock or alienate. He not only attacked the Darwinian determinists, he eschewed the psychopathological topics that later interested him immensely, the gray areas between consciousness and unconsciousness, the subjects of the Lowell Lectures of 1896–1897. He built a theory of a culturally acceptable, normal consciousness before considering the vagaries of mental abnormality.[21] There was a conventional propriety within his genius that insured his popularity. And James took a giant step in 1878 toward making his new psychology seem both scientific and respectable.

CHAPTER NINE

A Wandering Piece of Property

I must frankly confess to you that I am more unsettled than I have been for years.

To Charles Renouvier
JUNE 1, 1880

I am nothing but your wandering piece of property, and my one sacred duty is to see that my journeyings do not impair me, but pass me back to your arms worth more than I was when I left them.

To Alice Gibbens James from Nuremberg
SEPTEMBER 14, 1882

ALMOST EXACTLY one year after announcing their engagement, Mr. and Mrs. William James became parents. William's mother reported the birth to his brother Robertson: "A son and heir was born to [William] on Sunday afternoon at 5 o'ck. The mother had a *very* suffering time all day, but has been doing well ever since. The boy is a fine perfect child weighing about 9 lbs, and so far promises to be a great comfort to his fond parents." William's fatherly innocence amused Mary; she felt that her son's naiveté resulted from his scientific perspective. "Will is very funny about the boy. He seems very surprised that he is a definitely formed human being, he seems to have had an idea that it would be a vague object at first, that would gradually develop itself into a human being. He pretends to take a wholly scientific view of the case. I presume he has been studying this infant . . . very intently, for he said this morning that he would

Henry James, Sr. and William's first-born, Henry James, ca. 1880.
By permission of the Houghton Library, Harvard University.

be sure he would know him among a hundred babies."[1] William was delighted with his son, named "Henry" after his grandfather and uncle, and his banter was in full swing. "My domestic catastrophe is now a week old," he wrote Robertson. The child "has a rich orange complexion, a black head of hair, weights 8-1/2 lbs, keeps his eyes tight shut on the wicked world, and is of a musical, but not too musical disposition." He admitted that "I have a strong affection for the little animal—and though I say it, who should not, he has a very lovely and benignant expression on his face."[2]

Several months later James was less flushed with parental enthusiasm. Alice was totally absorbed in caring for little Henry, and William began to resent the strictures the new arrival put on his time. "I find the cares of a nursing father to be very different from those of [a] bachelor. Farewell tranquil mind! It is replaced by terrible solicitudes about the color of diapers, the meaning of all that 'gulluping,' wonderings whether the darling is not either too hot or too cold, anxiety about the mother overexerting herself and a thousand more perplexities that fill the day and . . . kill the time."[3] James was a devoted father, and as the family grew and the children matured he took time to be both companion and adviser. Yet he also began to construct a domestic environment, with Alice's concurrence, that put Alice, her mother Elizabeth Webb Gibbens, and a housekeeper in charge of the children.[4] The joys, cares, and real tragedies of child-rearing aside—the Jameses would lose the infant Herman to pneumonia in 1885—William's most pressing anxiety throughout the 1880s would be the cultivation of his intellectual and professional life. His concern for Alice and the family never diminished, but his energy and interest were focused most on forging a scientific psychology, broadening his professional relationships, and articulating a world view.

Back at Harvard in the fall of 1879, he was restless and impatient for not having made much progress on his promised textbook of psychology. "College has begun," he wrote former student G. Stanley Hall; "I have three rather lowly graduates in physiol. Psych., 5 seniors in Renouvier, and about 30 juniors etc. in Spencer's 1st principles, and a lecture a week on Physiology. . . . My psychology hangs fire awfully, and my ideas are stagnant from want of friction."[5] By early 1880 James sought to escape to a summer in Europe and hoped he could meet Hall there, so that together they could "make psychological feathers fly." Meanwhile, he was working "more and more deeply into my pure phenomenalistic standpoint."[6]

James' decision to go once again to Europe went deeper than impatience with not having made much progress on his book, although that frustration was part of a larger dissatisfaction. With the exceptions of Hall, Henry Bowditch, James Jackson Putnam, Charles Peirce, Thomas Davidson, and Josiah Royce, a bright, young philosophical correspondent at the University of California, America offered little intellectual excitement. Royce would eventually supply a challenge, but in 1880 he and James were just beginning the relationship which would bring pluralism and monism into a classic dialogue. Psychology in the New World was still in the grip of moral philosophy, and fads like mind cure and spiritualism held the popular imagination.[7] Moreover, the Harvard philosophy department was dominated by the quasi-religious perspectives of Unitarian metaphysician Francis Bowen and ethical philosopher George Herbert Palmer. Charles Eliot had been instrumental in getting psychology transferred to the philosophy department in 1876, a move which James enthusiastically supported. In 1880, however, what would soon become the golden age of philosophy at Harvard, with James, Royce, Hugo Münsterberg, and George Santayana, had not yet arrived.[8] Even the Metaphysical Club which had stimulated James in the early seventies had disbanded. He confided to Royce that "our Philosophic Club here is given up this year—I think we're all rather sick of each others' voices."[9]

William had complained about Harvard to Robertson a full year before leaving for Europe. "Here in the College Faculty there are men who are perfect nullities mentally, but who are needed, and are most valuable here for that very reason—they will do in a deadly serious way, what can't be done by genuises at all—they know all the past votes of the faculty, all the hours of recitation etc. etc., and have no horizon beyond. When one finds one's self beaten and judged a failure by the standards such beings use, one naturally feels like smashing something."[10] James had even considered teaching part time at Johns Hopkins, although more to supplement the family income than for a superior academic setting.[11] He told Charles Renouvier that "I must frankly confess to you that I am more unsetttled than I have been for years."[12] But he was not ready to sever even partially his ties with Harvard. Instead, James' restless mind reached toward the Old World as he continued to advance professionally in America.

By 1880 James had published three major articles in two prestigious Euopean journals, *Mind* and the *Journal of Speculative*

Philosophy. These had been well received, particularly by Charles Renouvier and the notable English philosopher Shadworth Hodgson. In "The Sentiment of Rationality" James had characterized Renouvier and Hodgson as "the two foremost contemporary philosophers."[13] Now William longed to meet the men who had in turn praised his work. In June, William was in London visiting Henry, but he was poised to befriend first Hodgson and then Renouvier.

When they met, the Rugby and Oxford-educated Hodgson was at the top of the English philosophical world. In 1880 he had just been elected president of the Aristotelian Society, an office he would hold for thirteen consecutive years. Like William James, Shadworth Hodgson had criticized both the British empiricists ("associationists") and the German idealists, the former for their too-easy clarity and the latter for their vagueness. Hodgson emphasized the relationships between qualities rather than their intrinsic properties. For example, he pointed out that color and shape had distinctness only when relationships were drawn between them. Like James, he repudiated atomism and turned metaphysics toward the study of relations. Hodgson was also occupied with the problem of space, one that interested William greatly in the late seventies and early eighties.[14]

William was much impressed, finding Hodgson "charming in the extreme," if "bashful" and somewhat absent-minded. James tried valiantly but with mixed success to draw him into Renouvier's perspective: "Despite the fact that he rejected R's two most important levels, the finiteness of the world, and the free will, he says enough . . . to make R. the most important philosophical writer of our time—you can't think how it pleaseth me to have this evidence that I have not been a fool in sticking to R."[15] Hodgson helped confirm James' conviction that neither empirical or idealistic explanations were adequate to describe reality. Moreover, to bring the Englishman Hodgson, the Frenchman Renouvier, and himself, the American, into an intellectual alliance gave powerful international support to his own thoughts and reputation.

Though meeting Hodgson was a highlight of his London stay, William also managed to make the acquaintance of several other notables at the Athenaeum Club. Among them was the now aged philosopher Alexander Bain, whose criticism of John Stuart Mill and theory of an active will had impressed William in the 1860s.

James also liked Croom Robertson, the young editor of *Mind* to whom he had sent "The Sentiment of Rationality," and with whom he continued to correspond until Robertson's premature death in 1892. And he lunched with a group that included the great Herbert Spencer, James' intellectual foil throughout the seventies. He reported to Alice that conversation with Spencer was "rambling and trivial," perhaps because William "avoid[ed] the subject of philosophy carefully and the great Herbert seemed quite content to gossip and quote Mark Twain."[16]

When James left for the continent in July, he had became personal friends with a major English philosopher and the editor of that country's best metaphysical journal. He hurried to Heidelberg, where he met Hall and talked psychology. William found Hall "a remarkable being. Such a simple craving for truth and boundless power of acquisition with such absolute modesty."[17] But Hall's talents and stimulation notwithstanding, William was looking forward most to meeting Renouvier. After resting in Switzerland in late July and early August, he caught up with the other great contemporary philosopher at the Uriage-les-Bains health spa outside Grenoble.

Renouvier's influence on James extended back to the late sixties, when in deep personal and intellectual crisis, James had referred favorably to *Essais de critique generale,* a four-volume work published between 1854–1864. William's memorable attempt to break the grip of determinism had borrowed freely, perhaps desperately, from Renouvier's second essay. In a limited sense, his sixties crisis was a young man's struggle to find a philosophy of action.[18] James had tried to promote Renouvier's philosophy at Harvard through conversation and course work. Yet Bowen and Palmer paid little attention. In 1876 William had complained in a letter to Renouvier that "two *professors* of philosophy, able and learned men, . . . hardly knew your name!!" He had vowed that before long "the name of Renouvier will be as familiar as that of Descartes to the Bachelors of Arts who leave these walls."[19]

Renouvier, unlike Hodgson, was not closely associated with academic philosophy. Born in 1815, he had been educated at the Ecole Polytechnique, specializing in mathematics and natural science and absorbing the broad theorizing of Auguste Comte. During the period immediately following Napoleon's demise, Renouvier became a socialist. After the coup d'état of Louis Napoleon, however, Renouvier retired from politics and devoted himself completely to philosophy. In 1867, along with close friend and collaborator Francois Pillion, he began to publish a metaphysical journal, *L'Annee philosophique.* The monthly propagated Renouvier's perspective and came to publicize James' views as well.[20]

Renouvier criticized philosophers who attempted to make too-fine distinctions between appearance and reality, and thereby erected a foggy transcendental ideal. He felt that after all metaphysical analysis, there was a finite and irreducible reality that could be mathematically expressed in cardinal numbers. Individuals were real and existed within a pluralistic world in which each could choose within the welter of experience; they were rational, distinctive beings who were free to act on their beliefs. Although James ultimately could not embrace Renouvier's entire metaphysics—particularly the notion of a finite universe—he was much attracted to the Frenchman's attack on determinism and on Kantian idealism. Renouvier's emphasis on moral responsibility in individual decisions was especially important to James, for it was living proof that determinism was not omnipotent.

With the exception of Scottish realism, the only well-established, respected intellectual choices were between empiricism and idealism, between Darwinism and divinity.[21] James, the still largely unknown American, was casting psychological and philosophical stones at the intellectual Goliaths of the western world. It would have been considerably less trouble for him to have thrown his lot in with one side or the other; most American thinkers of his era were elaborators rather than forgers of new traditions. Since James' intellectual ambitions were greater than those of most of his contemporaries, his anxieties, which must have included fears of failure and ultimate obscurity, were also more acute. How Hodgson and Renouvier received him was critical to the success of his own intellectual and professional mission.

James had an emotional crisis, in fact, while vacationing in Switzerland shortly before seeing Renouvier, with whom he would spend only one day. On the second anniversary of his wedding, William confessed by letter to Alice that the previous night he had "knelt down as we have done together and prayed for you Alice and for myself. I can be better to you than I have been, more of a rest, less of an anxiety, if only I get a *start.*"[22] His frustration climaxed three weeks later, after returning one evening to his room at the Rhone Glacier Hotel feeling "enraged at my eyes, and at the nature of some English people I had been eating near, etc. etc.

I lay awake for a long time and my rage finally vented itself on myself for my unmanliness. . . . A sort of moral revolution poured through me."[23]

He glanced out the window and beheld the mountains and the Milky Way, "with big stars burning in it and the smaller ones scattered all about, and with my first glance at it I actually wept aloud, for I thought it was *you,* so like was it unto the expression of your face—your starry eyes and the soft shading of your mouth." He exclaimed, "I am not crazy dear, only I had one of those moral thunderstorms that go all through you and give you such relief . . . and nature, God and Man all seemed fused together into one Life as they used to 15 or twenty years ago." He thought the experience was "really a physical one—a break up of the old worn out condition and pouring in of new strength."[24]

More than once James would lapse into an associative trance from which new moral resolve and creative energy would emerge. These episodes invariably followed periods of academic tension and/or professional anxiety. And they always occurred away from Cambridge, Harvard, and Alice. The release into a unity of God, man, and nature came when James was struggling to find scholarly recognition through building a metaphysical road between empiricism and idealism, between an isolated atomic clarity and vacuous, shapeless monism. James conflated Alice, natural beauty, and moral rectitude into a sense of rebirth that allowed him to deflate anxieties as he continued to pursue his unique intellectual path.[25]

BY SEPTEMBER 1880 James was back at Harvard, where he began the fall term as assistant professor of philosophy. But his intellectual production slowed. After producing six major articles in 1878 and 1879, he published just four over the next two years.[26] By the winter of 1880–1881, domestic developments threatened to upstage intellectual life. The Jameses had been living in Alice's mother's home on Garden Street in Cambridge. But in the spring of 1881 with Alice pregnant again, William rented a house on Quincy Street, next door to President Eliot's residence. The family now had more room for little Henry and the expected child, even with the continued presence of Elizabeth Gibbens, Alice's mother. Indeed, the William Jameses seemed an extended family, as William reported to Thomas Davidson that "Mrs. Gibbens and her daughters [Margaret and Mary] will form a common menage with us."[27] William's parents and Catharine Walsh—Aunt Kate—now resided on Mount Vernon Street in Boston, along with William's sister Alice, whose health remained poor.

For several years Mary Walsh James had suffered what Aunt Kate called a "functional disturbance" of the heart. In late January 1882, Mary contracted pneumonia, weakened rapidly, and died before the month ended. Her funeral brought all the James children together for the last time. Mortally ill Wilkinson came from Wisconsin. A homesick Henry had left London for Boston in early January, unaware of his mother's desperate condition. Robertson, troubled and a heavy drinker, was residing in Cambridge but would soon depart for a complete rest at Fayal off the Portuguese coast.

Unexpectedly, Alice James seemed to benefit from her mother's death; it appeared to Catherine Walsh "to have given her new life, those poor nerves having apparently found their long needed stimulus in the tremendous sense of responsibility which had fallen upon her."[28] Alice now stepped into Mary's place to care for Henry Sr., who, as events proved, would stoically refuse aid. Henry Jr., on the other hand, seemed immensely depressed. He wrote Thomas Sergeant Perry that his mother had been "the sweetest, tenderest, wisest and most beneficent being I have ever known."[29]

William, however, remained silent in the period immediately following the funeral. There was, in fact, little to suggest that mother and eldest son had been particularly close. One scholar has even argued that Mrs. James adversely affected William's self-development.[30] Mary James had been able to see through her son William's youthful deceptions, particularly those relating to his health, much more astutely than had Henry Sr. She had also resisted her eldest son's wanderlust, which tended to strain the family budget. Alice Gibbens James, however, believed that the death of her husband's mother had a more profound effect on him than first appeared: "Though they may not realize it . . . their mother's death has been a physical strain upon all the sons—at least the three that I see [William, Henry, and Robertson]. And her going has so changed everything; broken up the house and carried away so much kindness and friendliness that the wonder seems that anything should go on as usual."[31]

Eight months later, in Vienna, William observed "a peasant woman, in all her brutish loutishness" who precipitated a gush of sentimentality for motherhood: "All the mystery of womanhood seems incarnated in [the peasants'] ugly being—the Mothers! the Mothers! Ye are all one! Yes, Alice dear, what I love in you is only what these blessed old creatures have; and I'm glad and proud, when

I think of my own dear Mother with tears running down my face, to know that she is one with these."[32] Whatever the form of William's grief for Mary James, her passing probably did not touch William as fully as would the death of his father. Perhaps it was because his mother had not been intellectual that she had better understood William's tendency to exaggerate anxieties and health problems. William's paean to motherhood suggests that Mary was to be extolled for her embodiment of the universal characteristics of her sex rather than for her lasting uniqueness, hardly the most poignant expression of loss.

But whatever William's feeling for Mary, Alice was right: it was a wonder that things "should go on as usual," and indeed the domestic scene changed rapidly. Wilky returned to Wisconsin and died the next year. Even as she cared for her father, Alice James came more and more to depend on the woman who would soon be her live-in friend, Katharine Peabody Loring. Alice herself had only ten years to live and soon moved with Katherine to London. Robertson's recuperation at Fayal did not cure his alcoholism and later in the eighties he would be institutionalized. Henry returned to the ambience he had learned to love in London, to write novels and even a play. With his beloved Mary gone, Henry Sr. gradually lost interest in living, ate less and less and then nothing at all. In June Alice Gibbens James delivered another healthy boy who would be christened "William," and whom the father dubbed "William Jumbo Jr."[33]

The birth of a second son, however, did not root James in Cambridge. As the once cohesive James family lost its remaining basis for family solidarity, William planned to remove himself from wife, babies, ailing father, and from America as well.

WILLIAM'S NEED for yet another European trip arose from his continuing belief that his professional-intellectual future could best be nurtured there rather than in America. Then, too, his health was not good. He told Thomas Davidson in April that there was a debilitating relationship between his eyes and legs: "With my bad eyes and the conservation of energy between them and my legs . . . when I tramp I can't read and when I read I can't tramp."[34] His father noted in May that William had trouble with boils and diarrhea, although he seemed in good spirits.[35] The good spirits probably resulted from learning that his application for a year's leave of absence at half pay was accepted. James had

Mary Walsh James, William's mother, in middle age.
Courtesy of Henry Vaux, and by permission of Houghton Mifflin.

convinced young Josiah Royce, who felt isolated on the West Coast at the University of California, to be his temporary replacement at Harvard. It was a move which gave Royce the inside track to a permanent position at Harvard.[36]

James never intended to take his family abroad. He confided to Royce that "I am going abroad without my wife because I can't afford to take her, and because she and the babe will thrive better without travelling, and her mother and sisters will be filled with happiness at getting her to themselves again."[37] He would have preferred to have Alice make the trip without the children. "I . . . veer more and more toward a combined invasion by us both of that continent. All sweetness would be gone from it after six weeks with thee at home."[38] William may have been concerned about the sexual deprivation; he was concerned about it even before departing for Europe. Despite the presence of her sisters and mother, Alice wanted to hire a nurse to care for the baby. William agreed, but asked, "is there no way of her staying without absolutely separating us at night. I think that must and can be managed."[39]

All was managed well enough for James to embark from Quebec in early September. He had decided to ship out from the Canadian city, since the steamship lines advertised 1,000 miles of smooth water on the St. Lawrence river and the Gulf of St. Lawrence before entering the open Atlantic. Hence, one could avoid the "mal de mer," and it would be only "five days from land to land."[40] He had never forgotten his sea sickness aboard the *Colorado* in passage to Brazil nearly two decades earlier.

Although he avoided becoming ill, he was soon suffering pangs of separation from Alice. Being apart unleashed insecurities that could be quelled only partially through writing romantic letters. William depended upon Alice. She had, as much as anyone or any philosophy, saved him from morbid introspection and metaphysical obsession. Yet his professional ambition made simply settling down at Harvard intolerable. He needed to be abroad to make progress on his psychology; he needed to travel into the heartland of Europe and talk with the men who were pioneering in psychological science.

But to be gone a full year seemed at first too much to bear. "You are to me such a perfect haven or refuge from the rocks and labors of life," he wrote to Alice in passage. "I sat on the berth with your letter in my hand, and there came o'er me a sense of my premarital existence and the little sheet of paper seemed to *signify* such an immensity more than what it *was*." But Alice was apparently having her own woes as well. "Dear don't relapse into that state of self abasement!" he scolded. "It is simply pathological, and I am sure due to your nursing state." Although William was acutely sensitive to Alice's moods, he tended to take his own periods of depression much more seriously, dimissing hers as trivial or due to female complaints. For all of his psychological insight, he could be the typical Victorian male chauvinist. He exulted in Alice's "heroic motherhood," and admitted "how little credit I gave you for it."[41]

William could not resist abasing himself, however: "I am nothing but your wandering piece of property, and my one sacred duty is to see that my journeyings do not impair me, but pass me back to your arms worth more than I was when I left them." He reflected, "As I look back, it seems as if I had from that first night at Lake Placid been nothing to you but a source of anxiety; now I will strive with might and main and become something else."[42] From Vienna he implied that his marriage to Alice was a saving element. "I feel your existence woven into mine with every breath I take, dear; and I am . . . full of pity for the poor creatures who have to live alone."[43]

Separation from Alice, from Harvard, and from America not only allowed James to emotionally recharge his marriage, but also to achieve a fuller grasp of his special destiny. He found himself as much in return to the Old World as in building a home and a career in the New.

Direction, however, was not to be found in an instant. He worried over where to go and whom to see, and over what itinerary would best serve his intellectual interests and his professional future, which depended upon the completion of his psychology. As he roamed from England to Austria to Italy to Germany and back to England, William shared with Alice his professional plans, states of mind, and impressions of national character. "I feel more strongly than ever, that the psychology is the first interest of the year, which if done will justify the whole undertaking, but if not done will make it seem as if nothing were done," he wrote. He momentarily decided to avoid the German academic scene because "I have been *challenged* by what Berlin offers in the way of enlarging my culture; but an age comes, dear, when a man can't rise at every fly, and run after every bit of game; I think my previous German culture, with what I shall read, will have hereafter to serve my turn." He admitted that "I shall lose by not going to Berlin the meeting of a couple of men, [Ewald] Hering and [Ernst] Mach, in Prague and [Wilhelm] Wundt in Leipzig, by the way. But that is nothing." James resolved "that the only way to enjoy the year as well as profit by it is to make a moral task of it."[44] In the end, however, he went to Germany and saw all the men he had decided to avoid.

Though he had planned to go from Vienna for a short stay in the Alps, bad weather and physical complaints changed his plans, and he headed south to Italy. He confided to Henry, who was still in London, that enjoyment of the continent had been "a good deal impaired by the entirely unexpected condition of my sleep, my eyes, and my back, which have rather turned traitors. . . . Breathe not a word of my complaints homeward. *They* must be transient."[45]

In Venice his eyes improved, and the picturesque surroundings stimulated a slumbering interest in art. He planned to return to Paris by way of Florence, he told Henry, for "I have some ideas of which an article might possibly be made, relatively to the evolution of schools of painting, and Florence would help me out."[46] As had Brazil in 1865 and Dresden in 1867–1868, Italy seemed to bring back into focus an earlier creative field. And as in Dresden, James again juxaposed the modern and classic. "I hate to leave the glorious pictures, which one has to see in such an infernally unsatisfactory way," he wrote to his father. "They ought to be erected into a circulating library to which one might subscribe and have a masterpiece a month in his own house throughout the year." James experienced a sense of loss as he visited the galleries and felt that "the cheap young Italian swell has gradually stolen over my soul. . . . If anything can make one a fatalist it's the sight of the inevitable decay of each fine art after it reaches its maturity." He had thought of "scribbling something about it for the *Nation*" but decided against it. He was disheartened by the coarseness of tourists in St. Mark's Square, who were "Cads of every race, and to the outward eye hardly anything but Cads."[47]

But patrician disgust with modern aesthetics could not turn him away from modern science. James had marked out an ambitious professional future that demanded professional interaction. To properly complete his psychology he knew he needed to visit Germany, for it was in Germany that psychological science had made the greatest advances. Yet he resisted: "I renounce Berlin for this simple reason that my great trouble in Europe will be to pass the evenings. They will be very long, with no open fires, few friends to visit, etc. Moreover the Psychology is the one thing that justifies my trip, and what I should find philosophically in Berlin would take the form of rival and conflicting interests [rather] than of helps."[48]

This was weak rationalization at best. James was a social animal; he made friends easily and he spoke fluent German. What was unsaid was obvious: he had not established intellectual contacts in Germany. Who there knew of William James? He had published no book on either psychology or philosophy that would make him known to "rival and conflicting interests." He would be in the heartland of Old World science and metaphysics, the world of men that he had both admired and criticized. He would be stepping into that world not as a young man in his mid-twenties, who could afford to read Schiller and Goethe, ponder sculpture, and write a creative diary. James would enter Germany as a forty-year-old Harvard professor who was under pressure to write a scientific psychology; a man who did not have the stature of a Helmholtz, or even of a Mach or Wundt. Still, if he was to advance professionally, he had to bury insecurity. He did.

By the end of October, William wrote to Alice from Trieste Station of "passing into the land of sternness and Morality, after poor Italy, whereof the charms like that of the Virginia weeper in October, [are] not good for me every day of the year." He complained that "my mind has been like a stagnant lagoon, my circumstances ditto, and yet I've lain awake steadily since my first night."[49] From Prague he exclaimed, "I do love the atmosphere of Germany better than that of Italy. There is no doubt about it."[50]

In Prague he made what he had lacked in Germany—a professional ally. Shortly after arriving he befriended a young psychologist-philosopher named Carl Stumpf. Stumpf had studied law at Wurzburg but had turned toward other interests after coming under the influence of a forefather of phenomenology, Franz Brentano. In 1879 Stumpf became professor of philosophy at the University of Prague, and when James met him he was finishing the first volume of his classic work on auditory psychology, *Tonpsychologie* (1883, 1890).[51]

Like James, Stumpf had largely rejected Kant's categories and Hume's reductive empiricism. Reality for both men was largely a matter of perceiving relations; individual sensations were not real until they took their place in a relational world. Psychologists who believed that vision was an innate quality not originating in experience as sense perception were at that time called "nativists," and Stumpf had developed a nativist view of space perception that James had highly praised in his own work on space.[52] Further,

Stumpf was not bowled over by the Darwinists. In his view they had not eliminated teleology. Life's origin still required explanation. As we have seen, James had emphasized in the Lowell Lectures that evolution was simply a theory, derived, like all other theories, from consciousness. James later acknowledged Stumpf's importance in *The Principles of Psychology:* "Stumpf seems to me the most philosophical and profound of all the writers; and I owe him much."[53]

Carl was only thirty-four when they met, six years William's junior. It was perhaps easy for the younger man to like an older American with similar interests but with fewer publications. Conversely, perhaps James was comfortable in the presence of a young German who seemed to have the right approaches to so many of the matters that occupied his own mind. Whatever their attraction, Stumpf acted as James' guide in Prague and introduced him to two older, internationally recognized scientists, Ewald Hering and Ernst Mach. James wrote to Alice that Stumpf "insisted on trolling me about, day and night over the length and breadth of Prague, and that Mach, Professor of Physics, and genius of all trades, simply took Stumpf's place to do the same." In between tours James attended lectures. "I heard Hering give a very poor physiology lecture and Mach a beautiful one," he reported. Mach impressed him greatly: "Mach came to my Hotel and I spent 4 hours walking and supping with him at his club—an unforgettable conversation. I don't think anyone ever gave me so strong an impression of pure intellectual genius."[54]

Ewald Hering was a physiologist who along with Mach had been invited to join the prestigious Viennese Academy of Sciences. Both men were considered among the elite of European scientists, just behind giants such as Darwin, Galton, Pasteur, and Helmholz. Hering had given a famous lecture in 1870, "On Memory as a Generalized Function of Organized Matter," that suggested a psychological basis for heredity. And he was Josef Breuer's mentor, and through Breuer indirectly influenced young Sigmund Freud's perception of biological inheritance.[55] Like Stumpf, Hering was a nativist on the space question and had done vital work on double vision. When William told Alice that Hering had "cleared some things up for me," he was probably referring to vision.

Ernst Mach, however, stimulated James the most. The Austrian physicist and philosopher had studied in Vienna and became professor of mathematics at Granz in 1864. Only four years James' senior, he had held the chair in physics at the University of Prague since 1867. His broad intellectual scope attracted William, for Mach had ranged from physics and optics to the psychology of perception and aesthetics. In addition, he had a deep interest in the philosophical basis of science and felt that the division of science into various branches—physics, chemistry, physiology, and lately psychology—was largely artificial and mostly a matter of practical convenience. His mind, like James' had ranged freely from field to field. William reported excitedly to Alice that Mach had "read everything and thought about everything, and has an absolute simplicity of manner and winningness of smile when his face lights up that are charming." Mach was what William in fact was himself, a genius of all trades.[56]

At Aussig, on the way to Dresden, James decided to stay in touch with Stumpf. To Alice he reflected that the Germans "are not so different from us as we think. Their greater thoroughness is largely the result of circumstances." Magnanimity toward Germany and German academics, however, was not simply typical Jamesian capriciousness on national characteristics. After encountering three first-rate German thinkers, he had gained considerable confidence in his own abilities: "I found that I had a more *cosmopolitan* knowledge of philosophic literature than any of them, and shall on the whole feel much less intimidated by the thought of their like than hitherto."[57]

From Dresden he asked Alice to send him "*all* the copies of Feeling of Effort, of Sentiment of Rationality, of my Journal of Speculative Philosophy articles and a dozen each of the Atlantic and the Unitarian Review article." He explained that "swapping one's articles is the great way to get things early and surely from other men."[58] It was also a way to circulate one's point of view and reputation. Once sure he could compete successfully with the first rank of continental thinkers, James entered the market with gusto. Americans were the world's greatest salesmen, and James proved no exception.

But as his German stay lengthened, William became less enthusiastic, probably because he failed to continue to make the personal contact that he had made with Stumpf and Mach. In Berlin he "heard Helmholtz give the most idiotic lecture I ever listened to . . . on the simple subject of gravity which Mach treated so beautifully in Prague," he wrote. Nor did Berliners at large provide alternative pleasantries:

> As for Berlin and its helmeted race of inhabitants, what shall I say? I've got them all mixed up with the carnivorous animals at the zoological garden where I spent so much time, so that I can hardly tell which is which except that the carnivors have perhaps more suavity of manner—less prominence of the helmet element. I have hardly heard a moderately polite answer given to anyone since I've been here, except by the hotel porter, and he's paid expressively for it, magnificant creature that he is. Berlin's tone is almost exactly the same as that of the U.S.A. and I'm very glad I've already decided to have little to do with it and go to the mellower clime of Paris.[59]

James lingered a few days longer to see Hermann Munk, a German authority on the brain and the theory of localization that linked certain sensory functions like vision and smell with specific brain locations. Munk had studied the physiological questions that James had considered in the Lowell Lectures, questions he would again address in the the second chapter of *The Principles of Psychology*, "The Functions of the Brain." But the meeting proved only moderately successful: "Yesterday I went to the Veterinary school to see Munk the great Brain vivisector. He was very cordial and poured out a torrent of talk for 1 1/2 hours, though he could show me no animals." Munk did offer "one of his new publications and introduced me to Dr. Baginsky . . . [an] authority of the 1/2 circ. [semicircular] canals whose work I treated supercilously in my article. So we opened on the 1/2 circ. canals and Baginsky's torrent of words was even more overwhelming than Munk's. I never felt quite so helpless and small-boyish before, and to this hour dizzy from the onslaught."[60]

When William met men with narrower interests than Stumpf and Mach, men who had carved out specific areas of scientific expertise, he felt less confident—and with good reason. Munk and Baginsky knew more than James did about the physiology of the brain and ear. William would become more and more opposed to what he regarded as blind expertise in "narrow" sciences. This opposition was probably rooted in the knowledge that he had, as he put it in the Lowell Lectures, "spread myself thin." James was naturally interdisciplinary, and scientists who were less expansive would later charge him with interests inimical to true science.[61]

Things improved little for William in Leipzig, where he met Wilhelm Wundt, Europe's leading physiological psychologist.[62]

"Wundt," James observed, "has a more refined elocution than anyone I've yet heard in Germany. He received me very kindly after the lecture in his laboratory, dimly trying to remember my writings."[63] A week later, writing to Alice while on the train to Liege, Belgium, James elaborated on Wundt. "He made a very pleasant personal impression on me, with his agreeable voice, and ready tooth-showing smile. His lecture also was very able and my opinion of him is higher than before seeing him. But he seemed very busy and showed no desire to see more of me than the present interview either time." Wundt did nothing "to the gain of my ease," and encouraged "the loss of my vanity." After this cool reception, William underscored the superiority of his Prague contacts over those of Berlin and Leipzig: "Dear old Stumpf has been the friendliest of these fellows. With him I shall correspond."[64]

Although James recognized Wundt as a pioneer in scientific psychology, he had little positive to say of him in correspondence—particularly after failing to make acquaintance turn into friendship.[65] Interestingly, Stumpf was also critical of Wundt. To both, Wundt represented the archtypical self-assured, arrogant professional scientist. Several years later Stumpf concluded that "Wundt leads students and some others to believe that the ever-repeated measurement of reaction-time marks the beginning of an entirely new 'experimental psychology' from which one can look back upon the old psychology only with scorn and derision. . . . How often already has not psychology been made 'exact' in this way, only to be led back again into the path—into 'psychological' psychology!"[66] William added his own invective about Wundt: "Cut him up like a worm, and each fragment crawls; there is no *noeud vital* in his mental medulla oblongatà, so that you can't kill him all at once." This was so because "he isn't a genius, he is a *professor*—a being whose duty is to know everything, and to have his opinion about everything connected with his *Fach*."[67]

James was hardly kinder to Wundt's American allies, Edward Bradford Titchener and James McKeen Cattell.[68] William's breadth, his broad creative canvas, found him more and more opposed to any scientist who claimed too much for one science; and conversely, he aligned himself with thinkers who kept fields open and relationships between them problematical. Wundt remained unimpressed by James' growing reputation as one of the world's leading psychologists. After reading *The Principles of Psychology*, he commented, "It is literature, it is beautiful, but it is not psychology."[69]

In Liege James paused to meet Joseph Delboeuf, a Belgian physician who would later criticize Jean-Martin Charcot's hypnotic suggestion technique at the Salpêtrière Hospital in Paris. Delboeuf had reacted to James' published work, writing a criticism of "The Feeling of Effort" in *Revue philosophique*.[70] James, in turn, had read several articles Delboeuf had written for the *Revue* on determinism and had asked Renouvier to comment on them. Delboeuf intrigued James by suggesting that nature provided cracks in a largely deterministic world, cracks which formed discontinuities that provided the theoretical possibility of freedom. Such freedom did not, according to Delboeuf, violate the prevailing conservation of energy theory, because natural freedom acted as a catalyst to transform the actions of existing energies. Although Delboeuf did not propose a growing universe (as Henri Bergson would later), he argued that reality was charged with novelty.[71] James was already drawn toward a pluralistic world view and would later equate an unfinished world with a growing one. It is little wonder that he enjoyed Belgium more than Berlin or Leipzig, telling Stumpf that "I found M. Delboeuf a most delightful man, full of spirits and originality."[72]

Before leaving Liege for Paris, William assessed his continental visit. "I haven't gained enormously from talking philosophy . . . except a little with Stumpf," he told Alice. "The total lesson of what I have done in the past month is to make me quieter with my home lot and readier to believe that it is one of the chosen places of earth. Certainly the instruction and facilities of our University are on the whole superior to anything I have seen. The rawnesses we mention with such affliction . . . belong rather to the century than to *us* . . . ; we are not a whit more isolated than they are here."[73] Having sampled the continental academic setting, William found it more wanting than himself and Harvard. He would continue to cultivate European contacts, but he never again exhibited a sense of professional-intellectual inferiority—anywhere.

James remained discouraged about the lack of progress on his psychology, the major rationale for the trip. "You see I don't know yet how my psychology is to go on here. I'm decidedly tougher and better than I was and that's a clear gain worth perhaps the money. But almost literally the only intellectual occupation I've found time and eyes for has been that of writing letters." He declared that "if in a month I don't distinctly see my way towards getting on with the psychology (which I know I *can* get on with at home) it will, I think, be my plain duty to go back," to return to Cambridge.[74] But suddenly all professional considerations were upstaged. In early December 1882, a letter from Alice caused William to rush at once from Paris to join Henry in London.

HIS FATHER was dying. Alice reported that "your dear father has had an attack of nausea and faintness which left him very weak." Although the immediate crisis had passed, the prognosis was poor, because the elder James had decided to die. "He has distinctly made up his mind not to live and without disease, that at 73 is a good reason for going." Henry Sr. had "said good bye to many of his friends but he won't tolerate what he calls 'sentimentality.'" Alice advised William against hurrying home, because although the old man spoke "lovingly and often of you and Harry . . . I am sure he would be greatly disturbed if he thought either of you were coming."[75]

William heeded Alices's advice, but Henry left for America, leaving his brother in the Bolton Street apartment in London to await reports on Henry Sr.'s condition. Since he had been informed that Henry Sr. had sunk so low that he would not recognize his sons, William decided, "I would rather not see him, but have my last memory of him as I bade him goodby." And indeed William's memory turned back the decades:

> As I lay in bed last night I had such tender thoughts of poor dear old father, of the way we had leaned on him and looked up to him all those early years, New York, Europe, Boston—of his indomitable spirits, activity, genius, indulgence, that my heart was melted within me, and his life seemed but one thing. I pray God if the end comes it may come easily and quick. But if there's a chance of my seeing him yet, and in his senses, I will post straight home.[76]

Henry Sr. lingered in a coma into the third week of December. William made his last communication with his father by post. The letter showed clearly that he met his father's passing, at least the event of his death, without tearful sentimentality. "We have been so long accustomed to the hypothesis of your being taken from us, especially during the past 10 months, that the thought that [this] may be your last illness conveys no very sudden shock. You are old enough, you've given your message to the world . . . and you will not be forgotten." William generously acknowledged his father's influence—perhaps too generously: "What my debt to you is goes beyond all my power of estimating . . . so early, so penetrating, so constant has been the influence." He promised to do after death what Henry Sr. had had so much difficulty doing during life—publish his writings. "You need be in no anxiety about your literary remains—I will see them well taken care of, and that your words shall not suffer for being concealed."[77]

The end came on December 18, 1882. "Your father died peacefully at half past two today," Alice related. He had exclaimed, "I am going with great joy"; his last words, "There is my Mary!" The attending family—the two Alices and Aunt Kate (Henry was still in passage)—gazed at "the grand old face grave with majesty of a great triumph. Such a *satisfied* look—as if all expectations, all questions were already answered. We kneeled and sat about the bed talking of things he had said, and glad that the desire of his heart was granted."[78]

Alice wrote to William that his sister Alice, who had nursed her father since her mother's death in February, wanted an autopsy, especially "an examination of the brain to discover if there was any disease there. She thought it would be a satisfaction to you." But Dr. James Jackson Putnam, William's friend from medical school days, had advised that this was unnecessary, since "death had come from exhaustion, the force of the whole system having been spent."[79]

William also learned from Alice that his father had bequeathed to him his "Philosophical books and writings, [and] made you his literary executor." Sadly, for both the surviving children and posterity, Aunt Kate had " 'destroyed and sorted' father's letters."[80] Henry was chosen executor of Mr. James' will, which divided an estate worth approximately one hundred thousand dollars.[81] The Mount Vernon Street house in Boston along with income from railroad stocks and bonds went to his daughter, Alice. Three properties in Syracuse valued at seventy-five thousand dollars were divided equally between William, Henry, and Robertson. The new income from these establishments provided each son with between seventeen hundred and eighteen hundred dollars annually. Wilkinson was excluded, because Henry Sr. had advanced him substantial sums since the ill-fated Florida venture, including a recent five thousand dollar advance. William and Robertson ignored Henry Sr.'s omission and offered their mortally ill brother a portion of their shares. After Wilky's death his wife, Caroline, inherited her husband's considerable debts—some eighty thousand dollars. Carrie was periodically at odds with the James heirs as she contended that her late husband had not received his fair share of the estate.[82]

William's last letter arrived too late. The only son to attend the funeral was Robertson, since Henry arrived a day late, and Wilky was too ill to travel. There was no small irony in the fact that when the elder James died, the apples of his eye, William and Henry, were busy pursuing in Europe the careers and international fame that had escaped him. Only a troublesome and obscure son, one who like his father had no profession, along with William's Alice, shared the final scene: "Bob and I stood beside the grave till the last sods were laid upon it," she wrote to her husband.[83] The absence of William and Henry showed how Europe and ambition, as much as death, separated the father from his favored sons, while the common clay of family brought only the failure, Bob, and the attending women, the sickly Alice, the motherly Alice Gibbens James, and the protective Aunt Kate, to his final resting place.

William immediately regretted his absence. "It is foolish but I do so wish now I hadn't left home," he confided to Alice. To have been at his father's side "would have symbolized the unfathomable relation in which he stood to me . . . and would have been one of those things whereof the moment gives a life-long satisfaction, and is incomparable with an other gain. How I pity poor Harry that he should arrive too late."[84] Regrets were transformed into recollection as James tried to capture, three thousand miles removed, but now separated by eternity, the legacy of a man so close, yet now irretrievably distant:

> Father's boyhood up in Albany, grandmother's house, the father and brothers and sisters, with their passions and turbulent histories, his burning, amputation and sickness, his college days and ramblings, his theological throes, his engagement and marriage, and fatherhood, his finding more and more of the truths he settled down in, his travels to Europe, the days of the old house in New York and all the men he used to see there, at last his quieter motion down the later years of life in Newport, Boston and Cambridge with his friends and correspondents about him, and his books more and more easily brought forth, how long, how long all these things were in the living, but how short their memory now is! What remains is a few printed pages, us and our children, and some uncalculable modification

> of other people's lives. . . . For me, the humor, the good spirits, the humanity, the faith in the divine, and the sense of right to have a say about the deepest reasons of the universe, are what stay with me. I wish I could believe I should transmit some of them to our babies.[85]

William knew well his father's strengths and weaknesses: "Unlike the cool dry thin-edged men who now abound, he was full of the fumes of . . . human nature," he wrote. Yet Henry Sr. had conceived "more than he could formulate, [which] wrought within him and made his short judgments of rejection of so much that was brought, seem like revelations as knock down blows." For all his profundity, he had lacked great gifts of articulation, gifts that William had in abundance. Now realizing that he would never again look into his parents' kindly, care-worn faces, William offered a benediction: "He and Mother are in the Wintery Ground! Good night, good and dear faithful pair! We will come soon enough."[86]

Significantly, Mr. James' passing became as much an occasion to note publicly the fame of the sons as the death of the father. The *New York Post* observed that the elder James' "latter years were much cheered by the success of his sons—one the well known novelist, and the other a professor at Harvard, who has made his mark in psychology and physiology, and is one of the most promising of the young explorers in the field of philosophy." The *Post* also reported that "neither of them unfortunately, was with him at his death."[87] The *New York Times* mistakenly reported—and pangs of remorse must have arisen in the brothers—that "his last hours were made easy by the presence of his two sons, Henry James Jr., the well known novelist, who came from England to attend his illness, and Prof. William James, of Harvard College, with whom the philosopher had been living for the past 16 years."[88] But inaccurate or not, the obituaries testified that both Henry and William now had public reputations that surpassed their father's. Indeed, the earliest *Post* announcement of Henry Sr.'s death was William's most popular press yet, since the paper recognized "Professor William James of Harvard College, whose attainments as a scholar have given him a Transatlantic reputation"—which of course was just what William had hoped the European trips would accomplish.[89]

CHAPTER TEN

Charting the Realm of Consciousness

Consciousness . . . will take the form of a crescendo, among whose "objects" new relations, new qualities and new "things" will with equal frequency be found, and in whose stream new "faculties" show themselves at work as fast as materials accumulate for their exercise.

Loose Notes on Cognition
ca. 1883

BETWEEN MARRIAGE to Alice Howe Gibbens in 1878 and the publication of *The Principles of Psychology* in 1890, James built the professional and intellectual scaffolding for lasting fame. In these twelve years he climbed through the professorial ranks from assistant professor of physiology to professor of philosophy.[1] In 1878 James was thirty-six, in 1890, forty-eight; he had traveled well into middle age. His position within the James family was irrevocably altered, as he became more husband and father than brother and son. His parents had passed away and he passed through the fathering of five children, one of whom did not survive. If the period before his marriage, fatherhood, and the death of Mary and Henry James was prologue, what remained was culmination.[2]

The passage applied to intellectual matters as well as to domestic life, for his big book, *The Principles of Psychology*, separated a man seeking an international reputation from one known in Europe and America as a great, pathbreaking psychologist. The publication of *The Principles* in 1890 distinguished a James seeking an original perspective from a James who had one. His classic book stood as the culmination of a creative process which had began as early as 1868 in Dresden. And many seminal ideas in *The Principles* first appeared in a flurry of articles published between 1879 and 1885.[3] The early eighties were crucial to his sculpting of a world view, or as he preferred, a "world picture."

James' intellectual and professional stature matured within an American social and cultural context in the midst of massive transformation. Whether one looked at cities struggling to assimilate hundreds of thousands of newcomers from the countryside and Europe, or to the incredible concentrations of energy and capital in industrial development, or to attempts by urban and farm laborers to gain the solidarity to countervail the rising financial and industrial capitalism, the nation was undergoing fundamental changes. A catalogue of signs of increasing modernity in both Europe and America during the 1878–1890 period would include the internal combustion engine, the electric light, the first skyscraper, the first affordable camera, the first execution by electric chair, and the initial psychoanalytic sessions, conducted by Josef Breuer and his young assistant Sigmund Freud.[4]

James' psychology and philosophy were significant intellectual signposts in this shift toward modernity. He constructed his psychological house at a moment in the American experience when distinctions between the past, present, and future were especially problematical—when the past was being replaced by something not yet fully formed, to some extent still shapeless, but, like William's own mind, rapidly building into a protean reality.[5] Within such a milieu his genius could formulate an intellectual perspective in terms of metaphors which seemed apt both to fellow intellectuals and to the public at large.

James became ever more conscious of his mission within this transformation, a mission to provide a truly realistic description of mental life. He had introduced in the Lowell Lectures what he believed was a new psychological standpoint; he had presented an alternative to an automatic, predetermined mind. He had found the Darwinian explanation of consciousness to be no explanation at all, insofar as "survival" became ipso facto consciousness' most critical function. The theory of evolution had presented an inadequate psychology masquerading as science. The chief spokesman for such humbug psychology was Herbert Spencer, whom James criticized repeatedly in classes, public lectures, and essays illustrating the inadequacies of the evolutionary view of the mind. Spencer was James' foil in the crusade for the concept of a creative, nondetermined consciousness.

Then, too, as much as James appreciated and incorporated the Darwinian view of nature, he could not live humbly in the shadow of Darwin's immense influence, content to extend the great scientist's views into psychology and metaphysics. In the 1880s, surrounded by and immersed in cultural and intellectual transformation, James built his own reality. His world was meant to stand on its own foundation, which would be *The Principles of Psychology.* Its bricks and mortar—to borrow a favorite Jamesian metaphor—were first assembled and crafted into lectures and essays which James later developed into his two-volume master work. Ironically, the classic psychology that insured James' lasting recognition helped unhinge traditional understanding of the mind, and made mental life an analogue of modernity's essense—ceaseless, limitless change.

IT WOULD be an exaggeration to assert that *The Principles of Psychology* broke upon the intellectual scene of the western world with anything like the force of the *Origin of Species.* Yet to interpret *The Principles* as a psychological ripple, merely a derivative of elemental Darwinian power, would also be a large mistake, for James never intended his psychology to be simply an evolutionary refrain. He would take an untraveled road between the empirical and idealistic highways, or as he would later put it, between "tough-mindedness" and "tender-mindedness."[6] William wanted to be an original thinker; to refine already established intellectual systems, whether in science or philosophy, did not sufficiently engage his creative ambition. The writing of *The Principles of Psychology* was the story of original genius articulating itself. James' creative powers emerged full force in the eighties. How did profound insights into mental life come to James? Fortunately, it is possible to catch him in the act of constructing a unique description of consciousness, one that presaged his epic work, *The Principles of Psychology.*

The circumstances of the writing of *The Principles* have recently become much clearer. In June 1878, publisher Henry Holt proposed that James write a psychology for an American Science series of which Holt was the editor. James signed a contract after telling Holt he would need two years to complete the book. It took twelve years. James' procrastination, his fits and starts, along with Holt's faith that eventually a book would appear, provide an interesting episode in author-editor relationships.[7] But within the attenuated process of writing the book lay the secrets of James' creative process as well.

James did not send a completed manuscript to Holt. Instead *The Principles* went to the publisher piecemeal, chapter by chapter. Proofs for the first six chapters were sent to James between April 1885 and December 1886. He probably sent Holt manuscript for the remaining chapters, which constituted the great bulk of the book, in 1889–1890.[8] A few years later, in the preface to his one-volume textbook *Psychology: Briefer Course,* James would remark about *The Principles* that "with a single exception all the Chapters were written for the book; and then by an after-thought some of them were sent to magazines."[9] It appears, however, that at least some of the material for several important chapters did appear first in journal articles in the late seventies and early eighties. William drafted, redrafted, and redrafted again—continuing to revise various manuscripts even while taking advantage of opportunities to publish. The book was the large project, but if he could get his ideas circulated as articles, so much the better.[10] In the late seventies and early eighties James was anxious to impress men like Shadworth Hodgson, Croom Robertson, and Charles Renouvier. Visibility through publication could not await completion of the book.

Thus while William's most sustained work on *The Principles* came after 1885, most of his original ideas for it were developed and appeared as journal essays before 1885. We know that the 1878 Lowell Lectures presented his understanding of brain physiology, the role of habit, and the functional design of consciousness. These views also appeared in articles such as "Brute and Human Intellect" (1878) and "Are We Automata?" (1879). James tirelessly criticized deterministic Darwinists such as Huxley, Clifford, and most of all Spencer in the above articles, as well as in "Remarks on Spencer's Definition of Mind as Correspondence" (1878). In "The Sentiment of Rationality" (1879) James also considered the role temperament and emotion played in creating various philosophies. And in "The Spatial Quale" (1879), he interpreted how eye and mind perceived space. By 1880, in "The Feeling of Effort," he had refined his theory of the reflex arc and ideomotor activity. A year later he published "The Sense of Dizziness in Deaf-Mutes."[11]

Yet in 1882 when he left for what became nearly a year's stay in Europe, James had not yet formulated a sustained statement on how the mind constructs sensible experience into cognitive consciousness.[12] Although he had established a physiological basis for consciousness and had shown how consciousness attended to its interests, he had not described how the mind comes to *know* anything at all; he had discussed mental functions, but had not described how consciousness acquired knowledge.

Charting the structure of consciousness was the great task that came to fruition in *The Principles.* And it brought James face to face with metaphysical problems that engaged him for the rest of his life.[13] Two essays that appeared in 1884 and 1885 respectively, "On Some Omissions of Introspective Psychology" and "The Function of Cognition," provided the first published descriptions of such seminal Jamesian concepts as the stream of thought, the feeling of relation, and the distinction between perception and conception—between acquaintance with something and knowledge about it.[14]

The germs of these two articles that presaged vital ideas in *The Principles* as well as James' later radical empiricism first appeared in a notebook entitled "Loose Notes on Cognition." The notes were probably composed in London in 1883, several months after the death of James' father and shortly before his return to America. During his stay there he was stimulated by discussion with members of an English metaphysical group, the "Scratch Eight," whose meetings he had been invited to attend.[15] With the distraction of Henry Sr.'s final days now past, circumstances encouraged creative accomplishment.

"Yesterday I was parturient of psychological truth, being in one of my fevered states . . . when ideas are shooting together and I can think of no finite things. I wrote a lot at headlong speed, and in the evening, having been appointed, gave an account of it—the difference between feeling and thought—at the scratch eight," he wrote to Alice.[16] We cannot be certain that "Loose Notes on Cognition" was the result of this productive moment.[17] Nevertheless, in considering cognition, James was certainly distinguishing between thought and feeling; there is a suggestive shooting together of ideas in these notes, and they anticipate how he would structure consciousness in *The Principles.*[18] James was already fashioning the rudiments of a psychology of a sensible, thinking mind—an original psychology, conceived seven years before publication of his great book.

THE FIRST task James set for himself in the Notes was to distinguish between what was being studied—consciousness—and the person doing the studying—the psychologist: "Whether it be the cognitive function of one of his own, or an imaginary subject's feelings that he studies, it is clear that the psychologist, during the time of study, stands outside of the feeling in question, as well as of its supposed object."[19] James wanted to avoid at the outset the psychologist's fallacy, the confusing of the subject's consciousness and its objects with the psychologist's interpretation of them from his own standpoint.

But James was also aware, indeed acutely so, of the great pitfall in making this distinction between the psychologist and his data. "It is equally clear that this difference of standpoint may possibly lead to a great difference between the ways in which the feeling appears to itself and to the psychologist respectively" (Notes on Cognition, folder 2). So before describing cognition, James made a judgment that eventually became an important assumption about the mind—that is, that the mind feeling itself working was not the same as the mind describing itself working. This dualism, the separation of the psychologizing subject from the inherent, natural sensibility of mental life, was maintained in *The Principles of Psychology*.

The distance between the objectivity of the psychologist and the subjective data did not make psychologizing futile. The scientist could still construct a real mental world. "The word *cognition* will be employed to denote the knowledge of realities exclusively, and not cover cases of merely apparent knowledge such as fictitious belief, error, or illusion. Reality thus becomes the warrant for our calling a feeling cognition," James notes. "But what becomes our warrant for calling anything reality?" James asks himself, and admits, "The only reply, is the faith of the psychologist" (Notes on Cognition, folder 2). One did not need to despair about the objective limits of psychological science if one *believed* in reality. James' refusal to equate science with skepticism made faith in reality essential.[20]

But faith in what reality? Science regularly altered the nature of reality. How could one have faith in a real world if the real world was an unfixed, ever-changing one? "At every moment of his life, [the psychologist] finds himself subject to a belief in some realities, even though his realities of this year should appear illusions in the next," James acknowledges. Then why believe anything a psychologist says, if reality cannot be permanently located? Because "we ourselves are . . . in the position of the psychologist; and we shall find our burden much lightened by being allowed to take reality in this relative and provisional way without being obliged to define it or justify our views of what its special determinations are. Every science must make *some* assumptions" (Notes on Cognition, folder 2). Provisional conclusions were necessary if the scientist—indeed, if anyone—was to attempt to decipher reality.

James was searching for a place to begin talking authoritatively, if provisionally, about cognition. On the one hand he did not want to claim too much for psychology, thereby constructing an illusory mentality, an artificial science; on the other hand he wanted to show that psychology could describe how the mind could think. He was seeking an opening, a beginning for his special understanding of consciousness:

> The Psychologist is but a finite and fallible mortal; when he studies the function of cognition, he does it by means of the same function in himself; and, knowing that the fountain cannot go higher than its source, he is prompt to confess that his results in this field are affected by his own liability to err. The most he claims is that what he says about cognition may be counted as true as what he says about anything else. If his hearers agree that he has assumed the wrong things to be realities, then doubtless they will reject his account of the way in which those realities become known. But if they agree with him about the realities, he hopes to persuade them also to accept his doctrine of cognition. On the whole his science is less exposed than other sciences to the attacks of metaphysical criticism. His assumptions are fortunately few,—*some* reality to be known, *some* feeling to know it, these are the only ones he need make at the outset. (Notes on Cognition, folder 2)

With these humble claims James is ready to illustrate how the mind came to know. He begins by imagining "a statue endowed successively with different sensations" so as to "discover how an intelligence may gradually be built up" (Notes on Cognition, folder

3). He borrowed this idea from the eighteenth-century French philosopher Etienne Bonnot Condillac, who in *Traite des sensations* (1754) had hypothesized a marble statue constituted as a human being, but devoid of all sensations—rigid and completely dumb.[21] Initially the statue's experience would be entirely subjective, as it would be incapable of interacting with an environment. Condillac had then proposed smell as the first sense bestowed upon the statue because odor was one of the simplest of sensations registered by the mind. James too would begin with the sense of smell, and with a very specific scent: "Let then the first sensation be that which the fragrance of a rose excites" (Notes on Cognition, folder 3).[22]

The first sensation was so simple that it had no duration or location. The statue had not even been "expressly gifted with attention." James notes that "were we tracing the actual historical evolution of consciousness from its simplest beginnings, upwards, we should be obliged to suppose some sort of motor reaction, like an explanation at this point." But he is not doing so, and his statue cannot respond "Lo!" or "Ha!" to the fragrance of the rose. In an evolutionary model of consciousness, "sensation and reflex-action proceed hand in hand. But our fiction of the statue has nothing to do with evolution, being a mere device for showing how many types of cognition can be imagined" (Notes on Cognition, folder 3). Darwinism could not explain a knowing mind. Moreover, here is an excellent example of James' broad range; Darwinian science simply lacked the speculative powers to adequately explain thinking.

James asks, "In what justifiable sense *can* a first sensation be called cognitive and said to 'reveal' any 'fact of existence' whatever?" And in asking this question he nears the bedrock of consciousness: "We are so close down to the roots of things in this inquiry, and the elements to be discriminated so few, that such questions as these sound at first like mere puns and shufflings of words, hollow *Tweedledums* and *Tweedledees*." But the simplicity is "full of relevancy," for the first key to unlocking the genesis of cognition is to "consider carefully in what sense a first feeling can be said to 'know' its own quality" (Notes on Cognition, folder 3).

An elemental consideration calls forth an initial definition of cognition: "Let then the word knowledge or cognition simply mean that function of a feeling whereby from out of its own passing existence it grasps something *other* than that passing existence itself." Or in the case of the statue and the rose's fragrance, "the fragrance should *be* something *other* than the mere existence of sensation." The statue would have "knowledge in the sense of *acquaintance* with an entity other than itself." The statue's acquaintance with the fragrance would not be the same as the sensation per se. If the statue experienced the quality of fragrance, say, as a recurring sensation, the statue would have knowledge. "The 'nub' of this vindication of the cognitive function for the first feeling, lies . . . in the discovery that the fragrance *does* occur elsewhere than there" (Notes on Cognition, folder 4)—that is, fragrance was experienced as being different from the statue, outside of it.

The discovery of cognition is not the statue's, however, but the psychologist's. "The first sensation itself could not make this discovery, *we* had to make it. The sensation for *us* was a datum judged to have certain relations with other data—cognitive relations in this instance" (Notes on Cognition, folder 4). James is scrupulously avoiding the psychologist's fallacy while at the same time saying that only the psychologist—in this case himself—could explain consciousness.

William wanted to uncomplicate the concept of mind and remove unwarranted assumptions and artificial constructions. His psychology would stand on a more naturalistic base; and the point at which the mind achieved cognition was critical to the picture of the mind it would create. Elsewhere James had insisted that empiricism had misled psychology by picturing the mind as filled with discrete mental elements, by atomizing consciousness—in short, by committing the sin of clarity. He had also attacked the idealists for erecting a fictitious mental entity, for synthesizing consciousness—for committing the sin of unity.[23] Between these theories of mental life lay the road toward *The Principles of Psychology*. And as James considered cognition, he searched for a realistic simplicity that avoided the pitfalls both of an unrealistic clarity and of an artificial unity.

He is careful in the Notes not to prematurely bestow self-consciousness upon his statue: "The statue in knowing through it[self] the quality of fragrance, has not been represented by us as knowing itself as well as the fragrance, nor as being in any manner aware that the fragrance comes to cognizance in a subjective act. In other words . . . not even the germ of a consciousness of *ego* [has] been developed." James favored a phenomonal description of consciousness, and he did not *want* to bring the idea of self-awareness into his picture of mental experience at this early point. By saying that a mind can achieve acquaintance through sensation, but without self-consciousness, James challenged the commonsense philosophical tradition perhaps best represented by the Scottish thinker Sir William Hamilton, a tradition that James considered "one tissue of sophisms from beginning to end" (Notes on Cognition, folder 4).[24]

> Knowledge consists in no inward form or manner of feeling the fragrance as "fragrance-distinct-from-me." It consists in simply the fragrance as distinct from not feeling it. What this "feeling" is can by us neither be explained nor explained away. It is an ultimate simple nature or genus of the existent, with which we find ourselves familiar, and which the psychologist must beg or assume at the outset, if he is allowed to go on with his studies. It is wholly vain to seek to resolve it or its cognitive functions into anything more general, such as the scheme or form of self-relation or self-reaction. (Notes on Cognition, folder 4)

The commonsense tradition—"a jumble of bad metaphors and bad inference"—maintained that "knowledgeless things, when they grasp and posit, do not grasp and posit themselves; therefore to know all a thing needs is to grasp or posit itself." Or, "the senseless doesn't grasp itself; but the knowing is not the senseless; therefore the knowing does grasp itself." This, James maintains, is a large mistake. "But such grasping is not what makes them acts of knowing. They are acts of knowing because they feel what they grasp, whether that be themselves or anything else" (Notes on Cognition, folder 5). James puts feeling at the base of knowing. Knowing did not presume that consciousness sought out an object. While formulating the basics of his psychology, James would not let an intentional consciousness create knowledge. Egoless, objectless feeling rules cognition at his psychology's foundation. James was seeking to demystify consciousness and establish a psychological naturalism which would eventually permeate *The Principles* and become indispensible to his world picture.

He is willing to admit, however, the theoretical possibility of objects knowing themselves; he would not close off the possibility of finding self-relating phenomena. "There may be modes of self-reaction occurring outside the realm of consciousness altogether. Roses and other senseless things may 'grasp themselves,' or 'posit themselves' in countless ways we don't suspect." But this possibility seems to jeopardize the credibility of his own position. "How then can we say [objects] are not self-related any more than we can say they are so,—self-relation being so utterly vague a term?"

His answer points toward a pluralistic cognition. "But if they are [self-related], how doubly absurd to lay down such relation as the exclusive differentia of either consciousness or knowledge!" An open question openly considered makes his approach to cognition more plausible. "No! The differentia of consciousness is material, not formal. It is simply to feel, not feel in this manner or in that. And the differentiation of knowledge is, when we feel, to feel something real" (Notes on Cognition, folder 6).

James is not equating feeling with knowledge, only proposing that feeling was a necessary *condition* for knowledge. "What we have established is that if there be a fragrance in the world and if there be a feeling of that fragrance, then the feeling is the knowledge, of the acquaintance type, but nothing more" (Notes on Cognition, folder 6). Sensation was on the ground floor of consciousness. James had posited a physiological beginning to mental life. He had followed the general evolutionary scheme of lower to higher, homogeneous to heterogeneous development. He would, like other psychological pioneers, relate the rise of a thinking mind to physiological science, to findings on the brain, and to measurements of sensation.[25] Nonetheless, defense of feeling as the elemental data of consciousness set James apart. To take sensible consciousness away from James would be tantamount to taking away his psychology: there would be no stream of thought, no self, no time or space, no emotions. *Felt* consciousness was the mortar that held his psychological house together.

HAVING BESTOWED upon the statue knowledge as sensible acquaintance, he proceeds in the Notes toward a more complex consciousness that possesses "knowledge about." How does feeling simply as itself, he asks, "grow into something more?" "Suppose that no sooner has the feeling of fragrance departed than another similar feeling succeeds it. Or suppose a different feeling to succeed, that of sweetness. Does the statue's knowledge necessarily increase?" No, he answers: "Certainly not knowledge about either of the separately acquired feelings we have supposed" (Notes on Cognition, folder 6).

Why not? Because "each [feeling is] shut up in its own skin, and cognizant during the moment it lasts of nothing but its own content." As Condillac had said, the statue "would feel each time like the first time, years would be lost in each present moment." Or, "in other words, a succession of feelings is no feeling of succession; a difference of feelings is no feeling of difference; a similarity of feelings is no feeling of similarity." Feelings under these circumstances are "absolutely disconnected from all others [and] nonexistent for all others." Only "an outside intelligence" could synthesize these separate feelings (Notes on Cognition, folder 6).

But how was this knowledgeable synthesis, this relating of separate feelings, to happen? "If a mind with any structure is to come, the second feeling must be a fact with an altogether different sort of an inward constitution from that of the first," James notes. The second feeling "must be quite complex, involving the previous sensation and its quality, plus the sense of its difference from a new simple quality, now perceived to be on hand." And this sense of its difference was a necessity. Without it, consciousness would be "mere repetition of sensations for a million years." The statue would be "no wiser at the end than he was at the beginning" (Notes on Cognition, folder 6).

Such a repetitious consciousness was unacceptable, for "it is impossible to see what use consciousness would be at all when reduced to such an incoherent rope of sand," James comments. He needs a picture of a functional consciousness. Here he would have to breathe evolutionary complexity into his statue. That meant bringing the statue into life, bringing to bear upon it a physiological understanding of the mind. It had been necessary to begin theorizing outside the life sciences in order to make initial distinctions between subject and object, between the self and the feeling. Now it was necessary to abandon the initial abstraction in order to maintain the provisional realism that his psychology had promised. Without skipping a beat, James could move from philosophy to physiological psychology.

"We are organisms of a wholly different type." he notes. "Our nervous systems are so modified . . . by each impression that falls upon them, that the feelings that accompany their changes are feelings of more and more complicated objects as time goes on." Often in our experience this complexity was obscured "by various bodily happenings, such a fatigue, sleepiness, illness and old age; and it is also true that we may lapse from a higher into a more sensational consciousness by an effort of the will." Nevertheless, James is certain that "if we abstract from these causes of variation," we would find a rule of consciousness in which "each succeeding phase represents something related to the objects of all the phases that went before." Mental life took "the form of a crescendo, among whose 'objects' new qualities and new 'things' will with equal frequency be found, and in whose stream new 'faculties' show themselves at work as fast as materials accumulate for their exercise" (Notes on Cognition, folder 6).

In describing the mind as a crescendo or stream of sensibility relating itself to the past and seeking the new within itself, James pictures a mind whose experiences were relational and ever growing. Physiological occurrences such as fatigue, sleepiness, illness, and old age merely mask the mind's natural energy. Interestingly, William often complained that these "bodily happenings" interfered with his own creative life. In a broad cultural sense, biological limitations seemed compounded by contemporary American problems such as the growing urban-industrial crowding and bureaucracy that impede natural American inventiveness. Like Emerson, James needed to believe in a free-flowing, natural mental world, one unfettered by extraneous circumstance. Such a world would promote genius, not convention; it would nurture originality, not tradition. An expansive age needed an expansive psychology.

But as he moves in the notes from consciousness as feeling to consciousness as knowledge, James is still reluctant to impose a relating self. "Self as a part of the content of feeling, discoverable by analysis therein, is one thing. Self as an agent or causal condition of feeling discoverable [by] outside observation, induction, or deductive reasoning is entirely different" (Notes on Cognition, folder 6). By not bringing the self into an explanation of the rise of knowledge, he is able to work more freely with the objects of consciousness, that is, to relate them to each other in the sensible mental stream.

For example, he next discusses "objects of our consciousness [which] have the prerogative of lingering" and those which "can be apprehended . . . in flight." Those objects which lingered tended to be substantive; those that took flight relational. When speaking objectively, we might call the substantive objects of consciousness "qualities and things." And when speaking subjectively, we might call them "sensations and syntheses of sensation." Likewise, objects that took flight in consciousness were objectively called "relations"; and subjectively they were called "perceptions of relations" (Notes on Cognition, folder 7).

In distinguishing between substantive and relational consciousness, James was fashioning another fundamental characteristic of his psychology. Consciousness contained far more than substantive objects, being constituted also through relational qualities, "prepositions, verbs, conjunctions and the like" (Notes on Cognition, folder 7). Mental life contained the nonsubstantive characteristics found in speech and grammar. James' verbal skill, his acumen for the nuances of language, loomed large in his creative power, and allowed uncommon insight into the development of mental structure.[26]

James uses the transitional qualities of language as the key to cognizing mental activity. Declensions, conjunctions, and prepositions were linguistic signs that consciousness was relational and inherently dynamic rather substantive and static. "There is every reason to believe" he observes, "that sweetness thought in one relation is subjectively 'felt' differently from sweetness thought in another relation, and that sweetness thought in any relation is 'felt' differently from sweetness taken abstractly,—even tho it be all the while the 'same' objective sweetness that is meant." Without the power to relate, the mind could not identify. "When thought in relation, there is no sweetness in the mind apart from the relations. . . . There is just one immediate object, 'sweetness-in-those-relations,' and naught besides" (Notes on Cognition, folder 7).

So how then was the initial fragrance bestowed upon the statue linked to the next sensation of sweetness! James imposes a relational necessity: "And since confessedly we somehow end by the idea of what can only be expressed in the barbarous linked-together phrase of 'sweetness-later-than-fragrance-and-other-than-it,' what objection is there to beginning with it at once?" He treats this mental linkage "not as the resultant of a lot of mental entities and agencies whose reality and properties introspection unbiassed by language can never verify, but rather as the immediate counterpart in feeling of a certain complex action of the brain" (Notes on Cognition, folder 7).

Since sensibile relations were immediately felt, no other constructions were necessary: "Let us then feel free to go on . . . and to talk as if the most complex objects might be presented to consciousness in the unity of a single feeling. Let us repeat that one feeling is what enables the statue to know the connexion between its young experiences' different parts" (Notes on Cognition, folder 7). A cognizing mind was a feeling, relating mind whose sensible and relational existence was immediately felt; and in just this immediacy the most complex consciousness possessed simple unity. Condillac's statue had allowed James to posit three qualities of consciousness that forever after characterized his psychology—feeling, relation, and immediacy. Taken together they went a long way toward constituting Jamesian consciousness.

CONSCIOUSNESS NEEDED felt relationships if a knowing mind was to develop. Large sections of *The Principles* discuss how relationship between mental qualities are built up. A relational process looms large in several chapters—especially in "Discrimination and Comparison" and "Association."[27] James does not describe the great panorama of psychological relations when considering cognition in 1883. But he does outline how the statue identifies "sameness"—"saying the fragrance felt now is the same fragrance felt then" (Notes on Cognition, folder 8).

He recognizes that the search for sameness might be more desire than fact: "The world may be a place in which the same thing never did and never will come twice." Nonetheless, "Our intention is constantly to cover the same." Given the theoretical possibility of sameness in the world, in psychology it should be called "the law of constancy in our meanings." He explains that "this would accentuate its specific character, taken as a merely psychic element; and justify us all the more in laying it down as one of the most remarkable features, indeed one might say the very backbone, of our subjective life" (Notes on Cognition, folder 8). Since psychologically we intend sameness, indeed, find sameness a necessity, psychology is obligated to explain it.

"Not all psychic life need have the principles of sameness developed this way," he observes. "In the consciousness of polyps and worms, though the same realities may repeatedly impress it, the feeling of sameness may never emerge." But with humans such unrelatedness could never exist mentally. "We, however, running always back and forth, like spiders on the webs they weave, feel ourselves constantly working over the same materials and thinking them in different ways. And this feeling of the self-sameness of an object filling two different times, forms a far higher and more valuable principle of synthesis for us than the mere sense of the continuity of time itself" (Notes on Cognition, folder 8).

Here was not only a psychological principle, but a biographical truth. For much of his life James had been creatively running back and forth between physiology, psychology, and philosophy to weave an original mental web. And he had felt himself "constantly working over the same materials and thinking them in different ways." This was a language that mirrored his own genius, the inner biography that was his own mind—consciousness charting consciousness.

Cognition advanced when consciousness grasped sameness: "With the belief that things recur in time, and that the same reality can fill both now and then, we have already made a good step onward toward the possibility of translating our experience into conceptual form,—which is the peculiarly human way of treating it, and which we must soon teach our statue, if he is to continue to resemble ourselves in any high degree" (Notes on Cognition, folder 9). When the statue sensed that the "now" fragrance was also the "then" fragrance, a simple conceptual consciousness existed.

The principle of "constancy in our meanings" would lead the psychologist to consider "the relations of reality to the 'objects' in cognition." Historically "such conclusion well needs be of the highest importance, for ever since [George] Berkeley and [Thomas] Reid wrote their respective works, these relations have formed the most hotly fought-over battle-fields of psychology." James wants to "consider a question which has always been one of the dark points in psychology. I mean the question of the manner in which, in cognition, the 'object' is present to the feeling that knows it" (Notes on Cognition, folder 9). But he would not only join the protracted fray; he would present his own psychology as a means of ending the conflict.

James observes that "all the words commonly used to express the relations between the subject and object of knowledge are metaphors derived from physical processes." Therefore, "When we say we apprehend a truth, or perceive a fact, or grasp an idea, or seize a meaning, or 'intuit' a relation, we are speaking symbolically." But to say was not the same *as* knowing. "And when we think we have in any degree thrown light on the meaning of the simple word 'know' by the use of such synonyms as these, we delude ourselves in the most pernicious way." Why? Because "the metaphorical process of knowing must then be itself explained, and its difficulties, besides being gratuitous, are as great in their way as those of the real process" (Note on Cognition, folder 9).

In order to unravel the mystery of cognition, it would be necessary to interrelate physical and psychological reality—a complicated task. And there was an even more perplexing hurdle: "The main source of difficulty in knowing how the reality is present to the mind in cognition is due to the fact the principle of sameness or constancy in our meanings plays in a curious way right into the hand of the fallacy we have already noticed so many times. The fallacy, namely, of confounding the psychologist's standpoint with that of the feeling on which he is making his report" (Notes on Cognition, folder 9). Dualism made a realistic description of cognition well-nigh impossible.

Yet, having acknowledged the difficulties in even trying to explain a knowing mind, James proceeds to do just that. "What does the psychologist's principle of sameness enable him to do?" he asks. "It enables him to say that same thing, reality, or truth can be cognized by ever so many different feelings in ever so many ways." For instance, "A sensational experience apprehends it in one manner, an act of conception in another. When thought of dimly and remotely, as when we simply feel the relations to it of a topic of which we are speaking, it is known in a third kind of way" (Notes on Cognition, folder 9). The mind felt and knew objects *pluralistically.*

> When in talking we name [an object] and linger for a distinct "realizing sense" of it to develop, it is present to the mind in very different guise from when we indicate it by a pronoun, and pass so rapidly on that no images of it arise. Again, when we are straining memory to recall it and it seems just twittering on the brink of consciousness, we think of it in a peculiar way, and one different from all the foregoing. When persons in a room are talking about it, who will say that their feelings are reduplications of each other? And who will say that the manner in which a constant reality comes before us does not alter between youth and old age? Nevertheless, as aforesaid, 'tis the same reality that appears masquerading in so many garbs. (Notes on Cognition, folder 9)

The mind's sensible stream presented objects now in one way, then in another, and in still another; there seemed to be no limit to the forms of an object presented in an activated consciousness. James is in effect de-objectifying consciousess, and substituting transience, possibility, and expectation. Hence, he names without naming, identifies without labeling, describes without accounting for. That was new in psychology, and that was genius. His mind could be satisfied neither with merely sensible objects nor with transcendental ones—only with those that flowed and fluxed as did mental life itself. Already in these notes he was managing to talk about something that was, by his own admission, impossible to talk about. He had found a way to deal linguistically with dualism, even though he had not eliminated it.[28] Here was a great American intellectual escape from traditional European approaches to describing how the mind knows.

James had outlined a picture of the mind that no tradition,

either empirical or idealistic, had drawn. Consciousness was sensibly continuous. Space and time were felt first and then symbolically constructed. Emotional life arose out of physiological changes, cognition out of sensation. All this was shown in conscious experience. True, consciousness evolved from the physiological processes of the brain, but consciousness functioned beyond the physical brain even as it reflected the brain processes. Here lay the problem of knowing. And James was far more interested in describing relationships between the brain and consciousness than in building a theory of cognition. He wanted a world of cognitive relations always in some connection with physical reality.

This was entirely fitting, given his refusal to stay within the confines of one field of knowledge. Art, physiology, psychology and philosophy were presented to his mind as concatenation, as juxtaposition. The "products" of knowledgeable consciousness were not products at all, but interrelationships. This was the natural way to describe the mind. James never deliberately tried to be interdisciplinary; he could think no other way.

CHAPTER ELEVEN

The Professional Coil

For thirty-five years I have been suffering from being a Professor, the pretention and the duty, namely of meeting the mental needs and difficulties of other persons, needs that I couldn't possibly imagine and difficulties that I couldn't possibly understand. Now that I have shuffled off the Professional coil, the sense of freedom that comes to me is as surprising as it is exquisite. I wake up every morning with it: What! not to have to accommodate myself to the mass of recalcitrant humanity, not to think under resistance, not to have to square myself with others at every step I make—hurrah!

To F. S. C. Schiller
MAY 1907

WHEN JAMES returned to Harvard in 1883 he did not come back to an institution with thousands of students, a myriad of professional disciplines, and a multimillion-dollar budget. The college was small, about twelve hundred undergraduates and well under a hundred graduate students. Annual expenditures at the end of the eighties were not yet a half million dollars, while the school endowment approached seven million. Charles W. Eliot, taking his lead from the president of Johns Hopkins, Daniel Gilman, was taking steps to make Harvard a research-oriented institution. Although Harvard had had a graduate school since 1872, Hopkins remained the country's leading graduate center until the late 1890s.[1]

One distinct sign of the rise of graduate education was the rise of faculty salaries. Both Eliot and Gilman distinguished between teaching and research by rewarding original research and publication. James illustrated the trend. Between 1883 and 1889 his annual salary rose from twenty-five hundred to thirty-five hundred dollars, since every year he published several articles in first-rate scholarly journals. After the appearance of *The Principles of Psychology* in 1890, he received four thousand a year. Between 1895 and 1899 his salary peaked at five thousand, thereafter declining to about half that figure as poor health made him a part-time instructor.[2]

For this remuneration James was expected to instruct both undergraduates and graduates in psychology and philosophy. Although he began his Harvard career by teaching anatomy and physiology from 1873 to 1876, by 1877 he had added an undergraduate psychology course, "Philosophy 4." Between 1877 and 1882 he moved more fully into psychology and philosophy. For example, in 1881–1882, in addition to physiology he taught "Philosophy 2," an introductory psychology course emphasizing Taine; another "Philosophy 4," this one on contemporary philosophy, concentrating on Mill's logic; and a graduate course called "Advanced Psychology." From the mid- through the late eighties he prepared courses on English empiricism and on logic, and one called "Questions in Psychology," which involved laboratory research for graduate students.[3] James clearly did not neglect teaching for research and writing. Throughout the eighties he shared with students the gathering force of his special psychological and philosophical perspective. Most of *The Principles* was written not on sabbatical but during years of heavy teaching duties.

Harvard's graduate department offered advanced courses over a two-year period of study. Normally the student then spent a year or two in a German university before returning to Cambridge to write a thesis and take comprehensive examinations. James was on the ground floor of the development of the Doctor of Philosophy degree at Harvard. He was G. Stanley Hall's mentor, and in 1876 the latter became America's first Ph.D. to concentrate in the "new" psychology. Three graduate students—Francis Ellingwood Abbott, Benjamin Rand, and George Santayana—received philosophy doctorates at Harvard in the eighties. With the exception of Abbott, who graduated before Royce joined the faculty, all took philosophy courses from James as well as from Francis Bowen, George Herbert Palmer, and Josiah Royce.[4]

But in contradistinction to his colleagues, James had a special task. It was one thing to offer advanced courses within a traditional discipline; it was quite another to originate them in one that was not perceived to be distinct from philosophy. Although psychology as a subject had been introduced to American professors through the writings of John Locke, and "faculty psychology" had been taught throughout the nineteenth century, it was not viewed or presented to students as an *experimental* science. James led American psychology from the faculty tradition to the experimental perspective, from viewing the mind as spirit whose activities were somehow divided into special units or faculties, to viewing the mind as a subject for natural science. To study the mind scientifically it was necessary to understand brain processes and examine consciousness through the direct experience of introspective observation. It was through these methods that James believed psychology should become a science. And in 1876 Eliot allowed him to introduce psychology as natural science into the philosophy department curriculum, hence beginning the historical separation of psychology from philosophy in America.[5]

Nonetheless, until well into the twentieth century, all psychology courses at Harvard were taught in the philosophy department. Further, in the 1880s psychology did not have the institutional and organizational structures which marked a distinct professional discipline. There was no national psychological organization, only one national journal, and certainly no production of significant numbers of graduates concentrating in psychology. Throughout the decade psychology was still generally viewed as a branch of philosophy. Indeed, one of the great tasks of *The Principles* was to clearly distinguish between psychological and metaphysical standpoints.[6]

So as James struggled to write what would become recognized as the New World's first "scientific" psychology, he introduced students to what he later called that "nasty little subject." He had them read Locke, Mill, Taine, Spencer, and Wundt; he worked with G. Stanley Hall in the tiny laboratory in the Agassiz Museum and later with a few others in a larger facility in Dane Hall; and in lectures he introduced students to psychological subjects such as sensation, perception, emotion, the self, and the will—subjects that he was busy casting into classic written form.

James initiated among Harvard students a consciousness of a science that had hitherto been considered part of metaphysics. Yet he never considered himself a scientific psychologist. He did not possess the patience for thorough laboratory work, a quality he had found lacking in himself as early as the Brazilian trip. Moreover, he never believed that psychology could overcome what seemed an ontological dualism between the mind of psychologist and the objects studied. Most fundamentally—and ironically, considering that James was credited with developing a distinct discipline—he could not narrow his multiple fields of creative vision into one area; in the end he could not accept the discontinuity between the various disciplines in which he was interested; he saw them as interconnected expressions of a greater pluralistic reality.

On the other hand, psychology could help solve mental problems that philosophy avoided or muddled. What was a sensation? What was an emotion? What was the nature of the self? James reportedly told a Harvard sophomore who had filled out his academic schedule with philosophy electives only, "Jones, don't you philosophize on an empty Stomach."[7] Psychology demanded that students be knowledgeable in physiology, that they be able to distinguish between sensation and perception, that they engage in introspection and observe the nature of consciousness. Such an education would check the tendency of philosophy to build abstractions, the elaborate fabrications that obscured rather than facilitated understanding of the mind and its relationship with the world.

Harvard's philosophy curriculum was transformed under James' leadership. He opened the door which would eventually allow students to pursue psychology exclusively rather than metaphysics and psychology together. His goal, however, was not the single-minded study of one discipline. He would have students read widely and see the scope of extant intellectual traditions. They would know Locke, Mill, Darwin, Spencer, and Wundt as they would know Emerson, Kant and Hegel. But in building between empiricism and idealism a third road, James wrote and taught a perspective, a center of vision that belonged in neither tradition. He made tangible in the eighties what hitherto had been much more abstract; he propounded in teaching and writing authoritative work that had no authoritative past. His master text, after all, was to be called *The Principles of Psychology,* suggesting by its very title that it would right the multiple errors, misconceptions, and myths that permeated western thinking about the mind. And he was making the challenge in a department that looked less and less like the one dominated in the 1860s and 1870s by the Unitarianism of Francis Bowen.

HARVARD'S PHILOSOPHY department in the eighties included, in order of seniority, Bowen, Palmer, James, and Royce. Bowen retired in 1889. He had seen his influence decline not only because of his age, but because his Unitarianism, like other religious orientations, had failed to check the advance of intellectual movements such as Darwinism, positivism, and Hegelian Idealism. Then too, Eliot's program to build a graduate school serviced by research-oriented faculty put Bowen in the position of defending the past rather than building the future. Bowen had no progressive educational plans. Philosophy, a discipline traditionally aligned with religion, was best studied informally by sincere, dedicated men; it was not something to be transformed and reorganized simply because the college administration succumbed to contemporary whim. When James returned from Europe in 1883, Bowen had been Alford Professor of Philosophy for thirty years. A careful scholar of Harvard philosophy recently concluded that as a historian of philosophy, Bowen had "no superior at Harvard . . . and he left his mark on his students and associates—Chauncey Wright, Charles Peirce, and William James."[8] Bowen's mark on James was faint, however, scarcely as distinct as that of the younger men with whom he associated including Wright and Peirce.

By the mid-eighties it was clear that with Eliot's blessing, other men—Palmer, James, and Royce, soon to be joined by Hugo Münsterberg and George Santayana—were building a different academic context. The young British philosopher and mathematician Bertrand Russell believed that Harvard philosophy under James, Palmer, Royce, Münsterberg, and Santayana was the best in the world. They were the stars of Harvard's golden age of philosophy.[9]

Yet these five men did not constitute a modern university department. Departments were not yet centers of specialized academic activity. Members of different departments associated freely with one another and often had common scholarly interests. Indeed, the social sciences developed under the umbrella of the philosophy department. Bowen himself taught political economy in 1871, as James did psychology from 1876. Palmer, James, Royce, and Münsterberg all had a social component in their thought that made them more or less transdisciplinary. Even so, a tighter organizational structure emerged. In 1891 Eliot organized disciplines into divisions. Philosophy was made the departmental division that included psychology. In the eighties departmental chairmen served indefinite terms; in the nineties they had a set tenure and began keeping records of departmental meetings. Of the five golden age philosophers, only James and Santayana escaped chairmanship duties.[10]

James found himself ever more drawn to the shy, brilliant Californian, Josiah Royce. Royce had taught James' courses when the latter was in Europe in 1882–1883. By mid-decade Josiah was a full-fledged member of the department. He and James became the friendly antagonists in a protracted metaphysical dispute over the nature of the universe, over whether the world was monistic or pluralistic.[11] Neither found his deepest fulfillment in overseeing and encouraging departmental development. Their common commitment to the life of the mind, coupled with Bowen's lessening influence, meant that it fell to George Herbert Palmer to assume the most vigorous professional presence.

Palmer was more than equal to the task. Intellectual limitations prevented him from making an original philosophical contribution and he preferred the role of critic, teacher, and departmental protector. After some initial qualms about Eliot's plans to upgrade graduate education he wholeheartedly supported them. Palmer excelled in placing students and in ameliorating disputes among colleagues; he became Eliot's chief advisor on matters affecting the department. He was an ideal institutional man, doggedly loyal to Harvard and its future. He once remarked, "Harvard University, to its glory be said, is enormously unfinished; it is a great way from perfect; it is full of blemishes. We are tinkering at it all the time; and if it were not so, I for one should decline to be connected with it."[12] For all of James' attachment to Harvard and his belief in an unfinished universe, he was not, as was Palmer, predisposed to concentrate his energies on educational "tinkering."

While James would say of Royce, "You are still the centre of my gaze, the pole of my mental magnet. When I write, it is with one eye on the page and one on you,"[13] he was typically more irritated than impressed with Palmer. Sometimes the irritation went beyond departmental matters. In 1888 both men were interested in buying the same house. William told Alice that "Palmer has claimed the Gurney House, but he got in after I did, and I doubt if his claim be regarded as the stronger. There is nothing to do with him but keep out of his way." But though James had made the first bid, the owner disregarded protocol and accepted the second offer. William appreciated Palmer's dedication and teaching ability but did not really like him. Indeed, after losing the Gurney house he found him "an insulting brute." "Yesterday the manner of his speech to me on Harvard Square was enough to make me swear never to go near him or address a word to him again," he wrote to Alice. He promised to "try to keep the resolution; but I suppose that, as usual, I shall relax and repent of it again."[14]

Conflict emerged not only because of personality differences but because the two men were almost the same age and had entered the department at approximately the same time, Palmer in 1872, James in 1873. When Bowen retired in 1889 Eliot had to decide how to bestow rank. Who was to inherit Bowen's Alford Professorship? Although Palmer technically had seniority, he had not distinguished himself with original research. And though James was clearly the superior scholar, he had one year less of professional service. Seeing Eliot's problem, James approached the president with a solution. "Saw President Eliot about [the] Professorship," he informed Alice. "He evidently was set *against Palmer's* having it, but I think I've probably convinced him to the Professorship of Psychology, which will be a very good thing for me, and for the College. They have just started one with a laboratory in the University of Pennsylvania with [James McKeen] Cattell as Professor."[15] Knowing that Eliot wanted Harvard to be American's first university of scientific education, William knew he would be concerned that a less prestigious institution already had a professor of psychology.

As it turned out, although James probably would have preferred having Bowen's title, Palmer received the Alford Professorship and James was made professor of psychology. "It is easy to see how Eliot hates to put Palmer in that position," he rationalized. "But the arrangement is infinitely better for me. It has made me pretty blue about my responsibilities to see how little work I have done the past month, and how badly I teach my logic; and to be responsible for the limited field of psychology, instead of the whole howling waste of philosophy is a great load off my mind." Significantly he vowed: "I shall *do* philosophy all the same."[16] The narrower scientific discipline, now reflected in his title, encouraged him to concentrate on completing his book. But he could never narrow the *center* of his intellectual interest to one field simply because it was officially recognized as an academic discipline.

James soon learned that he had inherited more than just the privilege of being called "Professor of Psychology." In founding a natural science he had acquired the responsibilities of a founding father. Consciousness of those responsibilities increased as he neared completion of *The Principles*. Overworked, and anxious that the protracted project be completed, he complained to Alice that "I am not one of those happy beings who can do many things at once, and the sense of insecurity during the day coupled with the lack of sleep at night have made me a complete do nothing."[17] He mentioned being disheartened to his former student G. Stanley Hall, a man highly conscious of the rise of the new science, and one who helped found the first national journal devoted to experimental psychology. Hall reminded James that he was now encircled by the professional coil he himself had created:

> It distresses me to know that you have even moments of discouragement. To this you have no right. You started this whole movement yourself and are the very best man in my opinion, in the world at the present time in your own lines and I only fear that you are working too hard on your book. The cause of Psychology in this country is more dependent upon you and your safe delivery of that book than upon anything else whatever.[18]

There were signs by 1890 that James felt his responsibility to develop Harvard's psychological laboratory. A form letter was circulated throughout the Cambridge-Boston area soliciting funds for upgrading the Dane Hall facility. Although James did not sign the letter, he almost certainly approved its contents:

> The Corporation of Harvard University, recognizes that in our day Psychology has become a distinct branch of scientific research, and following an example set by other institutions, has recently founded a Professorship of Psychology and has appointed Professor William James to fill the Chair. The system of psychological instruction now given at Harvard was introduced there fifteen years ago by Professor James, and at that time no teaching of a similar kind was to be had at any other university or college in America. . . . This year one hundred and seventy-five undergraduates take Psychology as one of their elective courses, and our graduate students come from all parts of the country.

The solicitation urged "that in order to give the new department, the efficiency which its importance deserves, an expenditure of at least three thousand dollars will be necessary." Funds were needed for new anatomical models and microscopes. And equipment was "required for the lecture-room demonstration of the physiology of the senses, especially those of sight and sound, and for a small but carefully selected library of such books as are indispensable for reference alike for student and teacher."[19] Three respected local citizens—John Fiske, J. B. Warner, and George B. Dorr—signed their support.

James found himself in charge of the institutional expansion of a new science. As Hall had implied, the completion of *The Principles* made him the paterfamilias of American psychology. On May 17, 1890 he shared the elation of finishing with Alice. "The Job is done! All but some paging and 1/2 a dozen little footnotes and the work is completed and as I see it as a unit, I feel as if it might be a rather vigorous and richly colored chunk—for that kind of thing at least." But his jubilation was not tempered with a promise to spend the balance of his career nurturing institutional psychology at Harvard. Rather he felt reborn, "now that this big job is rolled off my shoulders like Christian's miserable pack."[20]

Indeed, after spending twelve years writing a book which made large claims that psychology was a natural science, James denied that in it he had established the field's scientific authenticity. "As psychologies go, it is a good one," he wrote Henry; "but psychology is in such an ante-scientific condition that the whole present generation of them is predestined to become unreadable old medieval lumber, as soon as the first genuine tracks of insight are made."[21] But how could he make genuine tracks of insight if bogged down with laboratory work and responsibilities for teaching hundreds of undergraduates?

James had never assumed that the most original psychological thinking came out of the laboratory. Introspection seemed to him the natural way of describing the mind.[22] Moreover, his mind demanded continuous creative excitement. A finished science that scholarship would thereafter simply refine cramped the imagination, constricted creative powers. He wanted to keep psychological science in open juxtaposition with other fields—as we shall see, with psychical phenomena and metaphysics. After completing his landmark book, James largely rejected the professional-scientific authority that he had tried so hard to develop in the late seventies and eighties. He took steps to free himself from the institutional responsibilities his scientific pioneering had engendered. The mind, its shape, its relationship to the world, its destiny, continued to be James' deepest concern.

JAMES' FIRST step in loosening the professional coil would be to find someone willing to direct the psychological laboratory, someone to develop a research program for graduates, someone with the stature to attract the best students and keep Harvard psychology America's finest. Hugo Münsterberg was the man James found to replace himself in these areas. Münsterberg had also based his psychology on physiology. He had made professional waves with his book *Die Willenshandlung* (Voluntary Action), challenging Wundt's position that the sensation of effort came from the brain rather than the muscles. James naturally sympathized with anyone who had the courage to attack Wundt. He had befriended Münsterberg at the International Psychological Congress in Paris in 1889. Hugo was twenty-one years James' junior, ambitious and quite anxious to secure a first-rate academic position. The two men discovered similarities in background. Like William, Münsterberg had developed an interest in aesthetics, had studied medicine, and had then moved into psychology without abandoning metaphysics.[23]

In 1892 "M," as William came to call him, agreed to take a year's leave from the University of Freiburg to come to Harvard. James hoped the temporary appointment would lead to a permanent professorship. He enthusiastically reported to Royce that Münsterberg's "indefatigable love of experimental labor had led him to an extraordinary wide range of experience, he has invented a host of elegant and simple apparatus, his students all seem delighted with him, and . . . everyone recognizes him to be, as a *teacher* far ahead of everyone else in the field, whatever they may think of his published results."[24] After protracted indecision about whether to leave Germany for Harvard, "M" accepted a permanent appointment in 1897. He would direct the laboratory and graduate study in psychology until his sudden death in 1916. Later James learned, as did everyone else in the philosophy department and even President Eliot, that Münsterberg could be an officious, meddling busybody as well as a gifted psychologist.[25] But in the nineties the transplanted German was crucial to James' plans to pass the institutional leadership of Harvard psychology to more willing hands.

James continued to take an interest in psychological affairs as departmental leadership in the fast-rising discipline shifted to Münsterberg. The success of *The Principles* led him to prepare a simplified condensed volume, *Psychology: Briefer Course* (1892).[26] In 1894 he was named president of the new national organization, the American Psychological Association. He became engaged in a debate over the place of psychical research in mainstream psychology in the late nineties.[27] And even shortly before his death in 1910, after refusing because of failing health to accept the presidency of the proposed 1913 International Psychological Congress, James tried to settle a dispute among psychological contemporaries over the Congress' presidency.[28] He remained the symbolic leader of American psychology, even though experimentalists such as Cornell's Edward Bradford Titchener and Columbia's James McKeen Cattell felt he had left scientific psychology for the fad appeal of psychical research.

JAMES' ESCAPE from institutional responsibility was more apparent than real, however, for with psychological success and symbolic leadership came public demands. Americans at the end of the nineteenth century saw the new psychology as a boon for solving social problems and relieving personal anxieties. The nineties had been unusually turbulent as the nation faced a paralyzing rural and urban depression, agrarian radicalism, and a spate of industrial strikes. The decade was a watershed era in which old values no longer seemed to sustain Americans through social and personal crisis.[29] Great cultural turmoil evoked among the middle classes what one writer has recently called a "culture of psychology," in which the new science to some extent took the place of religion as the hoped-for modern solution to a host of modern problems.[30]

James saw that American teachers were particularly curious about the educational applications of psychology. In 1891 Harvard appointed an instructor of the history and art of teaching, a step that culminated in the establishment of a Division of Education in 1906. Both Royce and James gave numerous lectures to teachers throughout the nineties and the first decade of the new century. James' psychological reputation and academic position virtually assured demand for a teacher-oriented psychology textbook. In 1899 his *Talks to Teachers* was published, a book that combined psychology of particular interest to educators with inspirational and therapeutic lectures.[31] The professional coil had a public dimension that James found impossible to ignore.

One occasion in the summer of 1896 illustrates the bittersweet consequences of psychological fame. As usual James was "fagged" from the rigors of the academic year, but he had agreed to be a guest lecturer at a teacher's institute in Buffalo, New York. He also planned to make an impromptu stop at Chautauqua and another at Syracuse to investigate a case of double personality which he hoped to include in a series of fall lectures, "Extraordinary Mental States," before hurrying on to Keene Valley for some much-needed rest. The trip started out on a sour note: "I keep having the most awful feelings about my display of temper," he wrote to Alice. "It is awful for me always to hurry off from you, as I seem now so fatally to do, after giving you such a scene—but time will assuage this as it has all misunderstandings, and meanwhile you know that my violence was only the irritable weakness of tired nerves."[32]

After several days of lecturing in Buffalo he arrived at Chautauqua, America's leading adult education summer camp. His hopes for staying incognito were quickly dashed. "The authorities nosed me out pretty soon. Unable to say no I have agreed to give my Power through Repose thing," his letter to his wife continues. The "Power through Repose thing" was an early version of the 1899 essay "The Gospel of Relaxation," which he included in *Talks to Teachers*. Although the lecture went well and the crowd was large—"1200 or more"—he was exhausted, and made a promise he would not keep: "I don't want any more sporadic lecturing—I must stick to more inward things."[33] Lecturing not only sapped his energy, it detracted from serious thinking and writing, depleted his creative powers just as surely as did being tied to the laboratory or advising psychology students.

James also discovered a dimension to fame that made him uneasy. Some in the Chautauqua audience viewed him as a moral therapist of extraordinary power. One morning he breakfasted "with a methodist parson with 32 teeth." The minister's wife revealed that she "had my portrait in her bedroom with the words written under it, "I want to bring balm to human lives!!!!!" "It was" he exclaimed to Alice in disbelief, "supposed to be a quotation from me!!!" He reflected on the terrible goodness of addressing such people as the minister's wife. "You bet I rejoice at the outlook of removal from the burdens of public lecturing," he exclaimed. "Even an Armenian massacre whether to be killer or killed, would seem an agreeable change from the blamelessness of Chautauqua as she lies soaking year after year in her lake side sun and showers. Man wants to be stretched to the utmost, if not in one way then in another."[34]

The experience made an impact, for he elaborated on the blamelessness of Chautauqua in the essay, "What Makes a Life

Significant," also included in *Talks to Teachers*. "The moment one treads that sacred enclosure one feels one's self in an atmosphere of success. Sobriety and industry, intelligence and goodness, orderliness and ideality, prosperity and cheerfulness pervade the air." But such perfection suffocated James. He found himself wanting "to enter the dark and wicked world again" and to thereupon utter, "Ouf! what a relief! Now for something primordial and savage . . . to set the balance straight again. This order is too tame, this culture too second rate, this goodness too uninspiring." He craved to "take my chances again in the big outside worldly wilderness with all its sins and sufferings. . . . There is more hope and help a thousand times than in this dead level and quintessence of every mediocrity." He cried out in sarcastic desperation, "The flash of a pistol, a dagger, or a devilish eye, anything to break the unlovely level of 10,000 good people, a crime, murder, rape, elopement, anything would do."[35]

Yet was it really only Chautauqua that moved James to castigate the mediocrity of the educated American middle class? Was he not also upset with himself? After all, he had willingly prescribed self-help therapy for himself and others for years. Even *The Principles of Psychology* had posited a mind that gave the individual power to change circumstances, merely by shifting his or her attention. Indeed, James' science rested not only on defining an active, choice-seeking consciousness, but in recognizing such consciousness as moral. Further, many of his best-known essays played to the reader's emotional need for individual action in an emerging modern world ever more governed by forces seemingly beyond personal control. Their very titles promised therapeutic and morally upright solutions to perplexing contemporary difficulties: "The Sentiment of Rationality," "The Dilemma of Determinism," "Is Life Worth Living?," "The Moral Equivalent of War," and most of all, "The Will to Believe"—the quintessential American argument for the validity of individual choice.

The very people who embraced the popular James were the Chautauqua crowd he loathed. He realized that his public career not only detracted from doing "inward things"; popular lecturing also demanded that he perpetuate Chautauquan mediocrity in himself as well as in his audience. Over and over he vowed to cease and again and again he lectured, even after his health made such ventures hazardous. In 1905 after lecturing in Chicago on another popular topic, pragmatism, James wrote to Alice, "I had a queer sense of the absurdity of the thing I was doing . . . to these virgin natures to whom it never occurred to think things in those aspects at all."[36]

As James entered his last two decades, he would increasingly find his creative powers at odds with the social duty to give morally responsible lectures. He could leave the psychological laboratory to Münsterberg, but he could not resist the moral obligation to promote his psychological-metaphysical perspective. Yet to see into the world, to formulate reality was not necessarily compatible with protecting public morality. In one sense James was amoral; his originality led where it would. He also hated pretense in himself as much as in the Victorian middle class that embraced his psychological therapeutics. The tension between creative drive and moral responsibility increased as his public reputation grew.

SEVERAL WRITERS have noted James' need to abandon not only professional psychology but the whole structure of academic life. "There was something in William James which was profoundly opposed to the whole life of scholarship, whether teaching, research or the making of books," remarked colleague and biographer Ralph Barton Perry.[37] Assessing his recently deceased friend, John J. Chapman observed, "It has sometimes crossed my mind that James wanted to be a poet and an artist, and that there lay in him, beneath the ocean of metaphysics, a lost Atlantis of fine arts: and that he really hated philosophy and all its works, and pursued them only as Hercules might spin or as a prince in a fairy tale sort seeds for an evil dragon, or as anyone might patiently do some careful work for which he had no aptitude."[38]

James regularly complained about being overburdened with teaching duties, even early in his tenure at Harvard. "I have been disappointed during the past month after beginning the new year very well to find myself getting fagged and unable to do much work," he reported to Kate Havens in 1876. "With Science and the reputation of one's college hounding one on as they do its a sorry thing to feel all the time that you can't respond to the call for lack of strength."[39] Nor with the passing years did he develop immunity to academic rigors. At the end of the spring term in 1888 he protested: "Time goes, but I don't seem to do much. The fact is, I'm fagged and the quicker I get away the better. I'm interrupted every moment here by students, come to fight about their marks."[40] He could barely wait to flee Cambridge each June, and occasionally graded final exams en route and posted them back to Alice so she could deliver them to the college.

Being "fagged" at the end of the academic year, of course, is the usual professorial complaint. But James rarely made comments about enjoying teaching. By the late nineties, even before injuring his heart, he asked the Harvard Corporation to decrease his salary by $500 so that he might reduce his teaching load from eight to five hours for one term. He explained to President Eliot that "my working powers are below average; and the condition of my doing anything at all satisfactorily seems to be that I should not have too many duties to think about."[41] In December 1900, he first seriously considered resigning. Although by then health was the deciding factor, the litany was familiar. He explained to Corporation member Henry Lee Higginson that "there is a cumulative amount of nervous wear and tear involved in preparing and delivering lectures at the sound of the bell, through so many weeks of the year, which is great and far in excess of the intellectual output proper. I can work my small intellectual capital far more economically and with more profit relatively to the animal expenditure . . . by the use of the pen than by that of the tongue."[42]

It would be a large mistake, however, to assume that James loathed teaching and loved writing, especially writing books. The protracted composition of *The Principles of Psychology* was anything but pleasurable. At one point publisher Henry Holt, tiring of what he viewed as unwarranted procrastination, accused James of reneging on a promise to deliver the completed manuscript by May 1, 1890. Holt reminded him that "I have just seen a contract signed by you to give us that MS. June 12/80 and yet you, you brute revile me for being a demon!"[43] But James replied that the delay in finishing was even more intolerable to *him.*

> Your fatal error however has been in not perceiving that I was an entirely *different kind* of author from any of those whom you had been in the habit of meeting, and that *celerity,* celerity incarnate, is the motive and result of all my plans and deeds. It is not fair to throw that former contract into my face, when you know or ought to know that when the ten years or a little more from the time of its signature had elapsed I wrote to you that you must get another man to write this book for you, and that, as things were then going, I didn't see how I could ever finish it.[44]

What Holt had failed to recognize was James' almost constitutional inability to work *slowly*. "Celerity" or speed—the "shooting together" of ideas he had alluded to in 1882—was essential to his creativity. He hated to plod, yet plodding was what writing *The Principles* required. "I have written every page 4 or 5 times over, and carried it 'on my mind' for nine years past," he wrote to Henry.[45] This protracted, painstaking work—a labor that incorporated over 2,000 citations—went against the grain of James' naturally celeritous mentality. "I'm a 'motor,' and morally ill-adopted to the game of patience," he once remarked.[46] Finishing the book was a twelve-year struggle not only with the enormous literature of the rising new science; it was a contest waged against the core of his temperament.

Moreover, James took little satisfaction in the resulting product. To Holt he exclaimed:

> No one could be more disgusted than I at the sight of the book. *No* subject is worth being treated of 1000 pages! Had I ten years more, I could rewrite in 500; but as it stands it is this or nothing—a loathsome, distended, tumefied, bloated dropsical mass, testifying to nothing but two facts: *1st,* that there is no such thing as a *science* of psychology, and *2nd,* that W. J. is an incapable.[47]

But he was relieved, even amazed to be done, proud of conquering himself, if not science. "At any rate, darling," he wrote to Alice, "it does give me some comfort to think that I don't live *wholly* in projects, as pirations and phrases, but now and then have some thing to show for all the fuss." He who had written only book reviews and essays had fashioned a 1,414-page tome. "The joke of it is that I who have always considered myself a thing of glimpses, of discontinuity, of apercus, with no power of doing a big job suddenly realize at the *end* of this task that it is the *biggest* book on Psychology in any language except Wundt's, [Antonio] Rosmini's and Daniel Greenleaf Thompson's." "Still," he confided, "if it burns up at the printing office, I shan't *much* care, for I shan't ever write it again."[48] And he never did. There would be but one edition of *The Principles;* all future printings were made off the original plates.[49]

A RARE PHOTOGRAPH shows James leaning casually against Putnam Shanty at Keene Valley. It contrasts markedly with

William relaxing with friends at Putnam Shanty, Keene Valley, New York, the American place he loved most, probably in the late 1880s. By permission of the Adirondack Museum, Blue Mountain, New York.

most of the other photographs we have of him, formal portrait-like images depicting a distinguished, bearded gentleman in starched collar, bow tie, and suit. The Putnam Shanty James was comfortably attired and at ease. At Keene Valley William was no longer thinking, as he told philosopher F. S. C. Schiller, "under resistance." In the woods he was free to hike, observe, and imagine, while in Cambridge he was yoked to the teaching duties and writing projects that a professional career demanded.

James had chosen the academic road. He had been unwilling to follow Henry Sr.'s vocationless example; indeed, he had ambitiously sought academic fame. And he had taken on the added burden of public lecturing. But he consistently rebelled against the academic regimen and looked for opportunities to travel to Europe, or to vacation at Keene Valley and Chocorua. As a young man William seemed always en route from America to Europe or South America, and from one vocational interest to another. In fact he had experienced remarkable educational freedom. Even during periods of depression, as in Dresden in 1867–68, his mind roamed imaginatively. In this sense, his rebellion against what Perry called "the whole life of scholarship" may have stemmed simply from memory of the more relaxed, creatively expansive years he had enjoyed before taking up teaching, writing his book, and public lecturing.

Yet the deeper strain in James embraced Victorian "civilized morality," the obligation to carry on the moral customs of his age.[50] The strenuous life, the striving will, the professional and public duty were built into the turn-of-the-century Victorian ethic. The need to cast off civilization at Keene Valley was an antidote to an exceptional inner burden; exceptional because James' constitution, above and beyond the prevailing cultural code, was naturally irresponsible, impatient, and easily bored with convention. Academic work was unnatural. The emphasis in *The Principles* on establishing habits, the great importance given to choice and will, was, beyond its psychological dimension, a code to insure that *he* did not lapse into a professionally and socially unconcerned self. His creative products—books, essays, lectures—for all their deference to "individualism" and "freedom of choice," were not antisocial. Quite the opposite: James, as George Santayana saw, perpetuated the same "genteel" values he had so sarcastically attacked in his remarks about Chautauqua.[51]

James was not altogether conventionally Victorian. His uncommon intellect, uncommon education, and pathbreaking intellectual productions obviously set him apart. But one does not find in James the kind of society-be-damned perspective found in his great European contemporary, Friedrich Nietzsche.[52] The professional coil and marriage saved James from himself, but they also subtly set limits to his radicalism, kept his psychology on a social leash. As Perry and Chapman suspected, James had within him a great potential rebellion against all social and moral prescriptions. At Chautauqua he had spoken of "a melancholic patient" whose "mind is fixed as if in a cramp on this sense of his own situation." He urged people "to forget their scruples and take the brakes off their hearts and let their tongues wag as automatically and irresponsibly as they will.[53] But he rarely found it possible, except perhaps fleetingly at Keene Valley, to take his own advice.

CHAPTER TWELVE

Solidly Established

I stole out and lay on my back on the floor of the old shanty-piazza with the scene before me on which our eyes rested so often during that sad-sweet honey moon summer. . . . If we could only have foreseen then how solidly established we should be twenty years later, it would have done us good at that time.

To Alice Gibbens James
SEPTEMBER 1897

DURING THE years James was working on *The Principles of Psychology,* he not only became immersed in the "professional coil," but in the cares of the extended James family—Alice, the children, and his sister and brothers, especially Robertson. After returning from the European trip of 1882–1883, he settled into a pattern of care and crisis. He sought—or so it seemed—the safety, the circumscribed, comfortable life of the upper middle class college teacher—the professional rank, the house, the summer cottage, the praise of colleagues and public. All the while he fretted about the well-being of loved ones, about family finances, and about progress on the book.

His health and spirits were good when school resumed in the fall of 1883. Forty-one, alive with novel psychological ideas, a personal friend of first-rate thinkers in Europe, the husband of a devoted wife and two growing boys, James had every reason to be optimistic. All he had to do was write the promised psychology.

Within weeks, however, there were domestic complications. In late September Wilkinson disclosed that he had Bright's disease, a severe kidney disorder made worse because of a chronic heart ailment. Wilky wrote that he had "lost nearly all my appetite—so that it looks as if it would not be long before I shall peg out . . . getting no rest nor sleep without the aid of drugs."[1] William immediately left for Milwaukee. His arrival so excited Wilky, however, that James stayed only a few days. After a brief rally Wilkinson died, on November 23. His death brought William's role as elder brother under unexpected pressure.

Henry Sr. had not given Wilkinson an equal share in the estate, since the latter had been advanced thousands for the Florida plantation and a business venture in Wisconsin. Since Wilky's wife, Carrie, had a wealthy father who agreed to provide for his widowed daughter and children, William saw no need to provide extra income from Henry Sr.'s estate. Robertson James, however, thought this was callous and shortsighted. "I cannot tell you how sorry I am to learn from you of your attitude toward Carrie!" he wrote to William. Bob felt that Carrie had been snubbed by the James family, and he wanted to "make some amends to Carrie for what I consider has been her social treatment from the James' in Cambridge up to the time of Mother and Father's death." He also noted that Wilky had not only made no provision for Carrie or the children, but died "leaving a heavy debt of honor unpaid to her father."[2]

Robertson believed that Carrie was hurt deeply by the family's social pretensions. "I have reason to know that she felt the indifference very keenly. . . . During the nine years of Carrie's married life she not alone never received an invitation to visit Cambridge, but was actually repelled from coming when her husband tried to provoke some expression of willingness that she should be allowed to go." And he chastised William in a way that eventually made him reverse his decision. "*It seems to me that our sense of honor ought to be touched* and that we ought to be too glad to wipe out as far as we may and in as generous a manner as we can what seems to me at least a dishonorable and the only dishonorable chapter in the history of our family."[3] Robertson asked William to send his letter to the other Jameses, meaning to sister Alice and brother Henry.

The long-term problem, however, proved to be not the distribution of estate money, but Robertson James himself. Bob's excessive drinking and on-again-off-again marriage with Mary Holton James brought William into the thick of domestic tur-

ABOVE, *Alice James, William's sister, ca. 1873. By permission of the Houghton Library, Harvard University.*

ABOVE RIGHT, *a young Garth Wilkinson James ("Wilky"), ca. 1861—William's second-youngest brother, who was severely wounded in the Civil War. By permission of the Houghton Library, Harvard University.*

RIGHT, *Robertson James ("Bob") in his fifties, ca. 1900—William's youngest and most troublesome brother. By permission of the Houghton Library, Harvard University.*

moil—at a time when such distraction, with its inevitable emotional disturbance, hardly promoted his efforts to complete *The Principles*. Bob was not only the youngest James, he was the black sheep. He had felt himself a superfluous child, drifting in the wake of his two famous older brothers. And for him the war had not been the great life-transforming event that it had been for Wilky.

To some extent, William had unknowingly laid the groundwork in the seventies for trouble with Robertson. Bob had suffered some of the same symptoms of back problems and depression that had plagued James. And William had provided medical and moral advice, playing the role of the wise, experienced older brother.[4] Thus for years William had willingly counseled Robertson, leading the latter to expect his help during difficult personal circumstances. Shortly after returning from Europe William had shared with Bob his feelings on the death of their father: "The gap left [by] my father is even greater than I supposed it would be when away, and makes me feel sad. The age comes to all of us when we first feel what a fleeting hour life is upon this mundane stage and death ensuing after death keep deepening the impression till we ourselves are swept away. Life can never again seem solid to me as it did a year and a half ago."[5] Yet life was more solid by far for the older than for the younger brother. William was experiencing professional success and a fulfilling family life. Robertson was without a profession and domestically in limbo, as marital difficulties with Mary Holton James and a problem with alcohol destabilized his existence. Bob was a perplexing, irritating, and at times embarrassing problem for William throughout the eighties and nineties.

Biographers have generally assumed that William and Henry, because of their differing temperaments, experienced the most acute emotional-intellectual competition among the James siblings. Leon Edel has explained their relationship as paralleling the biblical rivalry between Jacob and Esau, but Gerald Myers has recently questioned whether they felt such a deep-seated animosity.[6] Whatever the truth about William's feelings toward Henry, within the James family there was a much more tangible and volatile brotherly relationship. From the early eighties Robertson lived in the Boston area and eventually settled in Concord. "Bob" was the topic of conversation in William's letters to Alice much more than was Henry. Indeed, Alice wrote to Henry that William had cared for Bob for "so long he has got the habit of it."[7] Whatever the differences between Henry and William, they were contained by distance; after the 1860s their relationship was essentially that of pen pals: they exchanged letters rich in insight and social observation, but the band between them was largely an intellectual one. Robertson, however, was on hand in the flesh and presented to both William and Alice the antithesis of the very life they were successfully building.

In October 1884, Bob showed up in Cambridge drunk. He had left Mary in Wisconsin, who subsequently wrote to William about an affair her husband had indulged in. "The moment I saw him I knew he must have got into his drinking habits again—for nothing else could explain his corpulence," James replied. "He soon . . . told me of the singular love affair to which you allude, and of his declaration of it to you. He seemed shocked and surprised that it should have 'crushed' you." His brother's aberrant behavior made William a marriage counselor:

> I don't well know what the future has in store for *your* relations with Bob, but I strongly think you need not apprehend a great deal from this particular complication. His mental life is entirely discontinuous, he shoots instantaneously from one mood to its opposite, and is the sport of every jest. He has declared his intention of becoming a teetotaler. Heaven knows how many days it will work. . . . Bob seems to me a mere hollow shell of a man, covering up mental disease. I shall always be tender with him, because I seem now to be his only friend here and it is good to have one refuge. He is his own worst enemy—pity it is that you should be buffeted about at his chariot wheel.[8]

Perhaps William saw in Robertson the living reality of what he might have become, the shadow of his own nature, the ups and downs that he had carefully controlled to nurture the moral and the professional life. Perhaps his brother's mental illness made him keep a surer grip on his own normality. Whatever Robertson represented, he continued to trouble William and take time away from his work on the psychology.

By the spring of 1885, Bob had moved to Concord and was paying Mary child support from his share of the estate. William wished "he were harder pressed . . . for then he would have to go back to work." What really bothered William was his brother's drifting, unproductive existence. "I believe that the longer he leads this aimless life, the more inveterate the habit will become and the worse on the whole it will be for his future," he wrote to Henry.

Robertson justified his affair by maintaining that the young woman had "sworn she would wean him from drink." William found Bob "the strangest mixture of extremely manly and unmanly qualities."[9]

Bob also caused concern because he wanted to sell his shares of stock held in a family trust—stock which William's money manager, Henry Lee Higginson, felt would appreciate.[10] This put pressure on both William and Henry to find mortgage money to replenish the capital. "I got 3 rather exasperating letters from Bob yesterday," William wrote to Alice. "I think it barely possible he may be going to have an attack of insanity. Of course I can never again consent to be responsible to him for anything connected with the estate, and wish to heaven I had agreed, two years ago, when he began to worry, to let him have independent dealings with the agent."[11] William had offered to lend Bob money, but his offer was declined.

A relieved William reported to Henry in early September that "the doubts about Bob have been solved by his signing . . . a deed making Warner [the estate lawyer] and me trustees of all his property. The poor creature is constantly beset by temptations of the most morbid kind . . . and finally made this proposition himself." But grave doubts about Bob's long-term prospects remained: "Though he has [had] no spree since Spring, I fear he is running down morally and mentally from so long a lack of regular occupation and from such a constant strain of emotional excitement. I doubt whether he will ever be got to face regular work again."[12]

For over two years Robertson seemed somewhat better, living in Concord, temporarily reunited with Mary. But in early 1888 he relapsed and William again found himself in the thick of Bob's troubles: "Poor Bob has had another crisis," he wrote to Alice, who had traveled to South Carolina with son Billy, who was suffering from asthma. "Mary telegraphed me yesterday A.M. He had spent Monday night in town, he knew not how, had cut and bruised his face, was in his most pathetic mood, praying to be confined, saying he was really insane, etc." Edward Emerson, the son of the Concord sage, urged that Bob be put in prison or a reformatory, but William thought that "the only good thing about the reformatory is that prisoners are forced to work, in other respects, I should think it would legitimately drive Bob mad."[13] Nonetheless, some kind of institutionalization seemed mandatory. William took the train to Hartford to visit a mental hospital catering to alcoholics, Butler Asylum. He was convinced his luckless brother was doomed "to spend his life . . . between two contrasted emotional states, the irascible before, and the pathetic after, his sprees."[14] On February 6, 1886 William accompanied Robertson to the Butler Asylum.[15]

"He had been so savage with Mary all the winter, and with me that a break up was desirable," James explained to Alice. It seems possible that Robertson may have either threatened or possibly even physically assaulted Mary. Finding a solution to Bob's difficulties was improbable, even though James maintained that "he isn't technically insane." There was but one truly acceptable alternative: "The only manly and moral thing for a man in his plight is to kill himself, but Bob will ne'er do that, I'm sure," William concluded.[16]

He felt that suicide would be a face-saving escape for Bob, a means of ending his own sufferings as well as the shame he brought to the family name. While personally radical, it would have been socially acceptable—at least to the Victorian sensibility—as a solution to a situation that in William's judgment could only get worse. But as William predicted, Robertson refused "manly and moral" suicide and continued to drift in and out of crisis. After leaving Butler Asylum he remained under a doctor's care.

Money matters complicated the tension and anxiety which surrounded the unpredictable Robertson. Following sister Alice's death in 1892, Bob was slighted in the estate settlement. Both Henry and William offered to equalize the bequests, but the prideful younger brother declined. William, however, felt that Bob's refusal was actually quite sensible. "*Practically* he can do without the money better than we can, for if our earning power stops . . . we experience a terrible tumble, whereas he can't possibly tumble any lower than he now is, having ceased earning many years ago, and his wants being in equilibrium with his receipts."[17] Besides, William knew that Robertson obsessively spent any extra money on liquor.

Early in 1895, William's Alice reported to Henry about "our profound immersion in [Robertson's] problem." William had just left with Bob in route to "Danville, a great sanitarium near Buffalo." She described the bizarre happenings which had precipitated her brother-in-law's internment.

> He came Friday evening very drunk. The cabby who brought him and lifted him onto the porch told me that

> he found him in Bowdoin Sq. "knocking about." After some difficulty the man got the name of our street and number and so brought him out. Billy [the James's second son] luckily appeared just then and got him to bed. William had been very nervous all day and I had urged him to go to his Club dinner in town, so he did not see Bob till morning. . . . Imagine our amazement therefore at finding Bob in the library next morning perfectly sober and strangely *sane*. . . . He asked me if I had ever thought that his mind was failing. I was so wretched about him I told him the whole truth; that it *was* failing . . . and that another six months of such racking himself could have but one end.[18]

William had informed Bob, who was now very willing to be institutionalized, that "inebriates are no longer allowed in the State Insane Asylums," but were interned in "*Inebriate Asylums*." And although Bob "never flinched" at the possibility of being interned in "a semi-penal institution where the men are all locked in together in a common ward," William had proposed the Danville Sanitarium—an alternative "poor Bob fairly clutched at."[19] Robertson was periodically a patient at Danville throughout the late nineties.

By 1897 Bob had become involved with a married woman, one Mrs. Schishkar. It is unclear whether she was divorced or simply estranged from her husband. In any case, William found out from Robertson's doctor that his brother was living with her. William reported to Henry that the physician had enlisted his aid in convincing Bob to leave her home, even though the former also maintained that the woman was Robertson's "only prop in a lonely world, with only drink and ruin if she withheld her countenance." Bob refused the advice of both his doctor and his brother, however, and continued the liaison.[20] His stubbornness convinced William that Bob had degenerated to the point that "I don't believe he can live many years longer at this rate."[21] For his part, Robertson accused William of betraying the affair to Mary, who was still in Wisconsin. William told Alice that Bob "left me saying he should no longer view me as a brother."[22]

In early 1898 Robertson was back at Danville, a place that Alice thought suited him; "it is possible that they may keep him steady, but I am less hopeful than I was," she commented. She told Henry that Bob had asked William for money to purchase shoes, "and as soon as William had gone he went to the village and got drunk."[23]

By the summer Bob was in Cambridge visiting Alice. "What a nice time we might be having in Cambridge 10 to 12 days hence, if Bob were not there," William wrote from Keene Valley. "Perhaps you had better come up to save yourself from him."[24]

In the last decade of both their lives the relationship between the brothers settled. Robertson achieved some stability in Concord, where he lived with Mary and their children. He had considerable artistic talent and had begun to paint in earnest; one of his works has been preserved in the Concord Public Library archives. In 1907 William concluded that "Bob and Mary . . . seem to have come into pretty smooth water after Bob's tempestuous life. He is certainly in better plight than for 30 years past, and is wonderful as a talker and perceiver."[25] But by 1907 William was desperately ill with heart disease, and had passed the point in his own life at which Robertson could upset him. They would die within weeks of each other in the summer of 1910.

For all of James' sympathy for human frailties, he met in Robertson the limits of his own tolerance. In modern parlance, Robertson made William up tight; he forced his older brother into the posture of a conventional citizen, a defender of social codes and professional propriety. Bob represented an interesting antithesis to the life William was so fervently living—a life of domestic stability, productive work, and wide reputation. He was a man who had avoided the professional coil altogether, a man who did not have a good marriage, steady work, or his eldest brother's social standing. Yet Robertson was closer to William in temperament and physical appearance than either Wilky had been, or Henry. One wonders whether William might not have thought on more than one occasion that there, but for Alice, was the man he could have become.

TROUBLES WITH Robertson, the outsider, were complicated by circumstances inside William's immediate nuclear family. For one thing, three more children were born between 1884 and 1890—Herman in 1884, Mary Margaret in 1887, and Alexander in 1890. An enlarged family made William considerably more concerned about financial matters. Though his Harvard salary increased regularly throughout the decade,[26] and dividends and rents from stocks and property provided extra security, James was by no means sanguine about his income.

William worried about the Syracuse property inherited from

Henry Sr., which needed renovations that strained the family budget. "I am just back from Syracuse. The necessary alterations will cost $2,500 . . . inclusive of 500 for new sidewalks which the city ordered us to lay."[27] By 1888 James reported to Alice that taxes and travel costs to Syracuse had all but absorbed any profit from the property; "Our debts at the moment are . . . almost enough to wipe it out."[28] Frequent trips to Syracuse caused more anxiety than modest rental income could assuage.

The Jameses also purchased two major properties. In September 1886 while vacationing in the New Hampshire White Mountains, William bought seventy-five acres of land on Chocorua Lake. The property had "fine oak and pine woods, [a] valuable mineral spring, two houses and a barn." He enthusiastically wrote to Henry that "two thousand five hundred dollars will give us the place in fine order. It is only 4 hours from Boston by rail and 1 hour's drive from the station. Few neighbors, but good ones' [and the] hotel a mile off. If this is a dream let me, at least, indulge it for a week longer."[29]

The Chocorua place would provide a summer haven for the growing family in a setting not so rustic as Keene Valley, and one much closer to Cambridge as well. William realized that the purchase identified the Jameses with the materialism of the successful Victorian business and professional classes. "Verily the College Professor must beware of becoming a ridiculous stick-in-the-mud old fogy if he despises too much the ways of the successful and enterprising business classes and nouveaux riches who keep such places afloat," he told Alice with a hint of mixed feelings about the purchase.[30] Repairs to the buildings cost more than expected and kept William traveling up to New Hampshire in the off season to inspect the improvements. He never did enjoy Chocorua as much as Keene Valley, though he well appreciated that having the place was a social consequence of being caught in the professional coil. James understood bourgeois pretensions, was often critical of them, but largely embraced the social scene of the upper middle class anyway. Not only did he not deny himself or his family the good life, he actively sought it.

In the spring of 1889, after several years of unsuccessfully trying to find an appropriate house in Cambridge, William decided to build a home at 95 Irving Street. "I have . . . been computing the *entire* cost of the house . . . It won't be less than $15,000, and may even get a little beyond it," he wrote to Alice. The house would be sizeable, with three stories, three fireplaces, and an immense, twenty-two and a half by twenty-foot library-study. To offset the cost James considered taking in a tenant. "It is mainly the square thing to do, and will save us much in the way of entertainments commanded by outside expectation, as well as give us two hundred dollars and upwards in the way of rent."[31]

The thought of paying for a fifteen-thousand-dollar house made William extremely uneasy. Yet he vowed not to let the purchase alter his plans to go once again to Europe:

> Now that the waves are lulled I can tell you that I had a terrible day yesterday with my financial thoughts. I got up at 4 and began footing up everything and could think of nothing else. We shall live in that house next year and pay whatever it comes to, with or without a tenant. The following year we shall run at Europe. I am calmed down again, only hope I haven't agitated you too much. The house itself is *noble*. I hate to relinquish identity with it.[32]

Alice, however, stilled her husband's anxiety by observing that the Irving Street house was indivisible from their married destiny: "It's very queer, this place—the burden of it—the inevitableness, and all its sweetness," she had written. And James agreed: "That is exactly what I've been saying to myself all day about our new place here. Its like marriage, and all the bigger things of life, they're burdens in one sense but they're inevitiable and they're sweet, E.G. the children!"[33]

These burdens had became painfully evident in July, 1885. The year had been filled with illness. In February Alice had contracted scarlet fever. Then in June, little Herman, or "Humster," who was born in January and named after Professor Hermann Hagen of the Lawrence Scientific School, caught the whooping cough, which he then passed on to his mother. Alice's mother, Elizabeth Gibbens, lived nearby and was able to help tend mother and child, but William found "everything . . . in a confused condition."[34] The baby worsened and died on July 9th. William described the loss to his cousin, Kitty Prince:

> Our Little Humster, whom you never saw since his first boyhood has just gone over to the majority. We buried him yesterday under the young pine tree, at my father's side. For 9 days he had been in a desperate condition, but his constitution proved so tenancious, that each visit of the doctor found him still alive. At last his valiant little soul

> left the body at nine o'clock on Thursday night. He was a broad, generous, patient little nature, with a noble head who would doubtless have done credit to his name had he lived. It *must* be now that he is reserved for some still better chance than that, and that we shall in some way come into his presence again. The greater part of the experience to me has been the sight of Alice's devotion. I thought I knew her, but I didn't, nor did I fully know the meaning of that old human word *motherhood.* Six weeks with no regular sleep, 9 days with never more than 3 hours in the 24, and yet bright and fresh and ready for anything, as much on the last day as on the first. She is so essentially *mellow* a nature, that when the excitement is gone and collapse sets in, it will be short and have nothing morbid about it.[35]

And indeed Alice not only fully recovered, but two years later was pregnant with her fourth child. For the father, the birth in March 1887 was both expected and unexpected. He wrote to his wife's mother, "This morning . . . came a *daughter*—the living image of her mar [mother]. I kept talking of it as a *he* and *him,* from force of custom."[36] Little Mary Margaret, named for her two maternal aunts and known as "Peggy," helped the family overcome Herman's loss. Still, ill health continued to cause alarm, especially when Billy developed asthma. Taking no chances with the capricious late winter climate, the family sent Billy south with his grandmother to Aiken, South Carolina.

William was ever more conscious of parenting during the period when he wrote the great bulk of *The Principles of Psychology.* Fortunately, Mrs. Gibbens and her daughter Mary helped tend the children. And William, with his classroom duties and regular summer travels to Keene Valley and then to Chocorua, was able to isolate himself enough from the domestic scene to make progress with his writing, especially after 1885.

As he neared completion of the book, he once again could not resist travelling abroad—even in his worry about the added pressure the trip would put on his indebtedness for the new house. The domestic burden could rest with Alice—where it had mostly been anyway—while he visited with Henry in London and attended the International Psychological Congress in Paris. "You are pulling the wagon now," he told her, "and pulling it so successfully that the old sentimental adoration is wrapped with a new sort of gratitude and veneration."[37] What William did not tell Alice was what was obvious: her skillful homemaking not only allowed him a hiatus from his children, but encouraged him to travel abroad once again without worrying that his absence would undermine domestic order.

THE TRIP to Europe in the summer of 1889 lasted scarcely a month, as James visited briefly with Henry in London and traveled on to Paris and the International Psychological Congress, after revisiting the old James family seaside resort, Boulogne-sur-Mer. Being abroad, however, whetted his appetite for an extended European stay, a sojourn that finally materialized in 1892–1893. The protracted strain of writing *The Principles* and the problem of finding a successor to run Harvard's Psychological Laboratory had taken their toll. "I am tired," he wrote to Henry, "and being expected to keep some dozen magnificent youngsters making discoveries [in the laboratory], on penalty of Harvard University losing all prestige, etc. you may imagine that the discord between powers and responsibilities rather weighs on my spirits, and makes me long more and more for the year abroad to set the balance straight."[38]

Then, too, although Peggy and the infant born in late December 1890—Alexander Robertson, nicknamed "Tweedy"—were too young to much appreciate Europe, the older boys, Henry ("Harry") and Billy, would benefit greatly. Indeed, Henry was approximately the same age William had been when Henry Sr. embarked upon the 1855–1858 trip. It was time the boys learned French and German and began thinking about prospective vocations. Henry seemed more intellectual and inclined toward literature. Ten-year-old Billy had already developed, as had his father, a talent for drawing. James was not as obsessed with educational experiment as his father had been; but the thought of combining his rest and his sons' education was nonetheless irresistible. On the eve of departure he wrote to his former philosophical mentor Shadworth Hodgson, "I have 15 months furlough on half pay, and *ought* to make a good thing of it. . . . The prospect of the possibility of a little unimpeded *reading* is most sweet. We shall probably go to Paris or its immediate neighborhood, and put the boys in school there."[39]

But several months before the Jameses departed, another death saddened William. His sister Alice had succumbed to breast cancer in London in early March 1892. William and Alice had been close throughout childhood. Alice adored her oldest brother, while William loved to confide in and gossip with his little sister, whom he variously addressed in letters as "Dear Sweetling," "Dear Child," and "Cherie de Soeur." Their relationship had become more complicated, however, when William married Alice Howe Gibbens. Alice James' biographer has suggested that William's marriage precipitated her nervous collapse in 1878; and for a time Alice Gibbens James was cool, if not hostile, toward her new sister-in-law.[40] Later, relations between the two Alices improved, especially after the deaths of Mary and Henry James. But with Alice living as an invalid in London and a married William pursuing the academic life in Cambridge, brother and sister had never regained their earlier intimacy.

Beyond distance and disparate ways of life, there were marked differences in temperament that became more pronounced with the passage of time. William was gregarious, active, and fundamentally optimistic; Alice was introverted, detached, and fatalistic. After learning of her death he remarked,

> Poor little Alice! What a life! I can't believe that that imperious will and piercing judgment are snuffed out with the breath. Now that her outwardly so frustrated life is over, one sees that in the deepest sense it was a triumph. In her relations to her disease, her mind did not succumb. She never whined or complained or did anything but shun it. She thus kept it from invading the tone of her soul. . . . Her life was anything but a failure.[41]

And William was right. Two years later, after reading Alice's diary, he praised the composition that was later recognized as a minor classic. "It produces a unique and tragic impression of personal power venting itself with no opportunity. . . . It ought some day to be published. I am proud of it as a leaf in the family laurel crown."[42]

Actually Alice had bequeathed to William something more tangible than her diary. She left her brother an inheritance of twenty thousand dollars, which was paid within three months of her death.[43] And if the windfall was not what made the proposed European trip possible, it at least greatly reduced the strain on the family budget.

So by late May 1892, the entire family was aboard the *Friesland* bound for Antwerp. As it turned out they went directly to Freiburg, Germany, where William visited Hugo Münsterberg and convinced him to come to Harvard and direct the laboratory for a year; his hope was, of course, that the German psychologist would stay permanently, a hope that was eventually realized in 1897. The Jameses spent the rest of the summer of 1892 in Switzerland, where the children learned French and William climbed the Alps. In September they traveled south to Italy and wintered in Florence. The next spring they returned to Switzerland. Their last stop was London, where for a few weeks Alice and William vacationed alone for the first time since their honeymoon in 1878. The family returned to America in September 1893.

In view of her generous legacy, Alice's death was a bittersweet event, quickly forgotten in the bustle of the European trip. William was soon immersed in the immediate trial of living with four growing youngsters abroad. "Why didn't you warn me against coming abroad with Infancy?" he asked Henry Bowditch from Freiburg. "Practically we have all lived, quarreled, cried and rolled about in one room for 5 weeks. There are signs now, however of better times ahead. . . . Meanwhile the combining of a *holiday* with novel and perplexing responsibilities about the education of one's children and the most intimate and incessant contact with their bodies and souls is an idea worthy of a lunatic asylum."[44] In fact William was not getting the rest he had looked forward to. Alice reported to Henry that "the boys miss the house routine and the little ones are quite upset by the change. William has not been sleeping and this adds to the difficulties of deciding on our winter place—whether it shall be a French or German one." By late June William had escaped to search for an apartment for the family in Switzerland. Alice, alone with the children, had second thoughts as well. "I sometimes ask myself why I brought them so far from home?"[45]

James' jaunt to Switzerland was the first of several solo excursions. After considering keeping the boys in Swiss or French schools for the winter they finally decided, with pressure from Alice, on the balmy clime of Florence. They were installed in early October in "a large sunny apartment, very ugly but clean and roomy with excellent beds. We have two sitting rooms far apart so William can use . . . his study and really be very quiet."[46] But within a month William was off to Padua and Venice where he once again considered art, as if reworking the Dresden scene and perhaps even remembering the first stirrings of his imagination gazing at Delacroix in Paris years and years before, or when he had enjoyed Italian painting with brother Henry during his 1873–1874 trip.

"I surrender to Italy, and I should think that a painter would almost go out of his skin to wander about from town to town," he wrote to Alice. "One wants to paint everything that one sees in a place like this."[47] And from Venice he indulged the old habit of recording his mullings over art. "I noted down a good many things about paintings, but I doubt whether I can ever get up my famous lecture."[48] James wanted to rekindle his aesthetic creativity, but there is no evidence that he was successful this time. His imaginative pendulum was now swinging toward metaphysics. Indeed, it is indicative of his growing philosophical interest that at a ceremony honoring the two hundred fiftieth anniversary of Galileo's death, James received the honorary degree of "Doctor of letters and philosophy."[49]

In February 1893 William left for Munich, where he planned to make arrangements for the boys' schooling. Back in Florence the next month, he complained to Josiah Royce that the "sweet rottenness of Italy disagrees with me, the air is so devitalized that I long to get away."[50] By late April, Henry was at school in Munich, while Billy was studying at Lucerne. The family was happily situated in the Swiss city. Alice confided to her mother that "I am myself thankful to get out of Italy. I am no longer young enough to enjoy the general decay and shabbiness and ever present poverty of the people."[51] She felt remorseful about pressing Florence on William, and reminded her mother from Munich, where she was visiting young Henry, not to tell William that she had "outgrown Italy." "Don't refer to it, for I would not worry William . . . it was mostly my doing choosing Florence."[52]

But if Alice was too old for the "decay" of Italy, she was not so venerable as to turn down the chance for a second honeymoon. By mid-June William and Alice were alone together enjoying London. "Here we are in such comfort and luxury as I have never known," Alice told her mother. "William takes such good care of me that I don't feel at all forlorn in the midst of these Superiorities."[53] And William wrote to Shadworth Hodgson that it was glorious "being off on a lark *together,* for the *first* time since our honeymoon."[54]

Nevertheless, after a year in Europe William was not refreshed, and Alice held herself to blame. "I know not, nor why you should not be able to see me depressed in spirits without holding yourself responsible," he scolded. "The logical ground of my depression has been to see that after spending all this money, I am about the same in health, and that the year has gone with very little reading done. *You* have nothing to do with it at all."[55] But William in this instance was remarkably insensitive to the burden Alice had carried ever since their "tragical marriage" compact. She had good reason for thinking that if William was out of sorts she must share a portion of the blame. After all, he had said repeatedly that she had saved him from himself; and she had from the beginning felt immense responsibility for his well-being.

After returning to Switzerland for the remainder of the summer, they booked passage on the *Cephalonia* and were back in Cambridge by the first week in September.

JAMES FELT strangely empty after the fifteen months abroad. Although Münsterberg had come to temporarily assume the responsibilities of the laboratory, he had now returned to Freiburg. The expense of the trip, the lack of any great intellectual stimulation, and the prospect of resuming responsibility for experimental psychology depressed him mightily. "I had a pretty bad spell, and know now a new kind of melancholy," he admitted to the English psychical researcher, F. W. H. Myers. He made numerous visits to a mind-cure practitioneer, a Miss Clark, who had greatly benefited Alice's sister, Mary. At first she seemed to help William as well, who claimed to be "enjoying an altogether new kind of *sleep,* or rather an old kind which I have been bereft of for so many years that I had forgotten its existence." Clark had also helped cure two intimate friends of his and convinced William that her powers should be exercised on a broader scale: "I should like to get this woman into a lunatic asylum for two months, and have every case of chronic delusional insanity in the house tried by her . . . I may possibly bring it about yet!"[56]

Yet after a few months mind-cure no longer helped. "I am much blocked by the miserable Miss Clark business," he complained to Alice, "no sign whatever of benefit, though I have been to her every day this week."[57] His troubles seemed a personal shadow of national woes, which included a financial panic, a broad and deep economic depression, and labor unrest. James' son Henry later remarked upon the "complete absence of effervescence" in his father's letters from 1893 to 1899 and the noticeably depressed tone of some observing, "W. J. was in fact overworking himself and overdriving himself. He continued to do so more or less constantly until he broke down in 1899."[58]

William was certainly hard at work, becoming every more involved in psychical research, and struggling, as we shall see, to formulate a metaphysics of pure experience. He entered another period of intense creative activity, but with the exception of *The Will to Believe and Other Essays in Popular Philosophy* (1897) and *Talks to Teachers* (1899)—both of which were largely compilations of public lectures—James published no academic book. Indeed, in the nineties his development of a public persona became almost a second career.

In the summer of 1895 he lectured to teachers in Colorado Springs. He both loathed and enjoyed his new popularity. "Cities poison me in the summer time. The country does too, but I stagger on," he told Henry, to whom he recounted his recent schedule. "I have been a fortnight in Cambridge, lecturing 8 days to a lot of middle aged school teachers. Three other days quite idyllic (save for bad hotel) at Norwich, Conn; and now after 3 days and nights on the train have been 6 days in this operatic place, where I have given 2 out of a course of six lectures."[59] Yet James enjoyed imparting knowledge of the yet novel discipline of psychology to teachers. He remarked to Alice about his lectures, "The great marvel of them to the school-marms seems to have been that they were *intelligible*[;] such a thing was not to be expected of psychology according to their previous experience." He admitted to Alice that "I have been wrung by conflicting emotions—the desire to get home and opportunity to see more of this magnificent country have been at war in my breast all night."[60] The country won, for he decided to stay on for several more days.

James' love for rural living showed up in the pattern of his summers. Generally he would end the academic year and rush to Chocorua or Keene Valley, then fulfill the lecture obligations, and end the summer in Keene Valley. It was the latter retreat that attracted him most, the one place he found contentment throughout the nineties. "Oh the sweetness of this place dear Alice—such a heavenly peace and happiness have flooded me since I have been here," he wrote after returning from the Colorado trip. "You *must* not be low spirited when life is so good and your *husband* so swimming with happiness of which associations with your holy girlhood form the central element. . . . I am writing this only to express my joy to you and sense of the unfading beauty of this spot. . . . The boys . . . must be Adirondackers."[61] He was blissfully at ease. "I have been happy, *happy, happy*!—with the unperishable beauty of this place, the place I know so well. Nature had made it for falling in love in, passing honeymoons and the like. . . . This place is soaked with memories of you. I wish you could see it again."[62]

Keene Valley restored his perspective. He could take day-long hikes, often alone with a book and his thoughts. If he wished, there was social activity at the Shanty in the evenings, when the Putnams, Bowditches, and other prominent New England families put on plays and joined in song fests. As sister Alice had seen years before, "William's panacea for all earthly ills [is] the Putnam Shanty."[63] And in some incalculable way the Shanty rekindled both his sense of romantic union with Alice and a feeling of moral simplicity, which were inseparable in his mind. In June 1895 he wrote, "I have a little shanty all to myself, and the cedar posts, the balsomic smell, the rain streaming from the eaves into paper mache pails set out to catch the water, did carry me back most forcibly to that July 1878 when we first truly made each other's acquaintance in the Putnam shanty all by ourselves. I began to feel as if I were living on a normal moral plane again. It is curious how Society poisons me in seeming to take me away from the real springs of my life."[64]

A year later he exclaimed, "The past caught at my heart in the woods—my old boyhood here . . . the living you and all—I don't know a more exquisite emotion. . . . We must never forsake each other!!!"[65] Alice, again, must have recognized not only the genuine admiration but her continuing burden as well.

In truth, despite the periods of depression, William experienced a sense of personal achievement in the late nineties, a feeling of having at last gained a steadiness and solidity. His children were fast becoming adolescents with only the seven-year-old "Tweedy" still clearly a child. His professional standing at Harvard was unsurpassed; and a yearly salary of five thousand dollars plus book royalties and lecture fees reduced the financial anxiety he had experienced after purchasing the Irving Street house. By 1897 Münsterberg had decided to accept a permanent Harvard position, thus relieving James of the nagging laboratory problem. "I believe I am getting onto a more steady and normal plane of work for a few years to come, with M—g responsible for psychology," he wrote from Keene Valley.[66] "I stole out and lay on my back on the floor of the old shanty-piazza with the scene before me on which our eyes rested so often during that sad-sweet honeymoon. . . . If we could only have forseen then how solidly established we should be twenty years later, it would have done us good at that time."[67]

In 1897 he was offered both the prestigious Aberdeen and Edinburgh Lectures, and finally decided on the latter. But Alice was upset because William had suggested to Aberdeen officials that Josiah Royce would be a better choice than himself. She thought that Aberdeen's knowledge that they were William's second choice would somehow lose him the Edinburgh opportunity. Alice feared that her husband's generosity prevented him from getting his fair share of public glory. To Henry she commented, "If ever a man was slow to take the just-reward of his labours, and quick to avoid recognition, it is my dear William. Bless him. And Royce has not even said, 'Thank you!' "[68] As James' fame spread, Alice found her husband too self-effacing. She understood that William was now without question the Philosophy Department's great man. And she quickly sensed that his reputation extended far beyond the Harvard Yard.

In May 1897, Alice found an opportunity to impress upon William the fact that he was a truly famous man. When asked to give a memorial address on the occasion of dedicating a St. Gaudens statue of Civil War hero Robert Gould Shaw, James was sincerely surprised by the popular reaction. The audience at Boston's Music Hall listened "with rapt attention from the first word to the last," Alice wrote to Henry. "The cheering was *tremendous* and long continued. Speaking of it to me afterwards William said 'Did you notice that applause, it looks as if I were popular.' He will never know how many people here care for him."[69] William's delay in recognizing his own fame was suggested by his banker, Henry Lee Higginson, who had also given a speech on Shaw at the Sanders Theatre. Higginson remembered James remarking, "As for our speeches, yours was infinitely the more impressive, being the work of an honest man, and not that of a professional phrase-monger and paid rhetorician. Those are BAD devils."[70]

But William did see that popularity provided extra income. The sales of the *Briefer Course, The Will to Believe,* and *Talks to Teachers,* plus lecturing fees, meant he could invest in his family's future. "I have got finance very much on the brain," he confided to Henry.

> I am looking towards a gradual withdrawing from teaching within the next ten years. I have managed to put by 2000 a year for the past five—I have put 500 into my house mortgage within a month, and although the work is hard and the family very costly, it keeps balancing on the right side. I am practically spending at the rate of 10,000 a year—

> a fearful thing of which the only *extravagant* item is *education.* The boys don't seem in the least to be spendthrifts, and never need pocket money. Next year I shall earn a 1000 less by stopping outside lectures and instruction at Radcliffe. . . . It overworks me, and the Gifford Lectures will be arduous.[71]

But James did not, indeed could not, stop lecturing. "The public here grows increasingly eager to listen," Alice told Henry. "William in two days had five invitations to lecture. I think they average one a day."[72] Promises to stop lecturing and the inability to do so became a nearly constant refrain for the next ten years. "I am *never* going to lecture again outside," he resolved in April of 1898. "That is a fixed point and we shall shape our existences accordingly. I wish most devoutly that I didn't have this Bryn Mawr thing before me."[73] In fact, James found it difficult to turn down the money. Upon receiving the official letter confirming the Edinburgh lectures he noted, "The emoluments are at the ridiculous figure of 'about 700£' for each of 10 lectures each or $350 per lecture. I never knew before how much my time was worth an hour!"[74]

THERE WAS a curious denial in James. He consistently refused to admit that he enjoyed being in the public eye. In 1898 he was called to testify in the Massachusetts state legislature on the issue of denying faith healers and mind curers the right to practice their persuasions as medicine in Massachusetts. James defended their therapies as being as legitimate as the more traditional ones. The issue was part of a long-standing contest between "regular" and "irregular" physicians in nineteenth-century American medicine. More broadly it reflected the increasing power of professionalism in national life, a trend that James himself had participated in and yet had criticized. William's friend James Jackson Putnam had also spoken out in favor of allowing irregulars to continue to practice even though he told William that he did "not have any great feeling of fondness for them."[75]

Yet there was another matter that generated William's concern besides the licensing issue. James felt that he had to defend himself to Putnam, who had apparently suggested that William found the publicity surrounding his role in the controversy (he had been quoted in the *Boston Transcript*) stimulating. "On page 7 of the Transcript tonight you will find a manifestation of me at the State house, protesting against the proposed medical license bill," he related to his friend. "If you think I *enjoy* that sort of thing you are mistaken. I never did anything that required as much moral effort in my life. My vocation is to treat of things in [an] all around manner and not make *ex parte* pleas to influence . . . a peculiar jury."[76]

The next day he wrote to Putnam that "the rhetorical problem with me was to say things to the Committee that might neutralize the influence of their medical advisors who, I supposed, had the inside track and all the *prestige*. I being banded with the spiritualists, faith curers, magnetic healers, etc., etc., strange affinities!"[77] James found that he now had a reputation to defend rather than to make. Aside from a real sympathy with mind cure and psychical research he did not want to be associated with quacks. But his public interest in the issue had raised just that speculation, a charge that had been growing among James' psychological contemporaries throughout the late nineties.

The kind of public image that James projected may be glimpsed in a 1896 letter of his to Henry, written after the conclusion of a lecture series to Chicago teachers. William enclosed "a report of my Chicago lectures."

> The chief benefit for those who had already studied Professor James's Psychology resulted from the removal of things they knew before, but had been thinking of lately, and from the personality of the speaker. This last source of pleasure and benefit is very considerable, for Professor James is an exceeding fine looking man, with a high forehead, iron gray hair and beard, blue eyes, a resonant and agreeably modulated voice, *immaculate attire and the bearing of a perfect eastern gentleman* [James' emphasis]. He spoke without manuscript, with only occasional reference to notes, illustrating his talks frequently by rapid crayon sketches of the necessary mechanism of the brain.

Although he exclaimed, "You see why I hate to go to Europe,"[78] the flattering if stilted image conveyed by this Chicago article perhaps suggests the side of these Chautauqua-like occasions that James could not resist.

James had his share of vanity. Traveling to places where public deference and applause made him feel special was seductive. He had long sought reputation as well as creative accomplishment. By the late nineties he was reaping the rewards of twenty years of becoming established. His moral responsibility to the public was real—but so were feelings of awe and personal importance in the public limelight. For the last decade of his life James was seriously ill, and he derived less and less pleasure from popular lecturing. But even then he could seldom resist the social scene that made him the center of the public's vision.

For all his love of nature, solitude, and Keene Valley, James was exceptionally gregarious. With a popular reputation he could travel, meet interesting people, be the man of the hour and get paid to boot. As had Emerson and Henry Sr. before him, James took to the lecture circuit and further enlarged his popularity as well as his income. When fame came, he accepted it; and, despite constant disclaimers, he sought an even wider audience in Europe as well as in America. It was not simply the aptness of his subject, his gentlemanly demeanor, and his intellectual power that made James popular. He pursued celebrity and sold his own show. He was part of the popular culture of the Gilded Age, and although he at times loathed its inherent mediocrity, at others he encouraged what he hated. By the late nineties William had created more than an experimental psychology: he had developed a demand for his person and for his therapeutic educational-psychological message, a demand that grew ever larger as he attempted to satisfy it. Moreover, his public increased their demands on his time in the very period during which he entered into his last great intellectual speculations.

CHAPTER THIRTEEN

Conquering the Realm of Consciousness

Isn't the difficulty this?—to get out of a solipsism without jumping a chasm?

Pure Experience Notebook
1897–1898

JAMES' NEED to loosen the professional coil, to leave laboratory psychology to Hugo Münsterberg, to extricate himself from heavy teaching duties, to spend more time in the Adirondacks, were not merely signs of an anti-institutional temperament; nor were they solely related, after 1898, to his intensified efforts to regain his health. Famous and reasonably secure financially, by the mid-nineties James found little at Harvard—excepting Josiah Royce—to challenge his intellectual curiosity. Chronic restlessness coupled with unending dissatisfaction with his contributions in psychology and philosophy predisposed him to other explorations, other arrangements of his intellectual fields.

After creating the realm of consciousness in the seventies, and charting its regions in the eighties—climaxed by the publication of *The Principles*—he now sought new theoretical ground. Rather than being content to put the finishing touches on a perspective already gained, he sought yet another creative world. Paradoxically, as his health declined, his creative desire accelerated. Having plowed, sown, and harvested in several fields, his mind, like his person, sought the wilderness places, the untracked, unnamed, and even undreamed zones. He stalked the region inhabited by "wild beasts of the psychical desert."[1] He tried to conquer the dualism his psychology and metaphysics had uncomfortably assumed. He gathered together waning time and energy for one last assault on the citadel of consciousness.

There has been increasing interest in the past decade in James' later intellectual work. Indeed, scholars writing introductory essays for the definitive edition, *The Works of William James,* have reconsidered *Pragmatism* (1907); *A Pluralistic Universe* (1909) *The Meaning of Truth* (1909); and *Some Problems of Philosophy* and *Essays in Radical Empiricism,* which appeared posthumously, in 1911 and 1912.[2] These works give ample evidence that the later James wanted to work through the *fact* and *function* of consciousness toward the *destiny* of consciousness. His labors up to around 1895 revolved around two major questions: what was the function of consciousness? and what was consciousness like? Now he speculated about such matters as where consciousness went after death; whether consciousness in fact existed; and whether the mind was operative within an open, pluralistic, growing universe.

These changed emphases exposed a religious strain in James that had never been far removed from his thinking. In a sense his speculative interests moved closer to those of Henry Sr. But long association with experimental psychology and academic philosophy checked the tendency to embrace full-blown mysticism, preventing him from lapsing into his father's simple faith and self-abnegation. Acutely aware of numbered days and fragile health, William strove mightily to explain fully how the mind related to the objective world. He was intellectually taunt, more fundamentally serious, even desperate, than at any period since the late sixties and early seventies. To picture him basking in an open, pluralistic universe sustained by pragmatic solutions to momentary metaphysical puzzles is to badly misrepresent his mind and mood.

SINCE THE mid-eighties, James had described consciousness as continuous—the stream of thought flowed, the mind sensibly felt experience as an unbroken unity. He had taken great care to show how the substantive elements of the mind did not stand in isolation but were always relational; the conscious world was felt as a seamless whole. Yet in the mind's experience there were obvious discontinuities. Death ended the sensible flow. Amnesia obliterated memory. Dreams and pathological states such as hysteria and split personalities appeared as interruptions, as aberrations in the continuous mental flux. These seeming exceptions to the ever-moving, ever-sensible stream of thought fascinated James. And he gave some attention to mental pathology in *The Principles*, considered the case for an unconscious mind, discussed the disassociated personality, and devoted a chapter to hypnotism.[3]

Yet he had not really brought psychopathological phenomena into the descriptive center of his psychology: namely, his view that the mind integrated mental states into the stream, or rather that the sensible stream *was* the mind, whether perceiving or cognating. His great objective was a description of normal consciousness. Unlike Freud, who built a psychology that worked from pathological to normal mentality, James proceeded in the opposite direction, from normal mental life to abnormal mental conditions.[4]

Nonetheless, James sustained a keen interest in the abnormal mind. He had dealt with aberration in himself and in the James family, particularly. The perennial Victorian neurasthenia and its not-much-outdistanced cousins, hysteria and melancholia, were much suffered, much treated, and much discussed. Indeed, prominent New England families seemed to suffer an unusually high incidence of mental illness.[5] Psychological problems came to be almost a badge of class identity; woe to the healthy individual who had not at least experienced nervous disease, or some other debilitation. James' category of "healthy-mindedness" would have held very little interest if readers of *The Varieties of Religious Experience* had perceived themselves as Saluses, glowing with health and free from all mental complaints.

New England intellectuals had taken a pronounced interest in the mental underworld, an interest that dated from the Puritan period and the clerical crusade against witches. By the nineteenth century, revivalism had occasioned waves of religious fervor with attending conversions that produced what James called "twice born" souls. Moreover, a kind of nether world had emerged filled with spirit communications—the rappings, séances, and mediums which bespoke popular interest in what James and others would put under scientific scrutiny. New Englanders were especially receptive to those who found psychical linkages between the living and the dead.[6]

James' publications on psychopathology date from the late 1860s. Early book reviews deal with hypnotic states such as sleepwalking and spirit writing.[7] Even before he received the M.D. degree in

1869, James was fairly well acquainted with scientific attempts to classify mental illness. He had studied insanity under Brown-Séquard while in medical school and continued to discuss it with fellow medical student James Jackson Putnam, whose own family history was rife with mental illness.[8] He read literature of pathology, such as Henry Maudsley's *Physiology and Pathology of the Mind,* during the darkest days of his melancholia of 1869–1870.[9]

By the early seventies James also began to "scientifically" observe evidence for a spirit world at the scene of its alleged presence, the séance. "Being desirous of 'investigating' Spiritualism, I went a few days ago to see a medium who was said to raise a piano in broad daylight," he wrote Dresden companion Kate Havens. "She was a deceiver performing the feat by means of her wonderfully strong and skillful knee." But the charade did not end William's fascination with the spirit world. "If I go on investigating I shall make an important discovery: either that there exists a force of some sort not dreamed of in our philosophy . . . or, that human testimony, voluminous in quantity, and from the most respectable sources, is but a revelation of universal human imbecility."[10]

James put off the great search for this new force in philosophy for eight years, though he continued to write reviews of psychopathological literature.[11] Then, while in England during the professionally oriented 1882–1883 trip, he became acquainted with the founders of a new organization, the Society for Psychical Research—Edmund Gurney, F. W. H. Myers, and Henry Sidgwick. The society wanted to objectively examine psychical phenomena that were usually either blindly accepted or just as blindly dismissed. As a leading scientific psychologist who was receptive to the psychical realm, James was a welcome ally. "What we want is not only truth, but evidence," William reported to Thomas Davidson, who had scoffed at the society's antispiritual bias.[12] By 1884 James had founded an American branch of the English organization. Ten years later he was chosen president of the parent society.[13]

In 1885 he discovered a medium who could not be shown to be a fraud. Elizabeth Webb Gibbens, William's mother-in-law, visited a Boston medium who reportedly had astounding powers—not only of calling back the departed, but of relating details about family life no outsider could possibly know. Alice joined her mother, and together they convinced William to come observe the remarkable powers of Mrs. Leonora Piper. During the winter of 1885–1886, often accompanied by George Herbert Palmer (Royce scoffed at psychical research), James attended regular Saturday "Cabinet Séances." William found Mrs. Piper extraordinary: "My impression after this first visit, was that Mrs. P. was either possessed of super-normal powers or knew the members of my wife's family by sight and had by some lucky coincidence become acquainted with such a multitude of their domestic circumstances as to produce the startling impression which she did. My later knowledge of her sittings and personal acquaintance with her has led me absolutely to reject the latter explanation, and to believe that she has super-normal powers."[14]

James did more than attend séances. He compiled a questionnaire involving over seventeen thousand responses on trance and hallucination; he carried on an informal campaign to convince orthodox American psychologists to at least open their minds to psychical phenomena; and he published regularly in the *Proceedings of the American Society for Psychical Research.*[15] Psychopathological states such as split personality and hysteria seemed to him closely related to a medium's communication with her "control." The spirit world was a tantalizing field for psychological exploration, one with far more appeal to James than the conventional, unimaginative psychology of laboratory measurements.

The more James investigated psychical phenomena, the more he became convinced that the great obstacle to understanding was not the dissembling of mediums so much as scientific prejudice against taking such phenomena seriously. He wrote to Carl Stumpf, "I believe there is no source of deception in the investigation of nature which can compare with a fixed belief that certain kinds of phenomenon are *impossible.*"[16] And he was particularly incensed at the "brass-instrument" psychologists, the laboratory-oriented men who claimed sole responsibility for determining the scientific validity of psychical phenomena. He asked, "Shall the direct dealing of our minds with life, descriptive or non-descriptive, passively consent to be first falsified and mutilated and then . . . side tracked over to certain professionals called psychophysicists who claim to be the only persons licensed to give any account?"[17] In the nineties James was often attacked for having abandoned scientific psychology for the humbug of psychical research.[18]

There has been much interest in James' fascination with things psychical. Interpreters have linked it to his personal depression, to his father's mystical leanings, and to the New England proclivity for spiritual experience.[19] He was the first American to review the work of Freud; and his influence on James Jackson Putnam was crucial in inspiring the latter to become an American pioneer in psychoanalysis.[20] Recently, Eugene Taylor has reconstructed James' 1896 Lowell Lectures on Exceptional Mental States, thereby linking the normal psychology of *The Principles of Psychology* with the psychopathology treated in *The Varieties of Religious Experience*.[21] A psychical James has emerged alongside the philosophical and the more orthodox psychological ones.

Others have been bewildered, even embarrassed that James—a great scientific psychologist and practical-minded philosopher—was beguiled by the airy world of psychical research and the taint of psychopathology. After William's death, his son Henry tried to keep public exposure to his father's psychical interests at a minimum: "I would rather not use psychical research material as a [Harvard] manuscript exhibit," he wrote his mother in 1920. "It seems to be giving a sort of emphasis to something that has already been over-emphasized by the newspapers."[22]

Ralph Barton Perry devoted a short chapter of his James biography to William's psychical interests, interests secondary, in his view, to the philosophical ones. An open temperament made William receptive to a host of new phenomena, regardless how bizarre. And New England acquaintances encouraged forays into eccentricity and unconventionality. "He grew up in a circle in which heresies were more gladly tolerated than orthodoxies. Men like his father and his father's friends who were attracted to Fourierism, communism, homeopathy, women's rights, abolition, and spiritism were not likely to have any prejudices on mediumship, clairvoyance, mesmerism, automatic writing, and crystal gazing."[23] Perry's explanation is revealing: he *excuses* James' interest in psychical research—though he obviously disapproves—justifying it as typical Jamesian enthusiasm for unpopular causes. He treats psychical research as extraneous to James' central philosophical interests.

But Perry's interpretation contradicts James' own mode of thinking. Since William assumed that fields of thought were never isolated, but always in relation to each other, the meaning of one would invariably be revealed within the context of another; no field could be fundamentally extraneous. If all of James' intellectual interests were interrelated—though he might be emphasizing now one, now another—they did not present themselves to his mind in isolation, as exceptions, or as heresies, or as oddities to be dismissed for more serious-minded objectives. Rather, James would have asked, what was the meaning of psychical phenomena? And how did that meaning express itself in the context of another field? Lecture notes and notebooks written between the mid-1890s and 1905 suggest his answers to both questions.

JAMES' INTEREST in psychical matters peaked in the mid- and late 1890s. Between 1893 and 1898 he offered a yearly seminar on mental pathology. And in the fall of 1896 James delivered a series of lectures for the Lowell Institute at Huntington Hall in Boston's Back Bay. The Lowell Lectures on Exceptional Mental States presented eight psychopathological topics—"Dreams and Hypnotism," "Automatism," "Hysteria," "Multiple Personality," "Demoniacal Possession," "Witchcraft," "Degeneration," and "Genius."[24] While in the Lowell Lectures of 1878 James had been interested in establishing his authority as a physiological psychologist, and went on to describe consciousness in *The Principles,* he now presented a psychology of the unconscious or "subliminal" mind. The motive was partly didactic. He wrote to philosopher George Howison that although the subject matter was often morbid, he would offer "optimistic and hygienic conclusions."[25]

But the desire to help people with mental maladies was not the principal intellectual motive of the new series of Lowell Lectures. In the first, "Dreams and Hypnotism," James introduced a theme he would come back to time and time again. There was, he notes, a "common distinction" between "healthy" and "morbid." But in reality one could not make the distinction sharp. Not only was "no one thing morbid," but "a life healthy on the whole must have some morbid elements." Normal consciousness was only "a fragment of the whole mind." The mind in consciousness and unconsciousness, in dream and waking states, in illness and health, was "integrated." It presented a "field—focus—margin." And within the mind's field, "Margin *controls.*"[26]

By visualizing the mind as a field of focus and margin, James found a descriptive framework for interpreting psychopathology. In Lecture 1 he would develop the idea that in hypnosis, for example, "Suggestibility [is] due to the narrowness of field." And in cases of hysteria, which he would deal with in Lecture 3, "Conscious[ness] splits—the two halves shore the field." He would remind the Lowell audience in Lecture 4, "Multiple Personality," that "we are by this time familiar with the notion that a man's consciousness need not be a fully integrated thing. From the ordinary focus and margin, from abstraction, we shade off into phenomena that look like consciousness *beyond* the *margin*."

James was fully aware of recent attempts to explain and treat pathological mentality. "In the relief of certain hysterics by handling the buried idea whether as in Freud . . . or Janet . . . we see a portent of these new discoveries. The awful becomes relatively trivial." But he would caution his audience not to blindly accept or reject any one theory. He was uneasy with Freud's theory that "hysteria is obsession not by demons, but by [a] fixed idea of the person that has dropt down." For "to say that is one thing and to *deny any other ways* of phenomena is another." And the "other ways" he wanted to consider led into psychical research, a "portal . . . into which I said I would not enter." (Notes for Lecture 4, "Multiple Personality").

But of course James did enter. By positing the possibility that psychical transference existed, James moved discussion of psychopathology onto new ground, or in his language, into a field with a vast margin, an area that allowed fresh insight into pathological states.

James challenged the predilection in contemporary psychopathology for developing fixed classifications of mental illness.[27] An expansive mental field which incorporated new phenomena allowed him to juxtapose different historical interpretations of psychopathology, and in so doing conflate one with another. Hence, for Lecture 5, "Demoniacal Possession," he notes, "History shows that mediumship is identical with demon-possession." Psychical research encouraged remapping the mental terrain to reach an understanding of the mind free from both fixed theory and history. James escaped the belief that illness was still illness regardless of how one defined it. Understanding the shifting historical context of mental illness could dilute its terror, make it less special, and make psychopathological conditions nonspecific and relative within the open margin of the mental field. He helped move psychopathology away from defining mental illness toward describing mental experience. The mind considered as a field of focus and margin replaced pathological conditions with phenomenological descriptions.

IF THEREFORE one wants to describe the process of experience in its simplest terms with the fewest assumptions one must suppose 'fields' that 'develop' under categories of continuity with each other," James wrote in a notebook prepared for a 1895–1896 seminar titled "The Feelings." Thus he began the first of the three notebooks, dated 1895 to 1898, in which he would work toward a philosophy of pure experience, to be culminated by a fourth, entitled "World of Pure Experience," probably composed in 1903. The idea of a field of experience had provided theoretical ground for interpreting psychopathology in the Lowell Lectures. But James was not developing a philosophy of experience to better explain mental phenomena so much as he was looking for a better way to describe reality, in whatever guise it might appear. Conversely, if psychopathology or psychical research shed light on the semblances of reality, so much the better; "All the fields commonly supposed are incomplete and point to a complement beyond their own control."[28]

James realized that there was a cost to conceiving experience as continuous and open-ended. "We certainly have gained no *stability*. The result is an almost maddening restlessness." "But," he surmised, "we have gained concreteness." And indeed, in thinking about experience, James strove for example and illustration, making certain that the theoretical vagueness of "fields" never lapsed into incoherence. The problem was to be concrete and yet to discover a fresh, emerging reality. He would soon develop a special kind of concreteness which he called "pure experience" and "radical empiricism."[29]

William often jotted down problems to elucidate. The Feelings Notebook contains the following: "Our assumption of integral datum. Maybe it can't be made to work; but *Try*. Datum always complex. Former data figure as its parts. Don't start with present datum and say what it becomes prospectively; but treat retrospectively[,] the old datum given in the present one." He works through a personalized example of the datum of an actual field of experience:

> *Your* datum now is of me, and your having seen and heard the same me 5 minutes ago. The me as such shells itself out of your experience as such. The past me falls into one series, the knowing of me by you into another. The datum = *it* + me—two coordinate parts. The *it* part connects itself with past and denotes its. The me part with past mes. The question is: How does the whole datum come to be treated now as it, now as me? The it as then is to it as now as reality to a ghost.

The "field" was still not clear, however, so he drew a field and illustrated temporal relations within it. The notebook shows five small circles strung like beads equidistantly along horizontal and perpendicular lines which meet at a 90-degree angle. Another line connects the last circle on the perpendicular line to the farthest-most circle on the horizontal line, forming a triangle. The remaining circles are likewise connected, also forming triangles. The perpendicular line and its circles are labeled "The now datum," and the horizontal and its circles "The then datum." And James notes, "The triangles successively enveloping [the angle created by the horizontial and perpendicular lines] are the successive data" (Feelings Notebook). Thus he shows how the past and present are successively integrated as one continuous experience—juxtaposing different modes, the conceptual and the visual, to better explain his speculation.

But James had yet to make the remarkable assertion that allowed him to circumvent the dualism of *The Principles of Psychology.* By visualizing a field he brought to light the obvious. "The datum in itself and intrinsically considered is no more inner than outer. It becomes inner by *belonging* to an inner, it becomes outer by *belonging* to an outer world," he continues. Drawing a field explained how an inner and an outer datum were experienced: "[Datum] can be strung (in its intrinsic entirety) either on a vertical thread so to speak which unites it to associates that together with it make the inner world, and also on a horizontal thread with associates that together with it make an outer world" (Feelings Notebook). In other words, in thinking or supposing an outer world, one created *a relation* with it, a relation between an outer and an inner world. The datum of related experience was the ground upon which inner and outer reality rested. Within the relations of experience there was no *necessary* dualism.

James was far from satisfied. But he had suggested something critically important; that is, that experience when treated as data did not inevitably isolate subject from object. It was not necessary to bring the subjective and objective world together through transcendence or Hegelian synthesis.

The English philosopher Francis Herbert Bradley had suggested a way of transcending dualism in *Appearance and Reality* (1893), a book James had recently read, admired, and attacked.[30] In criticizing the empiricism of John Stuart Mill, Bradley had developed a philosophy of relations which avoided the usual atomic isolation of mental elements. But in his view, outer and inner relations could not be fully realized without a transcendental transformation that removed relational distinctions. Relations always produced contradiction, and where there was contradiction there was appearance, not reality. James, however, found reality in the datum. That was enough, even though he still needed to explain why the datum field of outer and inner relations held together. His view, his notebook acknowledges, led "to the question why don't immediate experiences break into units." Parenthetically he notes, "Reading related to these matters," and mentions "Bradley, Stumpf, Cornelus, Meinong, Baldwin, Richet, Reid, Hamilton, Avenarius, Royce, Hodgson, Spencer, Green, and Bain"—all of whom in one way or another were dissatisfied with dualism (Feelings Notebook).

He was now very close to explaining why experience did not dissolve into units. "A certain subjective stream goes on," he observes, using the metaphor of *The Principles*. True, "the collocation and sequences follow entirely different laws in the inner and outer worlds. A + B, two 'objects,' come together in my thought but as outer realities they are separated by the diameter of the globe." Still the great fact was that "the *presented portion* of the datum has pretty similar neighbors in the two worlds"—except that "the thought portion is freer" (Feelings Notebook). There was a semblance rather than an exact identity between an object as thought and an object as a material entity. Semblance provided an opening to continuing discussion of relationship in a field of experience. Relation did not eventually imply discontinuity or contradiction. Bradley was wrong.

Experience was simply *of* something in the objective world. *Relation,* the "of" between inner and outer, between subjective and objective, between consciousness and material things, was the immediate experience. James needed nothing more than relation to build reality. This was his datum. This was his field of experience.

> There *is* no stuff anywhere but data. The entire world (objective and subjective) at any actual time is a datum. Only within that datum there are two parts, the objective and subjective parts. Seen retrospectively, and as within the datum, the one part is to the other, so will the datum itself appear as the subjective part in the next datum which will contrast it with the objective part of its own content. [The world's] content goes on increasing without its bulk

> changing, as the landscape seen from the back window of a railroad train might, if new marginal (or physical) matter concentrically withdrew towards the center filling a constant space that stood there to represent the subjective fact. (Feelings Notebook)

James hypothesized reality as a field of relationship between past and present, between subject and object. He placed the participant in experience aboard a moving vehicle—the symbol of American progress in the late nineteenth century, the railroad train. The image may have come to mind as he journeyed by rail to Colorado in the summer of 1895—or during any other of dozens of train trips he took as a lecturer or on the way to Chocurua or Keene. As the train moved through space and time, parts of experience became less distinct and were lost in the constantly receding background, at the same time as new parts were gained; the world of experience could never appear static or comprehensive. As it passed it presented multiple standpoints, multiple relations, and endless transformation. As James pictured the world, margin and focus were in seamless, continuous, dynamic relation.

James was not above using a student's illustration to describe a field of experience. He transcribed it from "Cabell's examination book": "Our little datum at any given time is but a small fragment yet this fragment is surrounded by a fringe of symbols, like the visible area around a man walking in a fog which in the same way is ever changing, and so our fragment by intermediary data, might lead to any possible experiences." Cabell's imagery encouraged further description: "As the field alters and the older content shrivels it forms connexion in the new subjective value with the new objective content that marginally comes in. *That* was an appearance of *this* from an earlier point of view." And always in the alteration of a field came a meaning. "Around every field [is] a wider field that supersedes it, that yields more truth, the truth of every moment lying beyond itself. On my pluralistic scheme, the oneness of the universe comes from the parts overlapping" (Feelings Notebook).

By 1896 there was nothing in the universe for James that was truly discontinuous or beyond relationship. Certainly the troublesome dualism between the subjective and objective reality had crumbled into insignificance. It remained only to continue to speculatively explore the full significance of this new world as a field of relative experience.

As THE century closed, James was occupied with projects that had a more or less popular appeal—projects such as writing *The Will to Believe* (1896); researching and writing the Lowell Lectures on Extraordinary Mental States (1896); and composing "Philosophical Conceptions and Practical Results"—the University of California address that in 1898 marked the beginning of pragmatism as an original American philosophical movement.[31] He also began gathering material for the upcoming Gifford Lectures at the University of Edinburgh, later published as *The Varieties of Religious Experience* (1902).

Even so, his most intensive intellectual energies were concentrated on understanding experience. After 1895 consciousness was no longer the primary field of reality for him. James became less concerned about the mind per se and more concerned about experience—about transforming psychology and philosophy into accounts of the immediate stuff of experience. It is a large mistake to make pragmatism the irreducible core of James' philosophy. That much-publicized facet of Jamesian thinking was merely a way of dealing with the world of pure experience. As a scholar of Jamesian pragmatism recently observed, in it "experience is raised to the status of being the ultimate and only reality."[32]

In April 1898 James again began a notebook, recording thoughts on his new theoretical frame. "The whole *use* of the 'change' to pure experience is to see whether one may thereby solve certain problems which are *stickers* on the usual dualistic categories," he observes. He returns to "the idealist paradox—brain being a condition of consciousness, whose creature brain nevertheless is." This was the old puzzle of "psycho-physical causality," something he had broached as early as the Baltimore and Lowell Lectures of 1878 and had continued to consider while charting cognition. What was the relationship between the mind and the brain in a field of pure experience? He sketched a mind-body field:

> The content of consciousness has a centre, the self, corresponding to the body, which is the centre of the physical world. Where the body is the "here" of the self, when the body acts is the "now" of the self. Other remoter things are "theres" and "thems," so the field of consciousness is at all times systemized with reference to a focus of feeling whose position is in the body, and with which the other contents connect themselves in an ordered and shaded manner, as they become less and less important or near.

> As there are things constantly near and important for the body's action[,] theres with the body form the apperceptive system of me, contrasted as a whole with various not me systems.[33]

Mind and body formed in relation and yet out of relation with objective reality, the "various not me systems." "The whole point of the pure-experience hypothesis," James avers, "is that what figures or reports itself as the *same* stuff can figure over and over again, and they in diverse relations." The mind was presented with relations, the "stuff" of experience; and the stuff appeared pluralistically and continuously. "We strike thus upon the notion of a continuum. Is that perhaps the key? Does 'Pure experience' fall into discrepant continua each characterized [by] a certain dimension, so to call it, of relationship?" (Self–Pure Experience Notebook).

Posit a world of relations as a continuum of experience—in this case, relations between the mind and the body—and those relations would eventually reveal their distinctive qualities. "The chief difference that one sees between the brain world and conscious world is that the brain-world past and future as such are ineffective, and what ever happens has a present cause, whereas in [the] conscious world memories and future appear as incentives to activity." And then, "The sole point would be to show that the brain tendencies terminate differently on account of the representative character of present ideas." James probes the insight. "Can this be shown? Possibly the mere notion of representation involves it. A representation is a present idea that physically terminates in a remote reality. . . . We *call* it decision or selection, and the selection of the idea determines the selection of the terminus" (Self–Pure Experience Notebook).

The brain, considered only as a lump of molecules, was not the primary datum of experience. But in its representative quality and selective activity the brain became relative in an experiential field. "The brain is hardly ever given as a terminus. Its activities never so," James remarks in the same notebook. In 1898, as in 1878, James did not want discussion of the brain to end as simply discussion of physical matter.

Why not? Because his genius sought relations, rather than reducing qualities to physical matter or elevating them into transcendent essences. A field of experience was an open space in which to draw relations, just as surely as he had drawn them when creating landscapes on blank canvases while painting with John La Farge years before in Newport. In fact James uses paint itself as a metaphor to illustrate the relation between consciousness and its object. "[Let] oil, size, or water stand for consciousness while the object is made of the pigments held in solution."[34] The relations were as visible, as apparent in philosophy as in painting. One could describe reality as an experience continuum without erecting an elaborate synthesis or reducing the world to discrete molecules.

Always concerned with clarity, James describes his insight in the simplest way imaginable. "Make a violent contraction of the hand. You feel it, at the same time you see it, both times it is an object of sensation. Or you may feel it without seeing it, only thinking visually of it; or you may simply think of both the feeling and the seeing conceptually" (Self–Pure Experience Notebook). A contracted hand assumed various attitudes in experience which were not necessarily discontinuous with each other. Sensible and intellectual life were linked *as* a field of experience.

As James articulates a philosophy of experience he ponders the consequences of living with such a philosophy. "The great diff. between the phenomenist and the common sense view is that the latter gives *stable* elements, while the former is afflicted by a restlessness which is painful to the mind." The philosopher of pure experience "never gets out of the conception of flux or process: although it might well seem that all the actual found its place in the flux. For common sense the actually experienced is only a very small extract of the whole" (Pure Experience Cont'd. Notebook). Whereas common sense could separate experience from the rest of reality, pure experience for James *was* reality—and one experienced as painful to the mind. His special vision was self-inflicted. "Isn't the difficulty this?—to get out of a solipsism without jumping a chasm?" (Pure Experience Cont'd. Notebook).

By the late nineties James had discovered that the various perspectives associated with being an artist, a psychologist, or a philosopher were secondary to a primal encounter with experience, the "phenomenological" perception. James had not experienced whether to be an artist, psychologist, or philosopher as his essential intellectual problem. Rather he had always asked, in effect, "within a world of experience, how do I see?" And to ask that was to see linkages between art, psychology, and philosophy whereby each field fulfilled its own objectives and yet was functionally connected with the other fields. Such a perception was not only inherently interdisciplinary: it presented a constant pressure

William with his wife, Alice, daughter Mary Margaret ("Peggy"), and brother Henry at the 95 Irving Street house, Cambridge, in 1905.
By permission of the Houghton Library, Harvard University.

to avoid dead ends, whether dualism or self-centeredness; indeed, to escape from discontinuities and closures of whatever sort.

"My point must be to show that the beyond is part of the same continuum, whereas for common sense dualism, it is discontinuous, and separated by the epistomological chasm," James resolves (Pure Experience Cont'd. Notebook). Bridging the chasm was James' act of creating the world; and he had worked toward it for years as he related art to psychology, physiology to psychology, and psychology to philosophy. Separate disciplines and various vocations were the bridges he crossed and recrossed to enter the kingdom of pure experience.

How could he effectively illustrate the bridge? That still perplexed him. He refers to a setting familiar to contemporary museum visitors and exposition-goers. "In what we call a 'perceptual' experience there is perfect fusion of conceptual and sensible material, like the fusion of painted background and real foreground in one of those circular panoramas of battles etc.—when to inspection the line of division can not be made out" (Pure Experience Cont'd. Notebook). There were no joints between the real and the unreal in experience, any more than there were in the stream of thought. James was compelled to find the continuity between the real and the artificial in the panorama. One wonders whether another thinker—say David Hume—might have just as obsessively looked for breaks or disjunctions. But James' mind had to make experience a seamless scene of relationship.

Interestingly, he sought relationship and connection at a time when many Americans were experiencing the opposite, isolation and alienation. As an older America underwent the modern transformation, James sought to build a field theory that redefined the atomic and monolithic worlds of traditional philosophy. In his search for connection, James disclosed the lack of it in his own culture. In seeking to overcome discontinuity and "chasm," James joined a chorus of writers who were searching for the shelter of community in the face of modern dislocation.[35] In this sense James was not so much the philosopher of individualism as a great theoretician of relationship; there is a social implication to his metaphysics that has been largely overlooked.

JAMES DID not again take up the 1897 and 1898 "Experience" notebooks until late 1902 or early 1903. He was occupied with writing the Gifford Lectures during a period when his health had seriously deteriorated. Much of the time from 1899 to 1902 he spent in Europe seeking a cure for his ailing heart. But by 1902 he was once again engrossed in metaphysical thinking, which resulted in *Pragmatism* (1907) and *A Pluralistic Universe* (1909), as well as the posthumously published *Some Problems in Philosophy* (1911) and *Essays in Radical Empiricism* (1912). Each book further refined his world picture, and none is wholly reducible to the experience theme. A flurry of publication between 1903 and 1905 signaled a creative period—much as it had in the late 1870s.[36] As James had shifted his mind's eye from consciousness to experience, he had found a way around dualism; he now lived in a world of relations rather than in a stream of thought.

His 1903 notebook was called "World of Pure Experience," and it was to be a start on what he hoped would become a book on "radical empiricism."[37] It was closely related to five lectures on pure experience and radical empiricism that James delivered in August 1904 at Glenmore, the philosophical summer school near Keene Valley. Unfortunately the lectures were not preserved, but they, along with ideas in the 1903 notebook, were probably the basis of "Does Consciousness Exist?" and "A World of Pure Experience," two important essays published in the *Journal of Philosophy, Psychology, and Scientific Methods* in 1904.[38]

James continued to search for intellectual allies, and he found one in the French philosopher Henri Bergson, a thinker as influential for James' theorizing at this period as Renouvier had been several decades earlier. Nearly twenty years James' junior, Bergson was professor of philosophy at the College de France in Paris. Like James, he had broken away from Spencerian determinism, arguing that man experienced freedom of action directly. Bergson also took great pains to relate the body and the mind, thus presenting an alternative to the dualism William was so fervently dismantling.[39] Both commanded superior prose styles, filled with illustration and metaphor. In 1907, after reading *L'Evolution créatrice* (Creative Evolution), William exclaimed, "O my Bergson, you are a magician, your book is a marvel a real wonder in the history of philosophy, making if I mistake not an entirely new era in respect of matter, but unlike the works of genius of the 'transcendentalistic' movement (which are so obscurely and abominably and inaccessibly written), a pure classic in point of form."[40]

So under the stimulus of Bergson and in temporarily stable health, James once again thought through his new philosophical scheme. "By experience I mean . . . anything that can be regarded as a concrete and integral moment in conscious life," the

World of Pure Experience Notebook begins. "The word is exactly equivalent to the word 'phenomenon.' A phenomenon implies both something that appears and someone to whom it appears; and an experience implies both an experiencer and what he experiences." Admit this and traditional metaphysics no longer explained the world. "The essential consequence to remember is that, if experiences thus defined are minimal world-factors, absolute 'substances' in the old dualistic sense of 'material masses' on the one hand and 'souls' or 'spirits,' on the other can not be allowed to be real."[41]

James puts his world of pure experience into the center of philosophical debate. He wants to supercede the two major traditions. "There is a common tendency to philosophize as if some sorts of relation were matters of experience 'merely,' and as if other relations might belong to a more absolute order in which experience as a whole lies immersed." Here lay the absolute world of Royce and Bradley, where "the unity of the universe, they think is more profound, more true than its variety." In such systems unity overcame variety; the former was mistaken as the real, the latter mistaken as appearance.

On the other hand, "Empiricists . . . tend to make the disjunctive relations the more real. Hume, for instance, says that whatever we can distinguish from other things is as 'loose and separate' as if there were no manner of connexion between it and anything else." Connections were "mere 'feelings of easy transition' which 'custom' has engendered in our mind." The unreality of absolute monism and atomistic empiricism stimulates James' genius for metaphor:

> If the world be made of the stuff of experience, then it must be experienced throughout, just as everything in a picture must be painted, and everything in a story must be told. The picture it is true needs a canvas, and the words need a tongue, so some philosophies have thought that experiences also need substantial supports. But as painted fishes can live in a painted sea, and a storied palace hold a storied king and court, even so experienced connections are sufficient fasteners together of experienced terms, and transitions realized by us are the only relations possible between such things as our experience grasp.

Representations, whether painted or verbal, were as much in experience as anything else and needed no foundational "stuff." They were simply in the world next to one another, as in the panoramas of battles where "the real earth, grass and cannon of the real foreground connect themselves so subtly with their continuations on the painted canvas, that it is impossible to part them, and the whole scene shares in the effect of reality."

Such a world did not need consciousness any more than it needed material substance. "There is no such entity as consciousness, as thought stuff, and there is no such entity as either separable or inseparably combined thing stuff. There is only one original stuff, and this the undivided Stuff of what I shall call 'pure' experience."

Aware that to deny consciousness as real might easily be misunderstood, James explains: "To deny, as I do, that 'consciousness' exists inevitably shocks the hearer, and awakens opposition. But I only deny it to exist as an original quality or *nature* to which one aspect of being is determined." Consciousness, though not the primal stuff of experience, was functionally related to experience. "Pragmatically the function [of consciousness] is knowing. 'Consciousness' is invoked to explain the fact that things not only are, but get reported as known." Consciousness had always been functional in Jamesian thinking: it could stabilize an unstable brain; it could *act* this way or that; it could attend to any part of the stream of thought it chose. But his desire to break down the ancient distinction between subject and object calls for something even more radically original, a description that denied *all* existing categories—even those assumed in *The Principles of Psychology.*

He explains that when psychology became a natural science, it separated mental from physical phenomena. In fact, "This frankly dualistic or realistic standpoint is that of the author's own work, *The Principles of Psychology,* 1890." There, consciousness was a kind of "impalpable though substantive material in which the states of mind float, or of which they are framed." As such, consciousness was "a pure nonentity." "It has become too diaphanous to be a ghost of a gas even; a bare verbal symbol of the fact that experiences communicate; the last estate of the substantial soul in the process of evaporation."

Consciousness as an entity was gone, as the soul had gone before it. In its place was pure experience which could, however, still be described using James' great psychological metaphor. "If we look at our experiences with the simple aim of describing their success, we see that they form a stream which in important respects possess the quality of continuity." But consciousness had no reality outside experience, anymore than did a material object.

"There is no stuff but pure experience-stuff, and whether a given bit of this shall be treated as physical reality or as a conscious state depends entirely on the context in which it is taken. It is an affair of functional relations and resultant classification exclusively, to which the stuff in its immediate nature is indifferent." And having established these relations, there was only one answer to the question of what pure experience was: "If you ask 'what' it is, the only answer is '*that!*' "

In answering critics who did not believe he had effectively bridged subjective and objective reality he wrote, "My inner biography doesn't work either directly or indirectly on my mind, it *is* my mind." He believed that the description of what a mind was like could never *be* the mind—his, or anyone else's. "The Whole psychological analysis is after all only a retrospective interpretation of the total content of perception." Mind—the total content of perception—was actually only part and parcel of the "datum" or the real "stuff" of pure experience.[42]

Working through all his "fields," James came to recognize fully the phenomenal nature of his mind. He found a field of perceptions that related experiences as they came into it. Art, physiology, psychology, and philosophy had brought him here. He saw the world this way; and he thought this was the way the world was truly seen. This neutral "stuff" of experience was reality. How the mind used experience was another question, a value-laden one. But he had conquered consciousness: the stream of thought had all along flowed toward the bedrock of experience.

CHAPTER FOURTEEN

God Help Thee, Old Man

God help thee, old man, thy thoughts have created a creature within thee, and he whose intense thinking thus makes him a Prometheus; a vulture feed upon that heart forever; that vulture the very creature he creates.

Lines from Melville's Moby Dick, *recorded in James' Pluralistic Metaphysics Notebook*
ca. 1903

AS JAMES developed his ideas on radical empiricism, at one point he expressed the intention of using as "a motto for my book" first mate Starbuck's judgment of the crazed Captain Ahab of *Moby Dick.*[1] The lines never actually appeared as an epigraph in any metaphysical work, but that James would even consider using a quotation from an author whose work emphasized evil, madness, and death seems strange. After all, William James was an optimist who believed that despite life's tragedy, one could act, could make a difference. Why consider linking himself with a fictional creation so obsessed with thinking that mind destroyed heart? Why would a man disposed to hope for the best be attracted by devouring death?

Scholars have usually assumed that after James' time of troubles in the late 1860s and early 1870s he achieved a lifelong, if fragile, inner balance. They have noted that chronic neurasthenia continued and that an ailing heart plagued his later years. But they have taken for granted that after reading Renouvier in 1870, after receiving a teaching appointment in 1873, and certainly after marrying in 1878 he never again suffered deep melancholia.[2] A clear faultline has been seen separating the morbid James from the optimistic James. From the 1870s on, he remained exemplary, a life with a moral message, a legend fitting the Protestant tradition of crisis and redemption: the American intellectual, "twice born."

James himself was responsible for perpetuating the twice-born theme. His oft-quoted diary of 1868–1873 records feelings of melancholia, which reading Renouvier on free will overcame. And he explores the "conversion" antidote to "sick-mindedness" in *The Varieties of Religious Experience,* the work that includes the account, disguised but doubtless autobiographical, of his startling *"that shape am I"* existential fear (see chapter 6, note 72). Popular articles such as "The Dilemma of Determinism," "The Will to Believe," and "Is Life Worth Living?" also reject pessimism and encourage an optimistic philosophy of action. Moreover, James fashioned a psychology that found a scientific basis for the minds' freedom from deterministic bondage. Biographers have taken up his own cues, either out of sincere belief that this was the true James, or because they themselves have been caught in the twice-born theme so deeply embedded in the American ethic.

There was, however, another James—the genius, the original thinker, the man whose own mind called him to describe it. He was, as we have seen, obsessively compelled to theorize. Creative life entailed constant demands and pressures; the costs of this incessant mental labor were not all absorbed in the twice-born experience. When alone James was more apt to be in a state of unease, restless or depressed, than at peace and happy. As his perceptive friend John Jay Chapman saw, "There was, in spite of his playfulness, a deep sadness about James. You felt that he had just stepped out of this sadness in order to meet you, and was to go back into it the moment you left him."[3] Below the surface of the twice-born James, Chapman had located the still-burdened James, a man essentially untouched by the optimistic rhetoric he had done so much to perpetuate. From the late nineties until his death, the burden became more painfully visible.

I HAVE HAD an eventful 24 hours and my hands are so stiff after it that my fingers can hardly hold a pen," William wrote to Alice in July 1898 from Keene Valley. He had just returned from climbing New York State's highest peak, Mount Marcy. "I left . . . the lodge at seven, and five hours of walking brought us to the top of Marcy. I carrying 18 lbs. of weight in my pack."

That evening at Panther Lodge Camp, excited from the effects of the climb, and enthralled with the clear, starry night, he had "the most memorable of all my experiences." And in similiar vein to his efforts to describe his 1880 moonlight and mountain trance in Switzerland, William exclaimed that "it seemed as if the Gods in all the nature-mythologies were holding an indescribable meeting in my breast with the moral Gods of the inner life. . . . It was one of the happiest lonesome nights of my existence, and I understand now what a poet is." He predicted that "things in the Edinburgh lectures [on the varieties of religious experience] will be traceable to it."[4]

But James' exuberance was short lived. The Marcy climb irreparably strained his heart. William did not mention that during the climb he had added extra weight to an already substantial pack. Among the climbing party were two young women from Bryn Mawr College and an older Keene Valley enthusiast, Pauline Goldmark. The college students either complained that their packs were too heavy or the male climbers complained that the women slowed the ascent. In any case, sundry items were transferred to the guide's pack, and 58-year-old William, wanting to do his fair share, insisted that the guide transfer part of his load to him. "The girls probably had no understanding of the unwisdom of this for a man of W. J.'s age," his son Henry later observed. But when Alice learned of the added baggage, she was furious; Henry noted, "My mother, who was anyhow unable to find Pauline Goldmark sympathetic, could never forgive her for their absorbing W. J.'s guide."[5]

William kept to himself the immediate signs that his heart had been damaged. By August he had left Keene Valley for the University of California at Berkeley, where to a standing-room-only audience of eight hundred he delivered "Philosophical Conceptions and Practical Results"—the lecture that is considered the beginning of the American pragmatic movement. After lecturing, James could not resist climbing and camping high in the Sierras at Yosemite. Casually he mentioned to Alice suffering "some heart palpitation the 1st night from high altitude."[6] Back in Cambridge for the 1898–1899 academic year, James continued to have an irregular pulse and occasional chest pain. In November Dr. Jim Putnam examined his old friend from Harvard Medical School days and found valvular injury.

But it was difficult to know how much damage. Not only was there no electrocardiogram, but Wilhelm Conrad Roentgen's new x-ray technique was not yet widely used. The only practical way to diagnosis a physical injury to the heart was through auscultation, through interpreting the heart action by stethoscope. The diagnostic difficulty pointed to the shifting context of American medicine, a climate in which "homeopathic" and "regular" physicians not only argued theory, but competed for patients.[7] During his last eleven years, James would vacillate, occasionally with dire results, between experimental and more conventional physicians, attempting to find the correct diagnosis.

By spring 1899 William decided to seek medical consultation and rest in Europe. He had consulted Bowditch, "who seems to have better ears than others (beautiful creature that he is) and has defined my trouble more exactly," he told Alice. Anxious not to upset her, he condescended gently, "You would hardly understand the technical terms; and it is no matter," because Bowditch was "no more alarmist" than other physicians who had examined him. Bowditch advised going to the German baths at Nauheim, whose waters were "antirheumatic and antineurasthenic."[8] There he could consult the great German heart specialist Dr. Theodore Schott.

William's cares increased because he had promised ten Gifford Lectures for 1900 and had yet to write them. Exhausted from worrying and teaching, the mere thought of taking up a pen produced more anxiety. Bowditch urged complete rest, a prescription James heartily seconded. By June he had fled to Keene Valley. "In the woods one is morally safe. They do one's spirit a curious kind of good. I took the walk yesterday to Indian pass, very slowly—8 1/2 hours with not more than 3/4 hours rest."[9] Old habits died hard. He was walking—injured heart or not—all day long!

Alice informed Henry of his brother's condition and plans to come abroad. Although Putnam had assured her "that the disturbance was very slight and that with proper hygiene it need not increase," she worried about William's way of life. "Was ever man born of woman harder to take care of than William! Do you wonder that I want to get him away from all the frustration which wearies more than work, and establish if possible a well ordered plan of life," she said.[10] And when Henry wrote to William what seemed to the latter an overly sympathetic letter, James scolded: "Don't take it seriously—it doesn't menace either longevity or life—it only checks me in too rapid mountain climbing—a bad check to my personal self-consciousness, to be sure, for that resource has been my main hold on primeval sanity and health of soul."[11]

In fact not to climb while in Keene Valley was impossible. James admitted to Pauline Goldmark that weakened heart or not, he found himself trying again for Marcy's 5,433-foot summit. This time there was no starry evening trance. "Once a donkey, always a donkey; . . . after some slow walks which seemed to do me no harm at all, I drifted one day up to the top of Marcy . . . and converted what would have been a three-hour saunter into a seven hours' scramble emerging in Keene Valley at 10:15 P.M." He explained that "my carelessness was due to the belief that there was only one trail in the Lodge direction, so I didn't attend particularly, and when I found myself off the track . . . I thought I was going to South Meadow, and didn't reascend." He had been an "ass" and feared "we shall ascend no more acclivities together."[12] William's son Henry later remarked, "It seems unbelievable now that the doctors whom he'd consulted should not have forbidden much walking. But then, those mountains and woods were as irresistable to W. J. as a bottle to a drunkard."[13]

By August William, Alice, and Peggy were settled at Nauheim so that William could consult Dr. Schott and take the baths. Shortly after arriving he wrote Charles Eliot that "the great man here now is Th[eodore] Schott. The success of the treatment here in every sort of heart trouble seems very positive and extraordinary."[14] James did not get on well with Schott, however, whom he called "the oracular."[15] Schott was self-assured and scoffed at the value of the baths for heart patients. He told William to stop them and "walk harder than usual of late for two days," after which "he wanted to listen to the poor organ again to observe the effect," William told Henry. The doctor "prophecied that certain distresses which I now have in my chest on walking . . . would after several weeks disappear."[16]

William was more impressed with Dr. William W. Baldwin, an American living in Rome, who was recuperating from a heart condition of his own. He had first met Baldwin in Florence in 1892 and found him a pleasant experimentally oriented contrast to the more conventional Schott. Baldwin was using animal extract injections on himself. And James was so taken with his medical abilities that he recommended to Eliot that Baldwin be installed as "Professor of Hygiene" at Harvard.[17] What William kept to himself were plans to try Baldwin's injections.

Despite his lack of enthusiasm for Schott, William's condition improved enough for him to depart for London and a visit with Henry. Alice noted, "Dr. Schott said, 'I am pleased' and in answer to questions gave him the precious word that his heart had resumed the normal size. The painful constriction of the chest remains but Schott says that will disappear after some weeks."[18] But the pain persisted, and in London William became the patient of Dr. Bezly Thorne, whom he regarded as "the first English heart specialist."[19]

Thorne found his aorta enlarged. He urged rest and recommended that if the symptoms did not diminish he go back to Nauheim. All the while the patient gathered material for the Gifford Lectures and managed a little writing. But the event loomed the following year and he was finding it impossible to work steadily. Moreover, he did not feel well enough to even think about returning to Cambridge. He asked Eliot for an extended leave and notified the University of Edinburgh he would be unable to deliver the lectures in 1900.

In early December Thorne recommended that William move out of Henry's London flat to the Cotswold countryside near Great Malvern, where the climate was more salubrious. But the weather turned cold, and with Peggy enrolled in school at Harrow, Alice and William decided to go to Italy via Switzerland for the winter. An added incentive was Baldwin's presence in Rome. The Italian-based American physician had written to James describing in detail the benefits of Robert-Hawleys extract, a mixture of goat lymph, brains, and testicles.[20] But Alice distrusted Baldwin, and so William kept his plans to use the extract a secret. By early January 1900 he and Alice were in Rome. Shortly thereafter Eliot wrote that academic leave had been extended for a second year at half pay. The Gifford Lectures had also been postponed until 1901.

William and Alice stayed in Italy until the end of April. By then it was clear that William wanted to resume treatment at Nauheim. In May they spent a week near Geneva with the Swiss psychologist Theodore Flournoy, who had done much psychical research and in whom William found a kindred soul.[21] In May William wrote to Henry that he had located "a first rate neurological specialist named Widmer . . . who also knows all about Nauheim, having made many 'cures' there himself, and yesterday he came and gave me great consolation. The heart, whose sensibilities have been getting worse rather than better all this time, he pronounces (after a protracted physical examination) to be in no very bad shape."[22]

Yet after almost a year abroad, James felt no better than when he had arrived. And when Widmer suggested that there was no extensive somatic damage, William found another explanation for his discomfort. "The nervous element preponderates," he continued to Henry, "but as it is the result of the heart's state, he wants me to go to N.[auheim] immediately again, but to take the cure with great moderation and precaution." Widmer recommended one Dr. Abbey, since Schott was "overworked and slap-dash." Buoyed by the diagnosis and a new physician, William confided that "all this gives me immense relief, since I must say that this indefinite vague waiting has been a real crucifixion to me." Nonetheless, he admitting being "a thoroughly ill man nervously, and writing my lectures . . . is a curiously exciting and prostrating performance."[23]

In May William again bathed at Nauheim. But there was little improvement, so the Jameses returned to Geneva, where Widmer began electrical therapy. Alice had stayed with her husband throughout the European trip; she had endured much and was as anxious to find a cure as William. She had received letters from family friends urging that William seek out a certain faith healer in Paris. James had tried a mind curer in 1893 when he had suffered from depression and insomnia, and had attributed his temporary improvement to her powers.[24] The lack of much notable progress now made Alice's suggestion seem worth pursuing.

The results, however, were something else. "I made an ass of myself in letting that spider of hell the 'healer' touch me," he wrote to Henry. "10 days ago, it became evident that I had to try a lower station [lower altitude]. . . . And the invitation to a private house there, [after] so many months of confinement to hotel bedrooms, all came in at a certain psychological moment which made the acceptance of the invitation seem a relief from perplexity." But to his embarrassment, James found that the "faith healer" had given him something other than a still-ailing heart to complain about: "I confess I did not credit them with the power to produce boils. Altogether it has been a nasty job, and I shall never dabble in the like again."[25]

The boils delayed traveling for a few days, but by mid-August James was again at Nauheim. Baldwin was also on hand, as well as William's former student and good friend Dickinson Miller, who also had a heart ailment. For some reason, possibly Alice's insistence, Dr. Schott continued as William's official physician. "Schott came to see me at 6:30 and somewhat relieved the depression—Nauheim depression!—into which I had fallen," William told Henry. "He was very considerate and consoling in his manner, said he 'hated' people with half medical instruction like me. Told me I mustn't think of or ever feel my pulse, etc." He found William's heart "more normal than last year, but weaker, and need[ing] strengthening."[26]

Schott saw that James' constant second guessing of his doctors concerning diagnosis and symptoms could do little good and could possibly harm both heart and nerves. Yet William's medical knowledge and acute sensitivity to bodily and psychological changes made it impossible for him not to get involved. That James did not stand in awe of established, reputable physicians was not that exceptional. Throughout the nineteenth century there had been open, public criticism of doctors, who with radical blood-letting techniques and unrestrained use of mercury and arsenic had in not a few cases killed rather than cured. The growth in the numbers of homeopathics and other "irregular" physicians resulted largely from the failure of orthodox medical authority.[27]

New Englanders had been unusually receptive to medical and psychical innovators, and James was a case in point. In 1898 when the Massachusetts legislature had attempted to pass legislation requiring licensing of all practitioners, mind curers included, he had publically spoken out against regulation.[28] Knowledgeable and impatient with authority, frustrated and anxious, hopeful and experimentally inclined, he proved to be his own worst enemy.

He became more and more drawn to Dr. Baldwin. "Baldwin was very interesting," he wrote to Alice, who had gone to London to be with Peggy, who was unhappy in her school at Harrow. "He is taking Spadel baths again, and says they excite him very much, and some of his old nervousness and wildness come back. . . . He is himself taking large quantities of iodide of sodium, and understands its application. He has been studying arterio-sclerosis and heart all the year and probably knows a good deal more about them than he did; and on the whole I find his presence a comfort, since I listen to him and needn't discuss."[29] Baldwin's rejuvenation may have made William more concerned about his own condition. "The truth is that I have *no strength at all,*" he complained to Alice a few days later. "Yesterday I felt dreadful weak and seedy all day, and in the night . . . I had indigestion attack and diarrhea—no analogy to the bloody thing last May, and today am as seedy as if I had been ill a week."[30]

Rest and inactivity, along with the baths, Nauheim's great attraction, seemed to bring more misery than relief. By contrast, Baldwin seemed to have formed the correct therapy, if not a cure.

James' spirits rose in early September as he anticipated Alice's return to Nauheim. "The only thing for us to wish for now is that the adventure of the next eight weeks for both of us may pass quickly," he wrote. He had not made her feel comfortable about being in England. "But how differently [time] will pass for the two of us—you lapped in joyous excitement and multitudinous experience of a rapturous kind during most of it (I don't count in the voyages)—I feeling my pulse and counting the minutes, and trying to do 'nothing' until I see you again. Oh! for one good *talk* together ere you go! This being lectured by you every night 400 miles distance, and hardly being able to make a reply is not what I'm accustomed to, thank heaven."[31]

He also looked forward with growing excitement to starting Baldwin's injections—which held out the prospect of a miraculous cure for *all* of his ailments. Seven years earlier, feeling depressed and fatigued, William had injected himself with animal extracts, what he had described to Theodore Flournoy as "Brown-Séquard's famous injections."[32] The result had been little short of disastrous—an abscess from an unclean needle, and five weeks in bed. Now, however, he believed that Baldwin had so improved both the extract and the injection procedure that all danger had been removed and success was highly probable. In his letter letter of September 8, William finally confided to Alice:

> I think I may as well tell you what Baldwin's famous "remedy" is, as you must be mystified at my keeping it a secret. It is Robert-Hawley animal extract, something similar to what I tried when I had the abscess, only the result of much more evolution and experience, and in the hands of one of the best medical men in the U.S. You need not fear any more abscesses—that one was due to my neglect of antiseptic precautions in injecting, and this time the antisepsis is carried to an almost ridiculous extreme. The only need of a medical man is to control the proper dose by ausculting the heart, and to inject into parts of the body I can't get at myself. The statistics are marvellous in all kinds of degenerative troubles, and there is no *bad* result—the only unfavorable possibility being an absence of good ones.[33]

There had been astonishing cures. Not only had Baldwin completely recovered, but "Baldwin's old father was made 15 years younger in every respect, and my total complex of symptoms is just what it *ought* to relieve completely," he continued. Baldwin as well as another Nauheim physician, a Dr. Reiger, had diagnosed

William's problem as arteriosclerosis; and the injections would soften hardened arteries. The treatment lasted for three months, and James was anxious to begin as soon as possible: "You can well understand how it shakes me to be here within direct contact of it and lose all these precious days purely on account of Schott's pettiness of personal disposition. This is not a harum fad which I am *engoue'd* about on account of its novelty. It is the end of a long and legitimate therapeutic evolution which is at last in scientific hands, and bearing legitimate fruit."[34]

But Schott did not see James' condition as fundamentally sclerotic, and favored mild exercise. William admitted to Alice that he was infuriated not only with Schott but with himself. "If it were only possible to have begun 3 weeks ago, I dare say I should greet you in November an unrecognizable youth! But Schott's susceptibilities and my conscientiousness have stood in the way. I insisted on asking his permission." He had been too cooperative, too deferential; "When you are in the hands of a doctor, you must play your part in the farce of assuming his infallible wisdom. One can get candour out of a common man or out of doctor like Driver [the James family physician] or Jim Putnam, but out of one like Schott who counts as an authority, never!"[35]

Meanwhile, Alice was having her own problems. She had put off going back to Cambridge, but worry over Peggy's inability to adjust to British schooling now tempted her to take her daughter home. When William learned that his wife now wanted to depart for America, he exploded: "I confess I hardly grasp your state of mind, when you speak of waiting a month more at Rye etc, with me here, and *then* (!!!) going to America; when the whole trouble from my side, is that you didn't go 3 weeks ago, so as to be back in time to help me through the winter."[36]

Upset with Schott, himself, and now with Alice, James revealed to her his true reason for returning to Nauheim. "One of the reasons why I wanted to come here by myself was that I might talk it all over with him (Baldwin) and read the literature without having to battle with your anxieties." He had considered going to England where, without Schott's influence, he could begin the injections: "I feel as if my presence here were nothing but a way of being 'complimentary' to Schott. I have *no* confidence in his total judgment; and imagine there is no rational ground whatever, for his vetoing my return to England." In the end, however, he decided on following Baldwin to Italy. Consumed with the prospect of beginning Baldwin's astonishing injections, other matters were tabled. "Peggy's fate and that of the lecture will have to be disposed of in the fullness of time," he concluded.[37]

By November 1900, the Jameses were in Rome where, under Baldwin's supervision, William commenced the injections. From the Hotel Hassler, Piazza Trinita de' Monte, he described the routine to Henry:

> Invalidism breaks my day into so many segments that they crowd out more real life. My two injections make holes of 20 minutes at 10 and 6. Baldwin always spends at least half an hour during his visit. I come back to lunch at one, wherever I may be. From 2 to 3 I lie on my back and nap if I can. At 5 I practice Schott's Wider-Stand's gymnastik, and so it goes. This is the eighth day of the injections, and I can not doubt that they are taking hold.[38]

And his heart seemed to improve. Unfortunately, nerve trouble increased, since between injections, naps, and sightseeing walks he worked on the Gifford Lectures in earnest. To combat neurasthenia, he told Henry, he would begin another therapy. "I am about to recommence with electricity (which has always relieved the fag of college work at home toward spring) using a battery hired here."[39] Electrical therapy was used widely to ease nervous prostration. Physicians frequently compared the nervous system to a battery; the objective was to balance negative and positive fluids. Electrical current supposedly rejuvenated the nerve tissue, increased the absorption of oxygen, and even enriched the red corpuscles. Galvanization entailed placing the feet on a sheet of copper connected to a negative pole. The operator, either James or Baldwin, then attached the positive pole to the hand or a sponge and massaged the head, spine, and other areas, often including the stomach and lower abdomen.[40]

There was fifty-eight-year-old William James in a Rome hotel, receiving daily injections of goat extract from a physician who had only lately taken up studying the heart; doing Schott's "gymnastik" exercises to build up a heart whose debilitation no physician had definitively diagnosed; and shocking overtaxed nerves with a rented battery. Through it all he made remarkable progress on the Gifford Lectures, finishing seven of the ten by the end of 1900.

Whatever the therapeutic value of the injections, by April William was well enough to leave Italy for Switzerland. After brief stops at Lucerne and Geneva, the Jameses hastened on to London for a short visit with Henry. By mid-May they were in Edinburgh, where the long road to the Gifford Lectures finally ended. "Well! I made the plunge yesterday, and have come out unscathed," he exclaimed to Josiah Royce after the first lecture. "The audience was ultra-sympathetic, and laughed whenever I uttered a polysyllabic word. . . . It is a rubicon passed, and parts me from a hideous phase of existence."[41] A month and ten lectures later he wrote to Henry, "The deed is done! The lectures that have so doomfully o'erhung my consciousness for the past 2 years . . . have melted into the infinite ague of the past and I am relatively free."[42]

After resting at Nauheim from July to late August, William, Alice, and Peggy left by Leyland Steamer for America. "We have been away now more than two years, and the prospect of 'home' again I must confess excites me," he wrote to his old philosophical correspondent, Shadworth Hodgson. "With all its fatigues, responsibilities and complications, and be it ever so humble there's no place like it after all."[43] But the relief of return did not restore Williams health, even though his teaching load was reduced to a half course, "The Psychological Elements of Religious Life"—a rerun of the Gifford Lectures. The Cambridge social scene made sustained relaxation all but impossible.

In September he went alone to Chocorua, hoping to ease nerve fatigue. "My 'nervous prostration' keeps pretty bad—the symptoms are frankly that, no longer confined to heart or aorta. It is rather disconcerting; but it will probably yield to galvanism; and I am sure that in the meantime nothing can be as good for me as the extreme simplification and sedation of the place. . . . Europe is too much for one!" he wrote to Alice.[44] By December his condition and spirits had improved. "I am going on splendidly—days with feelings just like my old ones of pride and power and adaptation to the world's demands," he happily told Henry. "I believe it all to be due to those lymph injections, which I began [again] 6 weeks ago and shall take for three months."[45]

It was evitable, however, that although James had periods reasonably free from neurasthenia and angina pain, the search for better health continued. He had faced in Europe the bleak prospect that if miracle cures like Baldwin's injections and electrical therapy did not help, his health might never really improve.

SOMETIME DURING the two-year European stay—it is impossible to pinpoint precisely when—James composed a two-part notebook with the titles "Memorandum for the Gifford Lectures" and "Original Plan for a Philosophy." These scattered thoughts provide ample proof that his high expectations concerning Baldwin's injections were countervailed by darker thoughts, by a struggle to formulate a philosophy to sustain him through physical decline. These mullings are as central to James' state of mind during this period as are those he recorded in his diary during his melancholia of 1868–1869. And they indicate unmistakably that he suffered a depression perhaps even more severe. Indeed, James no longer had the youthful resiliency to assume that a philosophy of action could resolve or even check discouragement about his health.

As he composed the notebook, he read the lives of persons who had suffered and had somehow sustained themselves through religious faith and mystical experience. James was centrally concerned with the nature of suffering, his own and that of others. In the notebook he remarks, "In the act of groaning and travail one certainly feels as if one were in a very central, very ontological position. One feels as if all 'formulations' were secondary and superficial matters, relative to this groaning life."[46]

What was the ontology of suffering? How could one express it? "One feels as if no formula could exhaust the life, or be quite adequate to the mystery. And the religious life is far more at home in the mystery than is the intellectual life. Religion naturally expresses itself in more or less paradoxical terms. The lower and higher—What? *self!*" He reminds himself to "Collect a mass of Christian phraseology to illustrate the point!" The religious experience seemed to step into the travail of human existence and speak to the self in a way that made sense of the suffering.

Religion presented the world as dualistic—as flesh and spirit—but as a world whose division was resolved in mystical ecstasy; "A give and take between dualism and monism seems religion's native element. An estrangement; and a return and identification in which the former estrangement is a dynamic moment. The very word 'return' contains the mystery. Otherness 'swallowed up' in oneness, etc.—darkness 'overcome,' etc." James notes, "Find a lot of words like 'return' that have this dual inner suggestiveness."

James finds himself grappling in a different context with the same metaphysical problem that he had battled in the notebooks on pure experience: dualism. "Isn't the difficulty of a single smooth scheme uniting the subjective and objective worlds due after all to the pluralistic constitution of things?" he asks himself. Yet a pluralistic world did not mean that an ideal world did not exist. "No one doubts that parts of the world are ideal. . . . The religious problem is as to *what* these parts are and *how* far they extend."

But how did one discover these ideal parts, this religious pluralism? James had to find religious ideals in experience, but not just in the experiences of mystics and saints. His own life was filled with foreboding about his physical condition, a despair that might yield a religious—that is, an ideal—meaning. And here the notebook launches into a truly remarkable discussion James calls "the broader indifference." The meaning of religious experience becomes intertwined with his crisis; it becomes a meaning inseparable from coming to terms with illness and impending death:

> I find myself in a cold, pinched, quaking state when I think of the probability of dying soon with all my music in me. My eyes are dry and hollow, my facial muscles won't contract, my throat quivers, my heart flutters, my breast and body feel stale and caked. Laughter and cheerfulness even about other things than my own destiny are impossible. My mind is pinned down to the continual contemplation of annihilation which fills me with a kind of physical dread, none the less positive for not being very acute. My queries all revolve about myself. . . . I have forgotten, really *forgotten,* that mass of this world's joyous facts which in my healthful days filled me with exhultation about life, facts which are there still, wholly undiminished by my own paltry little fading out. The increasing pain and misery of more fully developed disease—the disgust, the final strangulation etc., begin to haunt me, I fear them; and the more I fear them the more I think about them. I am turned into a pent-in egotist, beyond a doubt, have in my spiritual make-up no rescuing resources adapted to such a situation. The little black centre of my field has practically obliterated for me all the effulgent spheres of light and life that lie about it.

The fear was so worrisome as to become obsessive. James could not find spiritual solace. Interestingly, he lapses into the language he had used in his theory of pure experience—the "little black centre of my field." He had described fields of experience as open and continuous, exhibiting a variety of focuses. But now these flexible fields with their "effulgent spheres of light" had faded into a fixed, lightless center. Proximity to death diminished one's vision and narrowed an expansive, open, pluralistic world. There was a relentless human fate that pitilessly disregarded the picture of reality his philosophy of pure experience had drawn.

James realized that his sadness "still had the sort of major chord quality that goes with every universal insight"; and "the queer privateness and individuality of forsakenness, the intense lonesomeness, of my own case were always an absent feature of my perception of the general fact of other men's mortality." What he ascribed to others he could not now ascribe to himself. "I thought it natural that they should sink into the grave with a noble pensive resignation. But what is natural for *me* seems a state of feeling far less noble." Perhaps he remembered Henry Sr.'s dignity, his willing turn from life toward death to complete his spiritual journey. Perhaps he recalled countless séances that took the distinction between life and death to be of no real account. Passing from one world into another might seem for some as natural as moving from daylight to night, from a waking state to a dream state. But the prospect of his own demise posed a *felt* discontinuity; it seemed to painfully contradict the continuity the world assumed in his eyes.

To rest with contradiction was impossible for James. He had to find a way to reconcile his self-centered attitude toward dying and death with other more "noble" attitudes. So in his notebook he redraws the field of experience. "Between the general view [of dying and death] which of course by anticipation always included my own case, and the actually realized presence of my own case, there is no difference whatever in the facts admitted. The only difference is in the moods aroused. Sober acquiescence in the one instance, paling and pining in the other." Dying and death happened to everyone regardless of how one felt them. The field was there, it was fixed. The felt objects within the field, however, were pliable, capable of many shapes. "These emotional colour tones are independent variables susceptible, according to the constitution of the individual[,] of combining with any intellectual content whatever."

How to best use the emotional contents of the incontrovertable fact of animal death—that was the vital question. How one approached the inevitable was a pragmatic matter; and for William, worried, unsure of a diagnosis, and frightened about an approaching end, the pragmatic escape from the field of death's black center occupied his imagination. On the back of the notebook page describing symptoms and fears is this remark: "A 'gentleman' is a man who cares nothing for his life." Next to this epigram he quotes the Roman stoic Marcus Aurelius: "Why weep for me, and not for the plague and death of all?" James was veering back toward the solution he had found to his depression of the late sixties: one had the power to choose an attitude, to believe one way or the other. All one had to do was to redirect, to refocus one's vision.

It, was, after all, a wonderfully Jamesian solution, this willful changing of the center of one's attention to accomplish a difficult task or to carry out a moral imperative. Even unto death there were ethical choices: "It is one active attitude or the other, the pining slave or the high-hearted freeman, and but the individual's 'so will I have it' to decide which it shall be." In his jotting William allowed his natural optimism fuller expression.

Yet what worked in 1870 was not so workable in 1900. The active will was likely to be broken during a terminal illness. "But Stoicism, with its muscles always tense, always holding its breath in an attitude which is ready to break down and at the last extremity always does so break down[,] has in this instability an element of weakness which religion in the more extreme sense has never felt," James admits. What worked to overcome a young man's melancholy would not do now. Religious resignation was a more realistic solution. For then, "Will is drowned in peace attained, death swallowed up in victory enjoyed. The hour of muscular tension is over, that of happy relaxation, of calm deep breathing, of a present with no possibly different future to be on one's guard against has arrived. The negative element is not ignored and forgotten merely, as by stoicism, it is positively expunged and washed away." "Will" and "muscular tension" had long sustained James, providing antidotes to the drift and idleness which had often threatened. Now, however, the religious solution tempted him—tempted as it never had before. A philosophy of pure experience, however resourceful, had ontological limits that religion surpassed.

BUT WILLIAM never truly succumbed to religious consolations; at least there was no indication of resignation in his notebooks. Why, given his acute awareness of the advantages of religious faith for those suffering a fatal illness, could James not believe? It would have been the pragmatic solution, a better one, as he admitted, than gentlemanly stoicism. But philosopher William Barrett has arged that James never spoke "from within faith."[47] Although sensitive to the religious experience, William passed through his own spiritual crisis without a true conversion; he remained on the outside looking in. James' indelible experimentalism prevented him from accepting a time-worn answer to the human predicament. Reality came to him as novelty rather than reoccurrence. Always William looked for the new experience; he expected unexpected phenomena in mysticism—spiritual variety rather than final resolution.

Unlike Henry Sr., who after his vastation embraced a philosophical mysticism that renounced selfish ambition, after his first melancholia William had embarked upon a self-fulfilling professional and creative career. Although he attempted to relax the professional coil in the last decades of his life, he retained powerful creative designs until a year or two before his death. The strength of his imaginative ambition—in part a result of his professional education—made it impossible for him to see the world from within a religious vision. William sought endless speculative expression, a preoccupation quite different from the pursuit of lasting spiritual peace. He could not cancel his theoretical ambitions in exchange for the nirvana religious faith promised.

James knew that like Melville's Ahab, "intense thinking" compelled him to continue to seek the nature of reality; he could not simply accept it as an indecipherable mystery, and humbly give up the metaphysical quest. His obsessively speculative mind and experimental temperament prevented him from experiencing the one profundity that could have helped ease the final crisis. James spent his last years globetrotting back and forth to Europe, delivering public lecture after public lecture, and of course, pursuing metaphysics. It was as if by moving, keeping busy, and remaining speculative he could evade the ontological reality religion met head on. The will to believe was not enough; neither was pragmatism or pure experience; and writing the Gifford Lectures on religious experience was not a substitute for faith itself. James' "religious" perspective, for all its sensitivity and insight, was not really a religious perspective at all. As William once said, "Although religion is the great interest of my life, I am hopelessly nonevangelical, and take the whole thing too impersonally."[48]

But he was only half right: his great interest was the speculative power of his own mind. And that power blocked his path into the religious kingdom even as his physical suffering and psychic anxiety urged him to enter.

CHAPTER FIFTEEN

Give Me A World Already Created

I've outlived the pioneer conditions, where you have to make "allowance" for everything, and dwell on an emergency basis, in perpetual hot water and frustration and unfinishedness, as in America. I once relish[ed] the sense of creation: Henceforth give me a world already created.

To Pauline Goldmark
SEPTEMBER 1909

BETWEEN 1902 and 1910 James continued his attempts to diagnose the source of his heart ailment and to find a cure. But he also worked heatedly on what he hoped would be a metaphysics textbook, the capstone of his intellectual endeavor. He sought cooperation between body and mind: would his heart hold out, so that the mind could continue to create? James' last years tightened the enormous tension between health and ambition, between relentless physical decay and relentless mental work. James' magnificence, his exemplary courage in the face of physical deterioration, shone through in his last years. So did his folly, his frequent hurrying to and from Europe, his unnecessary socializing, and his bizarre pursuit of a medical miracle. But in the end, although he was unable to complete his textbook of metaphysics, James left a fully matured world view: pure experience and radical empiricism; the pragmatic method and a theory of truth—the pluralistic standpoint.

Less than a year after returning from their two years abroad, Alice and William were again bound for England. William had

Alice Gibbens James, ca. 1905–1910.
By permission of the Houghton Library, Harvard University.

William James, ca. 1905–1910.
By permission of the Houghton Library, Harvard University.

promised to give the second installment of Gifford Lectures before leaving for Edinburgh the previous June. His general health seemed improved. "My nerve condition has continually grown more normal, under the use, as I believe of those injections," he informed Henry.[1]

Yet William understood that restoring the health of his nerves and reversing his degenerative heart disease was becoming ever more difficult. He wrote to philosopher Shadworth Hodgson that "I have known in the past three years how it feels to be ninety years old, though I am actually but 60. Our consolation has to be found in falling back upon the general. It is mean to complain, in one's own case, of that which all flesh—even the most decorative members of the species have to suffer; and after all the world is as full of youth and maidenhood as it ever was, if we would but realise the fact when our own beards are grizzling."[2] Nonetheless, he continued with the youth potion by hypodermic, Baldwin's lymph injections.[3]

It was almost impossible for James to act aged, to rest and take solace in rationalizing generalities. By late 1902 he had returned to the problem of pure experience and was working up a new course catalogued as a "Metaphysical Seminary: A Pluralistic Description of the World." "The lectures have come hard because they were bran[d] new and required a great deal of preparation, so that of late the subject has hardly ever been off my mind," he explained to Henry. He planned to include them "as part of my next book," and predicted that "it will be very technical and command no non-professional interest; but professionally I am sure it will be regarded as my most positive performance, and original in a sense in which I have never been original before."[4]

The lectures to which he referred developed the pure experience and radical empirical perspectives as well as the important but less foundational pluralistic and pragmatic ones. They would, he believed, be the crowning jewels of his creative work. He continued in his letter to Henry, "Since in philosophy opinions get tagged to names, it is by these forthcoming ones that my name, if I am ever spared to publish them, will survive and be a thing for my children to point to with pride.—You see how egotistical I have grown, and can observe (what I only the other day remarked to Alice) that avarice and ambition have become my consuming passions."[5]

But even as intellectual work preoccupied James, extraneous activities distracted and impaired his general condition. Visitors streamed incessantly to the Irving Street house. Further, he was asked almost daily to give public lectures. "Don't you *understand* me better than any human being ever can or will?" he exclaimed to Alice. "We shall work out splendidly if I can only find it possible to live more free from social complications. They are what really break me up and make me *ill*." Yet in the same letter he announced that "I have . . . to give 4 or 5 lectures at Glenmore at the end of August [1903]. They will spread my gospel somewhat, help me with the preparation of my book, and make me feel a little like Davidson!"[6] Anticipation of talking about pure experience at the philosophical summer school a few miles from his beloved Keene Valley was a social treat James could not turn down, social tedium notwithstanding.

By the fall of 1903 his nerves seemed revived, though his heart was ailing again, "for certain exertions I made early in the summer." Walking in the Adirondacks, even at a slower pace, did William little good, and teaching plus the Cambridge scene put him back on the social track that aggravated his symptoms. In October he invited diet and exercise crusader Horace Fletcher to stay at the Irving Street house.[7] William began a vegetarian diet and tried to masticate his food at least a hundred times before swallowing. If regularly practiced, chewing rather than bolting his food was supposed to improve his entire condition. Fletcher claimed to have been a human wreck at forty-five, but to have been utterly transformed through proper mastication. Although James did not long continue with "Fletcherism" ("I forget—and *bolt*"), he recommended it to Henry, who suffered chronic stomach complaints.[8] Henry "Fletcherized" for years and nearly ruined his stomach.

Fletcherism proved but one of several fad cures that James tried, attempting to combat a syndrome of symptoms that seemed particularly virulent during the academic year. As the fall term ended he felt so ill that he made a long-contemplated but deferred decision. "I beg to place my resignation in your hands," he wrote to President Eliot in mid-December 1903. "Poor health of late has much impaired my working powers, and I believe that a 'foot-free' condition will be better for me personally. . . . I trust that you will accept this resignation as simply as I offer it."[9] Eliot wrote back that after presenting the resignation to the Harvard Corporation, they had refused to accept it.[10] Their refusal was probably at the president's request. Eliot understood James' institutional value, his ability to attract first-rate students, and he tried to keep his star professor teaching as long as possible. In the end, James simply found it impossible to resign; the Harvard association ran too deep. But in 1903 something else probably convinced him to continue; he still had hopes of being cured.

Early in 1904 James contracted a terrible cold and traveled to Florida to recuperate. By February he had recovered enough to go to Chicago to give a guest lecture on pragmatism and meet the man who would carry that philosphy to lengths he had never intended. "Last night I spent with the philosophical department of this splendid university, which under Dewey, is doing great things for *thought*—strange that that long-necked and abstracted dreamer should really put his stamp on a genuine new school of philosophic thought," he wrote to Miss Goldman.[11]

The sudden attention pragmatism had generated surprised James, for it was not the bedrock of his metaphysics. " 'Pragmatism' never meant for me more than a method of conducting discussions . . . and the tremendous scope which you and Dewey have given to the conception has exceeded my more timid philosophizing," he explained to F. C. S. Schiller.[12] Along with Dewey, Schiller, a philosophy professor at the University of Southern California, had become an aggressive advocate for the pragmatic perspective. Dewey soon transformed pragmatism into a strategy to remold American education toward total cultural rejuvenation. Pragmatism would become the social philosophy of the progressive movement.[13]

James spent the spring and summer of 1904 preparing two key articles—"Does Consciousness Exist?" and "The World of Pure Experience"—for publication in the new *Journal of Philosophy, Psychology, and Scientific Method*. Brother Henry, making a rare excursion to America, enriched the Jameses domestic setting from June to March of the following year. Perhaps in part because of Henry's incessant reminders of the superiority of Old World civilization, William began planning another European jaunt. That he was restless for a change became clear in January 1905, when he again attempted to resign and take a temporary, part-time professorship at Stanford University—an action that Eliot again thwarted: "You see that I am unable to contemplate the severance of your connection with Harvard. I think it ought to be continuous and permanent, with such leaves of absence as you may desire, and with such amount of work and amount of salary as you may desire."[14]

Encouraged by these generous terms, William went ahead with travel plans. He wanted to see Greece, the birthplace of western philosophy, as well as to attend the International Philosophical Congress in Rome. James had corresponded with Giovanni Papini, an Italian philosopher who passionately embraced pragmatism as a philosophy of action to instill new spiritual awareness into Italian culture. Papini and several other militant Italian pragmatists popularized their views in *Leonardo,* a nonconformist review. They championed James, Bergson, and Nietzsche, all of whom, despite considerable differences, represented intellectual independence from traditional European thought and culture.[15]

In Rome James learned that he had a truly international reputation. To Alice he wrote, "This morning I went to a meeting place of the Congress to inscribe myself definitively and when I gave my name, the lady who was taking them almost fainted, saying that all Italy loved me. . . . So I am in for it again, having no power to resist flattery."[16] He read a paper—"La Notion de Conscience"—in French, a condensed version of his published article, "Does Consciousness Exist?" And he discussed philosophy with Papini and a "little band of 'Pragmatists' who carry on a very serious philosophical movement, apparently *really* inspired by Schiller and myself." William contrasted Italian philosophical congeniality with "our damned academic technics and Ph.D. machinery and university organization."[17]

But James had more than philosophical conversation on his mind. That evening he dined with Baldwin and arranged for more lymph extract.

By mid-May James was on his way home with short stops in France and London. So far health had remained tolerable, a fact he attributed to the stimulus of Europe itself. "In spite of fatigue and so much of occupation 'unreal' at my age, and in spite of one's disgust for table d'hote fare, it does do one good to come abroad. It stretches the mind to have seen Greece, and Italy again, and now this clean and civilized France—simply to realize their size as *facts.* It keeps you from drying up, it keeps you young," he insisted to Alice.[18]

But by the time he arrived in England he was complaining to Alice of insomnia and listlessness. He vowed to return to Cambridge a little early in order to avoid British social complications—"the Oxford Racket." Now he lamented that "I shall not come back in good condition to confront the Cambridge summer, and must immediately begin on the 'lymph' . . . so as to be in shape for the Chicago lectures."[19] James could not get off the social treadmill as one engagement loomed after another. Yet with the injections he felt able to continue living in a way that probably shortened his life.

Back in Cambridge by early June, he rushed on to Chicago for five lectures which introduced pragmatism to audiences of between five and eight hundred. He sensed the charade of peddling philosophy to such large and general groups. He admitted to Alice, "I had a queer sense of the absurdity of the thing I was doing, trying to inflict what for me has been a very gradual growth, a highly artificial state of mind, important only through its relation," its contradiction of the thought of other philosophers, on "these virgin natures to whom it never occurred to think things in those aspects at all."[20] Several days later, however, he seemed more satisfied, writing to Peggy that "my Chicago trip was a success—very simple and pleasant socially . . . and on the last day I had them all pulling on my string like one fish." The Chicago social philosopher George Herbert Mead "said he had the impression of having witnessed something 'historical.' You see I can brag as well as any one."[21]

In the late summer Alice contracted a severe cold which she passed on to William. The cold subsided as he began the fall semester. But William had arranged to give a series of lectures at Stanford in early 1906—an occasion which would bring him to California at a geologically auspicious time—and the cold left him apprehensive about the prospect of yet another long trip. "I have about got my strength back, but I look forward with some apprehension to the long pull in California which will begin in January," he wrote to Henry. "I think I lecture much better than I ever did; but I seem to be pretty unfit for all unwanted forms of exertion."[22]

James went first to Arizona, viewed the grand Canyon, saw Schiller at the Unversity of Southern California, and then continued on to Palo Alto. He initially found Stanford idyllic, and told Eliot, "I find this a most utopian place. All the higher essentials, and no superfluities or antiquities. The simple life at its best"—praise that must have worried Harvard's president, given James' recent intention of teaching there.[23]

Yet despite local charm, James soon soured on Stanford. He had contracted to lecture until the spring term ended in April. The thought of four long months without close friends or travel became intolerable. "I want you dreadfully to come and be my companion," he pleaded to Alice. "You've no idea how destitute I feel, all alone. But no matter—life's a battle!"[24] The only thought that cheered him was the pecuniary advantage. "I don't know when they pay one here, but if I can only get through these next four months and pocket the 5000 I shall be the happiest man alive."[25]

William's unease increased in mid-January when Alice failed to answer his daily letters. He eventually wired home, fearing an accident or illness. Alice had indeed been ill, having been in a rundown condition since the catarrh of the previous summer. Nonetheless she joined him in mid-February. By that time, however, William had another problem. "I am cutting my lecture today, being laid up with a regular attack of gout—unitarian gout!" he complained to Peggy just before Alice's arrival. "No family! no servant! unable to move!"[26] His heart condition may have caused this complication. The "gout" would return four years later, only days before his death, when impaired circulation resulted in swollen extremities.

William slowly recovered with Alice on hand to nurse him and converse. The days passed smoothly until April 18. Alice described the occasion:

> At 20 minutes past five this morning I was lying in my bed half awake, when the earthquake came upon us. I sat up and watched the pitching house, shaken as a terrier might shake a rat, creaking timbers and falling articles on every side. Then William came running in, saying "This is an earthquake—are you frightened? I am not, and I am not nauseated either!" . . . All our chimneys are down, our cosy sitting-room piled with bricks, the walls cracked every which way.[27]

Four months later, relaxing at Chocorua, James wrote Eliot that "if the earthquake hadn't closed Stanford a month too early last year I don't know how I should have got through the lecturing, it fatigues me so."[28] James' recollection was inaccurate, for his health did stabilize in March and April. But by September heart and nerves were bothering him as much as they had during the Nauheim stay in 1900.

"The furious and distressing wakefulness which it took two weeks to develop last year, began this year the first night," he wrote to Alice from Keene Valley. The renewed symptoms were so severe that he thought of canceling plans for the American place he loved most. The family had recently purchased land on East Hill several miles from the Shanty, and William had looked forward to building a summer place there. "It is a painful decision; but the symptoms are unmistakably the same: tightness of the head, incipient headache, running of the heart, and restlessness in bed." He sadly admitted that "I clung to the chimera of permanent connexion with the Valley—but good bye!"[29]

James returned to Cambridge more certain than ever that his health was the first priority. He was scheduled to teach a half course, "General Problems in Philosophy." In addition he was preparing eight Lowell Lectures on pragmatism that were to be delivered in early December 1906. With lectures to finish and facing growing infirmity, teaching had to be terminated. The decision could no longer be put off. "For a year past my infirm 'heart' has been giving me trouble again," he told Eliot, "and within the past fortnight something like the bad symptoms of seven or eight years ago have broken loose." Although agreeing to finish the fall semester, he wanted his resignation officially recognized at once, asking Eliot "Is it too late for it to take immediate effect now?"[30] The president now saw that James could no longer continue and agreed to terminate his thirty-five-year teaching career in January 1907.

The Lowell Lectures were a great success, perhaps the greatest in James' lecturing career. Famous in psychology, religious studies, and philosophy, James was unquestionably one of America's best-known intellectuals, a man with multidisciplinary popularity. He reported to Peggy, who was at Bryn Mawr, that "I gave my last Lowell lecture last night with great eclat, being called before the curtain. Tip-top audience, but Lord! I'm glad its over."[31] To Henry he exclaimed, "I had an splendid audience, the intellectual elite of Boston, and about 500 stayed it out." And he joyfully anticipated another ending. "I have only 5 more weeks of College lectures, and then free forever!"[32]

After those five more weeks James was presented with a vase inscribed, "Remembrance and Admiration from his last class in Philosophy, Harvard University, January 22nd, 1907." The Harvard graduating class contributed a silver-topped ink well. One class member remembered that "W. J. was really surprised—said something about premature obsequies to his teaching career . . . then he was cheered, and the class dismissed before the hour was over."[33] One student asked for an autograph, which he photographed and sent to Alice some years later. It read: "Wm James good bye Harvard! January 22, 1907."[34] William wrote to Peggy, "You will doubtless have heard of my farewell ovation last

Tuesday. A silver mounted inkstand and an athletic trophy in silver from the students—your mother more than ever believes me to be great."[35]

Leaving teaching, however, did not mean leaving audiences. A week after James' last class, he delivered the Lowell Lectures on pragmatism to his largest gathering ever, at Columbia University in New York City. "My last lectures had well over 1100, that being the 'seating capacity' of the room, and walls and other spaces being lined. . . . I enjoyed a *success* in my professional way," he wrote to his daughter. The final evening he dined at Hapgoods with "Mark Twain, Peter Dunne, and 4 other eminent litterateurs, very pleasant indeed."[36] "It was certainly the high tide of my existence, so far as *energizing* and being 'recognized' were concerned," he remarked to Henry.[37] For two weeks in New York William felt reasonably well; the glow of attention seemed a tonic—but not a lasting one.

He put himself under the deadline pressure again, this time promising eight lectures with the general title "On the Present Situation in Philosophy," to be delivered at Manchester College, Oxford in the spring of 1908. He explained to Henry, "Lecturing is profoundly odious to me, and the socializing will doubtless be excessive and fatiguing, but I dimly see my way to another little book which should be useful, it will do good to Alice, so I make the venture. . . . It will be my last adventure of the kind."[38]

The Hibbert Lectures at Oxford were indeed James' last formal lectures. Once again he and Alice found themselves on board ship, the *Invernia*, and arrived in England at the end of April. "I have been sleeping like a top, and feel in good fighting trim again eager for the scalp of the absolute," he informed Henry. "My lectures will put [the absolute's] wretched clerical defenders firmly on the defensive."[39] His spirits were soaring because the lectures were all but written. A letter to Peggy at the end of May shows him still optimistic; the lectures had been well received, and better still, he had just concluded the final one: "The ordeal is over! I have just successfully finished my last lecture . . . and feel as if I might begin to make my case."[40] The "case" was for a pluralistic universe grounded in the immediacy of pure experience.

The Jameses remained in Europe for five months. Alice divided the summer between Henry in Rye and Peggy in Switzerland, while William vacationed in England's Cotswold country at Patterdale. He rested, and he ambled through the charming rural environs. Despite a summer cold his health improved. "I am walking here a couple of hours daily, cool and inclined to mist and drizzle, but very fine scenery. There is no doubt that England is good," he told Alice.[41] To his "beloved boys" James wrote, "The 'country' in the American sense is the thing for me in my vacations, with an earth that you can lie on, a wild tree to lean your back against, and hold a book in your hand without reading it. . . . Oh but the fine shape the English nation is in, with everything apparently better than its counterpart is in America! . . . So few of the *unwholesome* human being[s] that [are] about so in America."[42]

By September, English superiority notwithstanding, James had tired of the Cotswold countryside. He wrote to Henry P. Bowditch that "I have myself been more than ready to return home and go to the shanty, for six weeks past, but the wife and daughter want to stay—heaven knows what for—and I am at their disposition. At 66 a man's own library, fireplace and work table are what he is most happy near. Vacations ought to be short and purely recreative."[43] But "vacations" at Keene Valley encouraged climbing, which irritated his heart. In truth James' rural English interlude was a last European recess from the debilitating effects of American work and relaxation.

After a stormy October passage that put Peggy to bed for eight days, the family arrived in Boston. William immediately went to Chocurua "to get the last short taste of the country's sweetness." After nearly a half year in England, return to the American landscape stirred his deep ambivalence about the nation. "No sound, no people, earth and sky both empty, and almost alarming in their emptiness. The elaborateness of English scenery, the simplicity of American—its hard to be torn so both ways by one's admirations, and the best policy is to think as little as possible about the contrast," he told Henry.[44] James had come back to more than just an empty American autumn; he came back to his own bleak autumn as well, for his professional ties with Harvard had been irrevocably severed. Vocational emptiness, the absence of even part-time teaching, may have produced a feeling of uselessness signifying that he was now truly an old man—a depressing prospect that the limelight of Oxford and the charm of the English countryside had helped repress.

By early 1909 James was seeing a local homeopathic physician. He admitted to Henry, "I am going daily for my arterial troubles to a semi-quack homeopathist in Boston. . . . No results as yet but exacerbation—which he protests is just what he expects at first."[45] As 1909 wore on, the pain caused by James' angina increased; he noticed "the continuation of my precordial pain" which persisted "in spite of the daily assurances of a semi-quack doctor whom I have been going to all winter." Nevertheless, he remained a patient, probably because "he has done me good in other ways."[46] Failure to cure an ailing heart did not translate into complete therapeutic failure. The positive thinking of an otherwise ineffective physician gave James hope.

He even tried yoga but found he lacked the patience to sustain the relaxation-inducing postures: "I am so rebellious at all formal and prescriptive methods—a dry and bony *individual,* repelling fusion, and avoiding voluntary exertion."[47] Still, James believed that illness often resulted from failure to find deeper levels of energy, and tapping such energies was an objective that yoga embraced. In 1908 his tiny book, *The Energies of Men,* had been published; it called for the awakening of untapped sources of willpower which habit blocked or kept hidden.[48] Despite the realization that he had come to terms with the inevitability of suffering and death, James could not believe that infirmity precluded significant rejuvenation. As he experienced the fate of all flesh, something deep inside refused to accept it as natural. Creatively, however, he began to anticipate the end, began to see the limits rather than the possibilities of his intellectual efforts.

He continued to think about what he called the "writhing serpent of philosophy."[49] He still hoped to finish his metaphysics textbook—a book that would be posthumously published as *Some Problems in Philosophy.* But after the Hibbert Lectures the great creative moments had passed: "Last summer in Europe made, I am sorry to say, a great cleft between my past and my future," he informed Pauline Goldmark. "I am willing to live hereafter as an old gentleman, taken care of, but taking care of nothing. I've outlived the pioneer conditions, where you have to make 'allowance' for everything, and dwell on an emergency basis, in perpetual hot water and frustration and unfinishedness, as in America. I once relish[ed] the sense of creation: Henceforth give me a world already created."[50] Such an atypical pronouncement reveals that James had turned a decisive corner: he understood that although he still had projects he wished to complete, there would be no more significant transformations in his world picture. The flame of intellectual ambition had flickered out.

He had in fact less than a year to live. In a few months he would flee the unfinished, creative New World for the imagined safety of the Old—seeking in his last flight not intellectual stimulation, but relief from terrible angina pain and dyspnea—shortness of breath.

DURING THE summer of 1909 James' symptoms worsened. In August he went to Bar Harbor, Maine to consult a Philadelphia homeopathic physician, Madison Taylor. Dr. Taylor prescribed calomel, or mercurous chloride, on the theory that disease resulted from "venous congestion," a condition in which the body filled with "bilious fluid." Calomel, taken in infinitesimally small doses, was thought to be the best bilious purgative, a substance which if correctly administered would evacuate the impurities. It was commonly given until the patient's tongue turned brown or salivation commenced. Calomel dosage, excessive bloodletting, and powerful laxatives and emetics all persisted into the twentieth century as remnants of purgative medicine, which was based on the theory that medicine should balance bodily fluids which disease had unbalanced.[51] William's former medical school professor, Brown-Séquard, had favored purgative medicine. So did Henry Bowditch, who frequently consulted with William on medical matters.[52]

Taylor convinced James that his ailing heart was at root a circulatory problem, one that would yield to laxatives and calomel. William reported to Henry that Taylor "advised an eliminative treatment (a la Karlsbad). . . . It has worked ill so far, indeed a fortnight ago I thought I was falling into the same nervous prostration (with cardiac symptoms) that I had ten years ago. But on stopping laxative waters, I have suddenly improved, and shall probably cultivate costiveness [constipation] for the rest of my life."[53] Although discontinuing homeopathy seemed to help, the symptoms never really subsided. In experimental desperation he turned elsewhere.

"I am going to try Xian Science on myself," he wrote Robertson, who had been recommending it for several years. "Do you know a healer named Strang, of 551 Boylston St. Boston? He has been recommended to me."[54] L. G. Strang was a follower of Mary Baker Eddy, the Boston founder of the Christian Science faith. Practitioners denied that illness resided in the body at all and urged positive thoughts to banish sickness.[55] William wrote to his son Billy's betrothed, Alice Runnells, that "I have been seeing Mr. Strang for 21 treatments. I like him exceedingly and I don't think he had any fault to find with the mental attitude I manifested towards his system. I think he found a willing patient." Unfortunately there was no change in his symptoms; a failing, however, that William did not trace to Strang and his method. "I think there is a certain impediment in the minds of people brought up as I have been which keeps the bolt from flying back and letting the door of the more absolutely grounded life open."[56]

Chest pain and shortness of breath were now nearly continuous. Nothing gave relief. Even resuming the lymph injections had made no difference. James told Bowditch that "I tried it again last summer, after a year's interval, with *absolutely no* result. . . . Either I have changed my idiosyncrasy, or my specimen was bad." He waxed philosophical, probably as much to help himself as to console Bowditch, who was also ill. "I don't think *death* ought to have any terror for one who has a positive life-record behind him; and when one's mind has once given up the *claim* on life (which is kept up mainly by one's vanity, I think) the prospect of death is gentle. . . . The great thing is to live *in* the passing day, and not look farther!"[57]

James failed to heed his own advice. For the final time in the spring of 1910 he left for Europe, going to Paris and then to Nauheim to find relief for what by now had become truly a tell-tale heart. The pain abated temporarily in February when a virulent cold forced complete rest. By then he was taking .004 grains of nitroglycerin three or four times a day.[58]

William planned to go straight to Paris to see a French physician, a Dr. Moutier, who specialized in curing arterially sclerotic patients, and then go on for rest and baths at Nauheim. But the trip was complicated by the knowledge that Henry was lying ill in London. Alice soon found herself caring for both Henry and William. Her son Henry later remarked, "My poor mother was at Rye, distracted and half mad with anxiety and doubts—acting nurse to Uncle Henry who clung to her . . . wondering whether W. J. really wanted to be alone for a while, and all the time frantic to get to Nauheim to look after him."[59]

The strain Alice endured during William's final months was immense, and she might have easily broken down in the crunch between brother-in-law and husband. But she had spent decades dealing with William's complaints, if not Henry's, and probably expected the crisis to work itself out, as had so many others. Besides, given her commitment to "tragical marriage," she may have felt this her test as much as theirs.

In early May, after a brief stop at Henry's Lamb House at Rye in Sussex, William proceeded alone to Paris where he consulted Dr. Moutier. "Saw Moutier yesterday," he wrote to Peggy, "who I fear can't do anything for me, says I have no marked hypertension (which is what he can deal with) and that it is mainly a nervous trouble over which he has no control."[60] Once again the psychosomatic explanation was confirmed, a diagnosis that James had hoped would respond to the French doctor's electrical therapy. But it was not to be. He told Henry Bowditch that Dr. Moutier had diagnosed a local cardiac vascular spasm as the probable cause of my anginord pain, and failed to touch it by his currents."[61]

Disappointed, James now looked forward to Nauheim, telling Alice, "The enormous experience those Nauheim men have of every sort of circulatory disturbance makes me wish for their opinion. . . . I mustn't passively go on getting worse like this without end."[62] Yet he lingered in Paris for ten days engaging in conversation with the American psychologist James Mark Baldwin; chatting with Henri Bergson; and staying overnight with philosopher Emile Boutroux and wife, who several years before had been house guests in Cambridge. The social activity fatigued him—especially the visit with the Boutrouxes: "Mrs. B's commonness is terribly in evidence . . . and her way of eating simply revolts me. . . . Tiresome and common to the last degree."[63] From Nauheim he told Alice, "Paris killed me."[64]

Once at Nauheim William put himself under the care of Dr. Theodor Groedel. He decided not to discuss any symptoms before being examined. "I did not wish him to be prepossessed by what I might say. I was quite ready to treat my condition as hysterical and try to bully it down, if he failed to find ample objective justification," he told Alice. But Groedel found the objective justification. "He finds all the trouble to be connected with the heart . . . , and arterial complication to be insignificant. The angina comes from the aorta, etc. etc. . . . Well there you have it!"[65] A few days later James viewed an x-ray of his "thoracic contents" and was alarmed that "the aorta looked to me extraordinarily big."[66] He had visual evidence of unquestionable somatic damage, the original diagnosis Drs. Driver and Putnam had suggested in Cambridge nearly twelve years earlier.

After over a decade of flitting from physician to physician, of lymph injections, of galvanic therapy, of baths, of homeopathy, of mind cure, James had the answer. "It is a comfort to me to have a definite diagnosis of aortic enlargement to explain my symptoms by, and banish the reproach of 'mere nervousness' by which I myself, no less than others, had been fain to treat the case," he wrote to Henry. "The baths here can't cure the condition, but can 'adapt' the heart to meet it better."[67] So he lingered at Nauheim taking a regimen of thirty baths; reading proofs of his last essay, on amateur philosopher Benjamin Blood; and writing dozens of letters, many urging Alice and the ailing Henry to come to meet him as soon as possible.[68]

In early June Henry and Alice joined William, and after an abortive trip to Geneva where William's heart could not tolerate the altitude, the three returned to London. By July Alice could see that her husband had much weakened, and he had fallen into deep despair. "I am perplexed and very anxious about William," she wrote to her mother. "He seemed to get steadily weaker not only at Nauheim but after, and at last he got completely discouraged about himself."[69]

Then unexpectedly during the first week of August, on the eve of their return to America—no European diagnosis or therapy remained—he rallied: "There is no happiness in life to be compared with the lifting of a great anxiety. . . . William is emerging from what we can all now see to have been an acute attack. Why he was so ill, why his breathing became so painful no one tries to explain. But he is gaining strength, sleeping without narcotics, eating better and beginning to take gentle exercise," Alice reported.[70] Actually, her husband had only three weeks to live.

William wanted to get home, and the prospect of doing so may have temporarily buoyed him. Once he left Nauheim, knowing now that his condition was unalterable and feeling extremely weak, he had only one remaining objective. "My dyspnoea gets worse at an accelerated rate, and all I care for now is to get home—doing nothing on the way," he wrote to Theodore Flournoy.[71] He and Alice decided to ship from London to Quebec, avoiding hundreds of Atlantic miles by going down the Gulf of St. Lawrence. From Canada they would travel by rail to Intervale, New Hampshire, where they would be met by their son Billy, who would then motor them the short distance to Chocorua. For the first few days at sea William sat in a basket chair, weak but fairly comfortable. But as they moved into the Gulf of St. Lawrence his condition worsened. By the time they met Billy at Intervale William's feet were swollen, indicating that his circulation had failed to the point where his fluids were beginning to accumulate.

William lived one week after returning to Chocorua. He took morphine continually and lapsed in and out of consciousness. At

2:30 in the afternoon of August 26, William James died in Alice's arms. She had just come to feed him milk. "Our William has gone," she wrote to close family friend Fanny Morse sometime that evening. "It seems strange not to send you his love and mine as always."[72] A death mask was made, and the bereaved family accompanied the body back to Cambridge by rail. Funeral services were held at Harvard's Appelton Chapel and the cremation at Mount Auburn Cemetery. The ashes were scattered in a stream near the Chocorua place. An autopsy confirmed Groedel's diagnosis. James had succumbed to the circulatory ravages of a swollen aorta and an enlarged heart.[73]

JAMES ENDURED a prolonged crisis between 1898 and 1910 that distressed him as much as his time of troubles as a young man—probably more. Whatever the basis for his earlier depression, it was slowly dissolved by a philosophy of free will, a teaching appointment at Harvard, and a loving companion. But nothing could really alleviate his heart condition. And William's heroic and yet pathetic attempts to correct what was not correctable probably abbreviated his life. Josiah Royce thought so. After conversing with Royce several months after William's death, John Jay Chapman noted that Royce "says James shortened life by fussing over his heart—not that he wanted to live, but that he thought it somehow a disgrace to have diseases and was always trying to cure himself. . . . If he had only let it alone and thought of something else, R. believes he'd be alive today."[74]

Royce had touched on a deep truth about his late colleague. For all of his genuine sympathy for "sick souls," James wanted to be healthy; health was the natural state and disease should yield to the proper therapeutics. Disease was not to be simply accepted as a part of the human condition; it was something escapable, something unacceptable within himself and, perhaps, socially demeaning as well—James always took pains not to appear ill in public. Nevertheless, he knew that his days were numbered, knew that his heart might never improve. Illness put a premium on time and spurred temendous creative effort. Paradoxically, as James fixed on his heart condition, he described experience as unfixed, a field of multiple relationships that might be altered according to pragmatic considerations.

By the fall of 1909 he had sensed, as he had written Pauline Goldmark, that his creative world was complete, that his pioneering efforts were past.[75] And when at last the firm diagnosis came, he lived only two months. Did James avoid those physicians who would have confirmed somatic damage? To have known for sure would have been a great relief, but the knowledge would also have denied James his characteristic picture of the world. There would have been few alternatives, few experiments, few surprises. Such an outlook was not Jamesian, was not moral, was in fact un-American, for it would freeze experience into an unalterable predicament rather than allowing the individual the freedom to try a variety of provisional alternatives. It was fitting that the final diagnosis came in Europe. But it was perhaps even more fitting that once James possessed final knowledge, he hurried home to die in America.

Epilogue

Strange to say, the present has blotted out the past, so that even these historic sites failed to give the feeling of reality thereto, and as everything has turned out so much less tragic than might have been supposed, my mind has kept spinning toward the future rather than lingering in the past.

To Alice Gibbens James from Keene Valley
SUMMER 1895

SO WHAT was the center of William James' vision? Imagine James at any point in his life—in Paris discovering Delacroix; at Harvard reading materialistic science and studying physiology; delivering the Lowell Lectures; writing *The Principles;* composing a philosophy of pure experience; sitting in a séance with Mrs. Piper; pursuing cures for neurasthenia and an ailing heart. Is there a center, a core, an essence, a fundamental, irreducible perspective? All the traditional answers have shed much light on James. He did believe in free will; he did believe that consciousness was a sensible feeling that actively selected; he did believe that truth "happened" to an idea, and that there were multiple truths and perspectives; he did believe in a open, growing universe. And James championed the moral life; and he sympathized with the underdog and the unpopular cause; and he supported a mental topography that expanded the boundaries of consciousness, while narrowing the distance between the normal and pathological. Yet while all of these statements tell much *about* James, they do not identify his vision; he remains elusive—

flexible, experimental, pragmatic. To paraphrase William's 1878 judgment on the physiological psychologists, we have a believable portrait, but one not to be believed.

Yet in the shifting collage of James' multiple interests and perspectives, there *was* one preoccupation that guided his creative efforts. No intellectual production, no family member, no friend or professional colleague, was as significant or interesting to him as the life of his own mind. In this sense, his deepest involvement was the effort to describe how *his* mind encountered the world. James was a master at introspection, at apprehending the fine movements of his own consciousness. This was of far more importance than how the world influenced his mind. The existence of an objective reality unrelated to his mind was a problem to be solved, rather than simply a fact to be acknowledged. Hence he was more comfortable in deemphasizing chasms between his mind and objective reality than in recognizing them. Although acutely sensitive, as we have seen, to the psychologist's fallacy of confusing his own stream of thought with someone else's, James fashioned a philosophy of pure experience which allowed him to bypass this problem. As an entity, his mind (and for that matter, all other minds) disappeared in the primal datum and flux of experience. But how did William feel his mental life centered in experience? Centered, James would say, in relationships.

What comes to mind when we try to visualize or imagine how James experienced relationship, how his vision was centered? He left a large clue when he described the seamless panorama—a scene with no break between the painted background and the real foreground. He thought in terms of connection; his mind's eye focused on relation, on conjunction, on continuity. There was nothing in the world that was not actually or potentially related to something else. A discontinuous world was impossible for James to imagine, because such a world would not be capable of being directly *felt*, experienced, apprehended.

Another telling clue is James' description of a landscape as seen from the caboose of a moving train. As the train traveled, the margin of James' visual field became the center, and the center became the margin. The shape of the world changed as perspective changed—inevitably and dynamically. His mind was shaped in perspective, and perspective was relationship in transit. James' vision was a tireless shaping of perspective, a vision that was constitutionally incapable of *not* reshaping relationships. James used body, mind, and imagined objects as a field of experienced perception and ceaselessly drew connections among them. All the rest—the will, the pragmatism, the theory of truth, the moral fervor—was byplay to this dynamic vision, this essential perspective.

Take what has been deemed original in James—the "stream of thought," "the will to believe," "pure experience": each was a sign of something else. The stream was sensible and flowing; the will allowed for action and choice; pure experience resolved dualism. But sensibility, choice, and resolution were all aids to expanding and articulating relationship as perspective. James was so caged within his relational vision that he could not help but continually reexpress—and in imaginative ways—what such a vision demanded. James was also profoundly dissatisfied with a completed metaphysics. But that dissatisfaction was inherent in his unending effort to conflate experience and relationship; to stop redrawing a world of relationships was to deny what he experienced as reality—something he was, of course, unable to do.

The relational vision contradicted James' moralism. From the mid-sixties he had cultivated the moral standpoint: "I must gain a profession and escape my father's superfluousness"; "I must justify marriage in terms of tragedy"; "I must develop a responsible science." Yet intellectually he understood as early as 1868 in Dresden that the achievement of the authentic creative grace he so admired in the Greeks was not dependent on the moral point of view. Indeed, the authentic creative life was free of moral constraints; the mind simply saw and naturally constructed a world. Although *The Principles of Psychology* anticipated radical empiricism and pure experience, James' dissatisfaction with his psychology was grounded in the growing conviction that in it he had nonetheless elevated consciousness into a scientific defender of morality: the mind, if correctly trained, would pick and choose the moral life. His unhappiness with dualism was intertwined with his sense that *The Principles'* theorectical structure was flawed *because* in it he had designed consciousness for "responsible" purposes. In the end he could not fully express his center of vision within the moral perspective.

On the other hand, a philosophy of pure experience simply made a world. Experience was itself and nothing more. Moral content was reduced to relational context. Values were obviously part of experience; but in pure experience values had no meaning. Meaning and truth appeared symbolically, and as practical matters could be dealt with pragmatically. But meaning and truth were not for James the bedrock of reality; they were not the core of his mind's felt experience.

James' mature speculations about the nature of reality are found in the notebooks on pure experience. These speculations often consist of a series of work scenes, relating body, mind, and object, for example, or the foreground to the background, or the margin to the center. Natural experience was body, mind, and object in scenic relationship; to describe experience was to visualize it as a canvas of felt relationships. Important Jamesian metaphors such as "stream" and "field" were indispensible elements of verbal-visual landscapes in which the sensible action of relating took place—continuously and naturally.

In one sense it is highly misleading to speak of the Jamesian perspective as open-ended, pluralistic, and tolerant. True, his multifarious world picture denied the two great nineteenth-century philosophical traditions, empiricism and idealism, on the grounds that neither was capable of encompassing the wealth of reality. But while acknowledging a diversity of other points of view, James embraced only his own vision. For all of his remarkable capacity to sympathize with differing temperaments and their role in fashioning a variety of perspectives, he could tolerate no other point of view as authoritative. For James there was only one supreme authority, the ability to visualize his own mind in relation to its experience. And once he had worked his way through his fields of experience—art, natural science, psychology, and metaphysics—to a philosophy of pure experience, his speculations reached a felt dead end. His world was indeed, as he remarked in 1908, "already created."

His pragmatism dealt with human problems. How does one act given these circumstances, given these potential consequences? What was possible, what was moral, what was truthful depended on contingencies that made a difference to people. That is one large reason why pragmatism, not pure experience, has remained to many the essential basis of James' perspective.

But James' mind was not content to fix its speculative gaze on a philosophy that was largely a strategy to facilitate discussion or encourage appropriate human action. Pragmatism had been useful in his own life as a means to further more fundamental ambitions. It had been especially vital in the sixties and seventies in helping him sustain his creative powers until they could be expressed in a form that satisfied his enormous desire to be original. Intellectual ambition compelled James to seek a world where, as he said of classical sculpture in 1868, "things never have any point. The eye and mind slip over and over them and they only smile within the boundary of their form."[1] Within the boundaries of pure experience there was indeed no "point," simply the "form" of relationship—the only boundary James recognized.

This valueless, relational space might be imagined as the field out of which symbolic creation emerged. Artist, scientist, and philosopher all worked within this essential phenomenal setting, within the relationships established between hand, mind, eye, and object. Their creative products issued from the elemental relational process. But James' genius was to *end* rather than begin his speculations with this descriptive frame of reference, the ultimate context, the pure experience. In fact his speculative life had been, at center, a struggle to find this descriptive scene, this state of natural, theoretical grace. Once he found it, there he stayed.

James' transformation of metaphysics from the symbolic to the relational was an important expression of the movement of the American intellect away from *all* traditional religious and philosophical worlds, toward a final secularization—a noncommittal world of fluid, adjustable relationships. It is the neutrality of relationship, its plastic, manipulative essence, its failure to translate experience into a discontinuous empirical or transcendent ideal reality, that marks James as perhaps the greatest philosopher of modernity. The world is context, pattern, connection, potential and possibility, all shaped relationally. Philosophically considered, this was the consummate Americanization, the final New World metaphysical melting pot.

This may seem too harsh. He did retain, as his letters and popular essays show, deep-seated beliefs in the sanctity of the individual and the value of the moral life—both of which have a traditional, even backward-looking quality. And his psychology had a romantic cast—a large role for consciousness, a stream for it to flow in, a place for the "self," for "will," for "emotions." For all of the functionalism of *The Principles of Psychology,* it was not behaviorism.

Yet for better or worse, James broke down the romance and pretensions of the mind more completely than any other turn-of-the-century American thinker. It was his philosophy of pure experience, his relational reality, that presaged the intellectual tone of behaviorism, among other modernisms. A Skinnerian experiment in operant conditioning is in an incontrovertable sense simply the observation of a context of behavioral relationships. For all their scientific and intellectual differences, James and B. F. Skinner speak the same language, of building a context out of relations. When James accuses the Darwinists of being unscientific about the mind, when he charges Hegel with vagueness, when he attacks Wundt as a "narrow" psychologist, he criticizes from the center of his vision. Their realities are not real because, for different reasons, their realities are not sufficiently relational.

James' intellectual legacy should be distinguished from the popular mythology about him. The James who defended will, choice, and action in the face of determinism and pessimism is the James who has persisted in the public mind. But the relational James, the James who tirelessly sought the essence of experience, is the James who has had the greater intellectual influence. In his wake, psychology and philosophy at Harvard, and in America in general, have developed along the lines of what might be called the technology of relationship; cognitive psychology, behaviorism, and symbolic logic are all concerned more or less with treating their disciplines *as* relational context.[2] While James attacked, manipulated, and discarded what remained of traditional western metaphysics, his successors have built on James' base. And for what?

Some eighty years ago James suggested the intellectual legacy of a philosophy of pure experience. "We certainly have gained no *stability*. The result is an almost maddening restlessness. But we have concreteness."[3] He might have added, "the concreteness of relationship."

Notes

UNLESS OTHERWISE indicated, citations of works by William James refer to the definitive editions in *The Works of William James*, edited by Frederick Burkhardt et al. (Cambridge: Harvard University Press, 1975—).

The following abbreviations of frequently cited books and manuscript collections are used throughout the notes:

JFP James Family Papers. The Houghton Library, Harvard University, Cambridge, Mass.

LWJ Henry James, ed. *The Letters of William James*. 2 vols. New York: Longmans, Green, 1920.

Perry Ralph Barton Perry. *The Thought and Character of William James*. 2 vols. Boston: Little, Brown, 1935.

VC Vaux Collection. Henry Vaux, Berkeley, Calif.

WJP William James Papers. The Houghton Library, Harvard University, Cambridge, Mass.

INTRODUCTION

1. William James, *Essays in Radical Empiricism and A Pluralistic Universe*, 2 vols. (Gloucester, Mass.: Peter Smith, 1967), 2:87.
2. See criticism dated May 26, 1900, in *LWJ*, 2:354–355. The student's name was excised by the editor, Henry James (William's first son), and she was referred to as "Miss S."
3. The exception is Howard M. Feinstein's *Becoming William James* (Ithaca, N.Y.: Cornell University Press, 1984).
4. See Gay Wilson Allen, *William James: A Biography* (New York: Viking, 1967).
5. For commentary on Feinstein's psychoanalytic approach, see ch. 2, n. 22 and n. 30.

6. See Gerald E. Myers, *William James: His Life and Thought* (New Haven: Yale University Press, 1986).

7. For the biographical importance of attempting to place oneself inside the process of creation, see Ira Progoff, *Life Study: Experiencing Creative Lives by the Intensive Journal Method* (New York: Dialogue House Library, 1975); and Robert Schott Root-Bernstein, "Creative Process as a Unifying Theme of Human Cultures," *Daedalus* (Summer 1984), 113:197–217.

8. See Cushing Strout, "William James and the Twice-Born Sick Soul," *Daedalus* (1968), 97:1062–1082; reprinted in his *The Veracious Imagination: Essays on American History, Literature, and Biography*, pp. 199–222 (Middletown, Conn.: Wesleyan University Press, 1981). For a discussion of interpretation of James' psychic crisis, see ch. 6, n. 50.

9. Through 1987, fifteen volumes of James' writings had been published in this series, including an annotated three-volume edition of *The Principles of Psychology*. See bibliographical essay for a listing of the separate volumes.

10. William James, Notebook, Seminary of 1903–1904; item 4506, WJP.

1. REPENT OF THE PAST

1. The James family genealogy is outlined in the James Family Bible; JFP. See also see R. W. B. Lewis, "The Jameses' Irish Roots," *The New Republic* (January 6 and 13, 1979), 186:30–37.

2. Quoted in Jean Strouse, *Alice James: A Biography* (Boston: Houghton Mifflin, 1980), p. 5.

3. By far the best discussion of William James of Albany's career and of his stormy relationship with Henry is Howard M. Feinstein, *Becoming William James* (Ithaca, N.Y.: Cornell University Press, 1984), pp. 25–36; but see also Harold Larrabee, "The Jameses: Financier, Heretic, and Philosopher," *American Scholar* (1932), 1:401–413.

4. For a discussion of the period's striking transformations, see Edward Pessen, *Jacksonian America: Society, Personality, and Politics* (Homewood, Ill.: Dorsey Press, 1978).

5. Henry James, Sr., "Immortal Life, Illustrated by a Brief Autobiographic Sketch of the Late Stephen Dewhurst," in William James, ed., *The Literary Remains of the Late Henry James* (Boston: Little, Brown, 1926), pp. 182–183. For an interesting interpretation of the sketch, see Howard M. Feinstein, "The Double in the Autobiography of the Elder Henry James," *American Imago* (Fall 1974), 31:293–315. Feinstein maintains that Henry created an alter ego in Stephen Dewhurst to fulfill character development that was blocked by his father.

6. Henry James, Sr., "Immortal Life," pp. 185–186.

7. Henry James, Sr. to Robertson James, undated; VC.

8. Henry James, Sr. to Robertson James, undated (probably in the 1870s); VC.

9. William James of Albany to Archibald McIntyre, undated (probably in the late 1820s); JFP.

10. Quoted in Jacques Barzun, *A Stroll with William James* (New York: Harper and Row, 1983), p. 13.

11. Henry James, Sr. to Catherine Barber James, July 24, 1854; JFP.

12. Howard M. Feinstein, "Fathers and Sons: Work and the Inner World of William James, an Intergenerational Inquiry," Ph.D. diss., Cornell University, 1977, pp. 18–19.

13. William of Albany's will is most fully explored by Feinstein in "Fathers and Sons," pp. 60–71, and in his *Becoming William James,* pp. 61–62.

14. Henry James, Sr., in *Literary Remains,* pp. 58–60. Strangely, although this "vastation" passage has been much interpreted, particularly in conjunction with his son William's psychic or spiritual crisis in the late 1860s, it has not been directly related to Henry's own accident and amputation.

15. For excellent discussion of Jacksonian revivalism, see Whitney Cross, *The Burnt-Over District: The Social and Intellectual History of Enthusiastic Religion in Western New York, 1800–1850* (New York: Harper and Row, 1965). In *The Iron of Melancholy: Structures of Spiritual Conversion in America from the Puritan Conscience to Victorian Neurosis* (Middletown, Conn.: Wesleyan University Press, 1983), John Owen King III argues that Henry Sr. underwent one of the last truly spiritual as opposed to neurotic conversions in the American Protestant tradition.

16. There is as yet no adequate full biography of Henry. See however Fredrick Harold Young, *The Philosophy of Henry James, Sr.* (New York: Bookman Associates, 1950); Steve Brown, "Henry James, Sr. and the American Experience," Ph.D. diss., University of Oklahoma, 1981; Perry, 1:3–79; but Feinstein is without peer on the elder James' youth. See *Becoming William James,* pp. 25–88.

17. Henry James, *A Small Boy and Others* (New York: Scribner's, 1913), p. 271.

18. For a lively and perceptive account of popular fascination with the oddities in Barnum's American Museum, see Neil Harris, *Humbug: The Life and Art of P. T. Barnum* (Chicago: University of Chicago Press, 1973).

19. Henry James, *A Small Boy,* p. 277.

20. See Henry James, Sr. to Ralph Waldo Emerson, August 31, 1849; JFP.

2. A HUNGRY EYE

1. By far the most detailed account of the stay is Gay Wilson Allen's in *William James: A Biography* (New York: Viking, 1967), pp. 32–63. Unfortunately we learn from it little about how a novel culture and new educational experiences may have affected the way James perceived the

world. In *The Thought and Character of William James,* Ralph Barton Perry sees the sojourn as a pattern-setter for what would become William's chronic wanderlust. He does not ascribe any fundamental import, however, to the fact that the "winter of 1856–1857 in Paris was a year of exposure on . . . sensitive and maturing minds," one during which "William's growing interest in painting . . . was richly fed." See Perry, 1:177, 183, and more generally, 169–189.

2. Henry James, Sr. to Catherine Barber James, August 13, 1855; JFP.

3. Henry James, Sr. to Catherine Barber James, September 25, 1855; JFP.

4. Henry James, Sr. to Edward James, November 2, 1855; JFP.

5. Henry James, Sr. to Catherine Barber James, November 30, 1855; JFP.

6. See Henry James, *A Small Boy and Others* (New York: Scribner's, 1913), pp. 314–315. See also Howard M. Feinstein, *Becoming William James* (Ithaca, N.Y.: Cornell University Press, 1984), pp. 108–110.

7. Mary A. Tappan to Henry James III, January 11, 1921; WJP.

8. See copy of vol. 1 of *LWJ* annotated by Henry James III in the Houghton Library, Harvard University. In *William James: A Biography,* p. 47, Allen mistakenly has William receiving the microscope for Christmas 1857.

9. Henry James, Sr. to Catherine Barber James, October 15, 1857; JFP.

10. Henry James, Sr. to Edmund Tweedy, September 14, 1856; JFP.

11. Mary and Henry James, Sr. to Catherine Barber James, December 24, 1857; JFP.

12. See David H. Pinkney, *Napoleon III and the Rebuilding of Paris* (Princeton, N.J.: Princeton University Press, 1958).

13. Ibid., p. 10.

14. Both Perry and Allen maintain that William discovered Delacroix in the winter of 1856–1857. But he almost certainly saw his work at the Palais de l'Industrie Exhibition in October 1855. Allen, without documentation, mentions that James saw the works of Millais in the Exhibition's British display (*William James: A Biography,* p. 36). Yet there is reason to assume that William also viewed the French painters. After all, he was in Paris, the art center of the western world; he had made a practice of visiting galleries and art exhibits in New York; and his budding interest in art would have led him to pay special attention to controversial French painters such as Ingres and Delacroix, especially when both had dozens of paintings in special rooms.

15. See Henry James, *A Small Boy,* p. 344; Allen, *William James: A Biography,* p. 41; and Feinstein, *Becoming William James,* pp. 110–112. Feinstein has also suggested that *The Bark of Dante* attracted William because Virgil represented artist William Morris Hunt, who by 1858 challenged Henry Sr.'s control over William's career. And Feinstein has further maintained that William's sketches of animals devouring animals and of animals devouring men were symbolic expressions of murderous wishes

toward the elder James, who blocked his development as a professional artist. See *Becoming William James,* pp. 124–235.

16. Howard Feinstein also notes the absence of written comments by William on Delacroix's painting. See Feinstein, *Becoming William James,* p. 112. Perhaps James believed "Delacroix a greater dramatist than painter;" see Jacques Barzun, *A Stroll with William James* (New York: Harper and Row, 1983), p. 264. Yet no one, if we can believe his brother Henry, touched William's imagination on the 1855–1858 trip as profoundly as did Delacroix. In *A Small Boy and Others,* Henry recalled William's "repeatedly laying his hands on Delacroix, whom he found always and everywhere interesting—to the point of trying effects with charcoal and crayon, in his manner" (p. 359).

17. Progoff, *Life Study: Experiencing Creative Lives by the Intensive Journal Method* (New York: Dialogue House Library, 1983), pp. 14–15.

18. Both critics quoted in Maurice Serullaz, *Eugène Delacroix* (New York: Harry N. Abrams, 1972), p. 158.

19. For Delacroix's important place in the development of modern painting, see Fritz Novotny, *Painting and Sculpture in Europe, 1780–1880* (New York: Penguin, 1978), pp. 151–171. Delacroix's use of color was revolutionary, and he was criticized for sacrificing line for color. Novotny concludes that "Delacroix and Cézanne were the two painters who delved most deeply into the mysterious connexions between colour and line" (p. 154).

20. William James, *The Principles of Psychology,* 2:819.

21. See William James, Index of Subjects: A Collection of Miscellaneous Notes (1864–1890); item 4520, WJP. The entry "Art vs. Nature" is dated November 25, 1864, when he was studying science at Harvard. It reads: "Faire un choix dans la nature—on en fait tres adrotement une loi; parce que les 3/4 du temps la nature se passe des contrastes. C'est donc par insuffisance qu 'on choisit, parce que les moyens de l'art sont bornes, et qu' il lui faut toujours sacrifier une chose pour en faire valoir une autre. Delacroix." I am grateful to Professor Juliana Thomson for the English translation of this passage in the text.

22. Feinstein interprets William's Delacroixian drawings as evidence that William barely sublimated his rage toward Henry Sr. for blocking an artistic career; Delacroix suggested visual means for expressing deep-seated oedipal conflicts and wishes. See *Becoming William James,* pp. 145–177.

23. The mind as experiencing and making the world is best encapsulated in the famous "The Stream of Thought" chapter and again in "The Perception of Reality" chapter of William James, *The Principles of Psychology,* 1:219–278, and 2:913–951.

24. Eugène Delacroix, *The Journal of Eugène Delacroix,* Lucy Norton, tr. (Ithaca, N.Y.: Cornell University Press, 1951), p. 249.

25. Progoff, *Life Study,* pp. 121–123. It is important to make a distinction even in the early stages of James' education between his interior,

creative life and the progression of his career and to avoid imposing a compartimentalism that did not exist in James' own mind. Exposure to science at the College Imperial and joy in receiving a microscope did not mean that William was becoming a scientist. Whatever that exposure amounted to, James' creative life at this time focused most strongly on art, on Parisian painting, and especially on the work of Delacroix.

26. See Burton Bledstein, *The Culture of Professionalism: The Middle Class and the Development of Higher Education in America* (New York: Norton, 1976).

27. Allen, *William James: A Biography,* p. 41.

28. Henry James, Sr., *Moralism and Christianity* (New York: Redfield, 1850), p. 59.

29. See Neil Harris, *The Artist in American Society: The Formative Years, 1790–1860* (New York: Braziller, 1973).

30. See Feinstein, *Becoming William James,* pp. 89–100. Feinstein explains Henry Sr.'s intellectual flipflops not only in regard to art but to religion, science, and philosophy in terms of an ongoing intergenerational struggle. William becomes the pawn in Henry's erratic moves to settle unresolved conflicts with William of Albany. But did William really take his father's shifting points of view on his career and intellectual development that seriously? Feinstein makes William's destiny seem very dependent upon his father's opinions and behavior, when William's genius may have made Henry's perspective irrelevant. The conflicts seem much more the father's than the son's; after all, they were not deep enough to thwart the full flowering of William's originality.

31. Henry James, Sr., *Lectures and Miscellanies* (New York: Redfield, 1852), p. 102.

32. It is interesting that although William took an interest in the vocational decisions of his children, he never actively opposed their choices. None of them threatened his own genius, and he could look upon them with the relative objectivity Henry Sr. and William of Albany had lacked.

33. For a discussion of Henry James, Sr.'s essential "Americanism," see Steve Brown, "Henry James, Sr. and the American Experience," Ph.D. diss., University of Oklahoma, 1983.

34. Henry James, Sr., quoted in Allen, *William James: A Biography,* pp. 37–38.

3. DETACHED AND SLIGHTLY DISENCHANTED

1. *Newport Mercury,* July 10, 1858.

2. *Newport Mercury,* May 21, 1859.

3. Phrase quoted in Virginia Harlow, *Thomas Sergeant Perry: A Biography and Letters to Perry from William, Henry, and Garth Wilkinson James* (Durham, N.C.: Duke University Press, 1950), p. 6.

4. Henry James, "The Sense of Newport," *Harpers* (1906), 13:43.

5. The observer was Frederick Law Olmsted. See Lloyd Robson, "The Hotel Period, 1844–1865," manuscript, Newport Historical Society, Newport, R.I., p. 140.

6. Thomas Wentworth Higginson, in Mary Thatcher Higginson, ed., *Letters and Journals of Thomas Wentworth Higginson, 1846–1906* (Boston: Houghton Mifflin, 1921), p. 225.

7. On Hunt, see Helen Knowlton, *Art and Life of William Morris Hunt* (Boston: Little, Brown, 1899); Bruce Howe, "Early Days of the Art Association," *Bulletin of the Newport Historical Society* (April 1963), no. 110, p. 12; and Howard M. Feinstein, *Becoming William James* (Ithaca, N.Y.: Cornell University Press, 1984), pp. 117–122.

8. Henry James, Sr. to Mrs. C. P. Cranch, October 18, 1858; VC.

9. William James, quoted in Percy MacKaye, *Epoch: The Life of Steele MacKaye, Genius of the Theatre, in Relation to His Time and Contemporaries,* 2 vols. (New York: Boni and Liveright, 1927), 1:76. On La Farge, see Royal Cortissoz, *John La Farge: A Memoir and a Study* (New York: Houghton Mifflin, 1911).

10. On MacKaye and Perry, see MacKaye, *Epoch: The Life of Steele MacKaye,* and Harlow, *Thomas Sergeant Perry: A Biography.*

11. Henry James, Sr., quoted in Allen, *William James: A Biography* (New York: Viking, 1967), p. 55.

12. Georges Cuvier and John Ruskin, quoted in William James, Geneva Notebook, 1859–1860; item 4495, WJP.

13. Georges Cuvier, quoted in William James, Geneva Notebook; and William James, Geneva Notebook.

14. William James, quoted in Allen, *William James: A Biography,* pp. 59–60.

15. Henry James, Sr. to Edmund Tweedy, July 18, 1860; JFP.

16. Henry James, Jr. to Thomas S. Perry, July 18, 1860; quoted in Harlow, *Thomas Sergeant Perry: A Biography,* p. 252.

17. William James to Henry James, Sr., August 19, 1860; quoted in Perry, 1:198–199.

18. William James to Henry James, Sr., August 24, 1860; quoted ibid.

19. William James to Thomas S. Perry, August 5, 1860; quoted in Harlow, *Thomas Sergeant Perry: A Biography,* p. 261.

20. Henry James, Sr. to Edmund Tweedy, July 24, 1860; JFP.

21. There is only one short reference to Hunt in a standard study of nineteenth-century American painting, while contemporaries such as Martin Johnson Heade, William Sidney Mount, and George Caleb Bingham each have complete chapters. See Barbara Novak, *American Painting in the Nineteenth Century: Realism, Idealism, and the American Experience* (New York: Harper and Row, 1979), pp. 125–164 and p. 247.

22. William Morris Hunt, in Helen M. Knowlton, ed., *Hunt's Talks on Art,* 2d. series (Boston: Houghton Mifflin, 1883), p. 63. Hunt included Velázquez, Tintoretto, Veronese, and Millet among the great painters.

23. For a discussion of the social context of nineteenth-century American painting, see Neil Harris, *The Artist in American Society: The Formative Years, 1790–1860* (New York: Braziller, 1966).

24. John La Farge, quoted in Cortissoz, *John La Farge: A Memoir and A Study*, pp. 117–118.

25. Perry has suggested that William thought his own painting mediocre and increasingly recognized that he was cut out for a scientific and eventually a philosophical career (1:200–201). Allen says in *William James: A Biography* that Hunt believed America did not value painters and that this discouraged William; he also argues that William developed neurasthenia and eye problems during his second stint with Hunt, thus making a change of occupation more pressing (pp. 70–71). Feinstein suggests in *Becoming William James* that neither Perry nor Allen effectively analyzes the decision; Perry does not emphasize that William never lost his proclivity for art, and though Allen correctly dates the onset of neurasthenic illness in the second Newport period, Feinstein maintains that William's eye problems did not begin until 1865 and the Brazilian voyage (pp. 140–141). He suggests that William's decision to continue painting had adversely affected his father's health and that William turned from self-expression to duty in an effort to please Henry Sr. William in effect murdered his painter self, which as an alternative ego kept reappearing periodically, often with "symptoms that plagued him for the remainder of his life" (pp. 144–145). But I would maintain that creatively the painter self never really plagued William because it was used therapeutically in other expressive mediums. To assume that William's creative life rose and fell in response to Henry Sr.'s wishes is not to fully grasp the *unity* of James' perception—that is, how no single career or profession could satisfy that perception. In the activity of his own creative continuum, or fields of experience, James forged his own way. Even in his late teens and early twenties the selfishness of genius was clearly evident.

26. See Leon Edel, *Henry James: The Untried Years, 1843–1870* (Philadelphia: Lippincott, 1953), p. 170, and George Fredrickson, *The Inner Civil War: Northern Intellectuals and the Crisis of the Union* (New York: Harper and Row, 1965), pp. 158–161.

27. Henry James, Sr. to Edmund Tweedy, October 1, 1861; WJP.

28. See Allen, *William James: A Biography*, p. 53.

29. Henry James, Sr., quoted in Edel, *Henry James: The Untried Years*, p. 171.

30. Henry James, Sr. to Robertson James, August 31, [1864?]; VC.

31. Henry James, Sr. to Edmund Tweedy, July 1860; quoted in Edith G. Hawthorne, ed., *The Memoirs of Julian Hawthorne* (New York: Macmillan) pp. 120–121.

32. Using Henry Jr.'s retrospective glance, Fredrickson argues in *The Inner Civil War* that William felt considerable guilt about not enlisting (pp. 158–161 and 229–238). Perry, on the other hand, assumes that William's "physical frailty" and immaturity ("I can see in William James no evi-

dence whatever of his having entered manhood in the decade of the 1860s") naturally kept him out of the army. See Perry, 1:202–203. In *William James: A Biography,* Allen suggests that neither Henry nor William were rugged enough for the life of a soldier (p. 71). James' most recent biographer acknowledges that it is not clear why William did not serve, but suggests that both parental resistance and health problems were the probable reasons. See Gerald E. Myers, *William James: His Life and Thought* (New Haven: Yale University Press, 1986), p. 31.

33. For an interesting picture of Newport on the eve of the Civil War, see L. A. Robson, "Newport, One Hundred Years Ago," *Bulletin of the Newport Historical Society* (July 1961), no. 106, p. 7.

34. *Newport Mercury,* April 27, 1861.

35. See Edel, *Henry James: The Untried Years,* pp. 173–183.

36. *Newport Mercury,* May 4, 1861.

37. Perry, 1:203.

38. Eric Foner discusses the increasingly ideological nature of Northern politics in *Free Soil, Free Labor, Free Men: The Ideology of the Republican Party Before the Civil War* (New York: Oxford University Press, 1970).

39. See Fredrickson, *The Inner Civil War,* for a sensitive discussion of how the war transformed deeply held opinions.

40. Wilky was especially immersed, even becoming frozen in the glory of his wartime experience. See Feinstein, *Becoming William James,* pp. 254–258.

41. See for example Van Wyck Brooks, *The Flowering of New England* (New York: Dutton, 1936), and F. O. Matthiessen, *American Renaissance: Art and Expression in the Age of Emerson and Whitman* (New York: Oxford University Press, 1941).

4. WHO MADE GOD?

1. On Eliot's role in upgrading scientific education and professional development at Harvard before and during his presidency, see Hugh Hawkins, *Between Harvard and America: The Educational Leadership of Charles W. Eliot* (New York: Oxford University Press, 1972); Samuel Eliot Morison, *The Development of Harvard University Since the Inauguration of President Eliot: 1869–1929* (Cambridge: Harvard University Press, 1930); and Bruce Kuklick, *The Rise of American Philosophy, Cambridge, Massachusetts, 1860–1930* (New Haven: Yale University Press, 1977), pp. 129–139.

2. Charles W. Eliot, *Harvard Memories* (Cambridge: Harvard University Press, 1923), p. 66.

3. Morison, *Development of Harvard,* pp. xliii–xc.

4. Stephen P. Sharpless, "Some Reminiscences of the Lawrence Scientific School," *Harvard Graduates' Magazine* (1918), 26:532–540.

5. William Dean Howells, *Literary Friends and Acquaintances* (New York: Harper and Brothers, 1900), p. 153.

6. William James to Mary Walsh James, November 2, 1863; WJP.

7. William James to Henry James, Sr., September 7, 1861; WJP.

8. William James to Henry James, Sr., September 16, 1861; WJP.

9. Ibid.

10. William James to Henry James, Sr., December 25, 1861; WJP.

11. Agassiz was nearly the same age as the elder James. He had the European sophistication and ability with languages that Henry Sr. lacked. William's admiration of Agassiz's lecturing virtuosity may also have evoked resentment, since the elder James considered himself a first-rate speaker. Yet Agassiz, like Mr. James, was critical of materialistic science and retained a religious if not mystical perspective.

12. The definitive biography is Edward Lurie, *Louis Agassiz: A Life in Science* (Chicago: University of Chicago Press, 1960); but see also Sharpless, "Some Reminiscences of the Lawrence Scientific School," pp. 532–540.

13. William was particularly impressed with Agassiz's considerable descriptive powers. Agassiz would influence a generation of American scientists, even though many rejected his notion that divine intervention was necessary for the creation of new species. See Lurie, *Louis Agassiz,* pp. 252–350.

14. William James to Henry James, Sr., September 16, 1861; WJP.

15. There is no biography of Wyman, but see Sharpless, "Some Reminiscences," pp. 532–540, and Morison, *Development of Harvard,* pp. 379–382. Wyman's chief ally against Agassiz was Asa Gray, a Scottish-Irish New Yorker who became professor of natural history at Harvard in 1842. Gray's field was botany and he probably had little direct influence on William, since the latter gravitated toward animal rather than plant science. See A. Hunter Dupree, *Asa Gray, 1810–1888* (Cambridge: Harvard University Press, 1959).

16. Edward Waldo Emerson, *The Early Years of the Saturday Club* (Boston: Houghton Mifflin, 1918), p. 423.

17. Charles W. Eliot to Henry James III, September 7, 1915; WJP.

18. William James to Henry James, Sr., September 16, 1861; WJP.

19. William James to his parents, December 25, 1861; quoted in Perry, 1:211.

20. William James to his parents, September 7–8, 1861; quoted ibid.

21. William James to Mary Walsh James, quoted in Allen, *William James: A Biography,* p. 75. Allen does not precisely date this letter, but it was probably written in September 1861.

22. William James to Alice James, March 1, 1862; quoted in Perry, 1:212.

23. Charles W. Eliot to Henry James III, September 7, 1915; WJP.

24. William James to Alice James, October 19, 1862; quoted in Allen, *William James: A Biography,* p. 90.

25. Charles W. Eliot to Henry James III, September 7, 1915; WJP.

26. See William James, 1862 Notebook and 1863 Notebook; items 4496 and 4497, WJP. Hereafter these notebooks will be cited parenthetically in the text.

27. William James to Henry James, Sr., September 16, 1861; WJP.

28. On Peirce, see Murray G. Murphey, *The Development of Peirce's Philosophy* (Cambridge: Harvard University Press, 1961); Philip Wiener, ed., *Charles S. Peirce: Selected Writings* (New York: Dover Publications, 1966); and Kuklick, *The Rise of American Philosophy*, pp. 104–126.

29. Peirce may indeed have fueled William's inquiry into father-son relationships, for despite his brilliance, Peirce was never to achieve his father's professional prominence. Charles could not have known this at the time, for his career looked relatively bright. Yet the underlying pattern of viewing himself as an outsider was already in place. For Peirce's protracted problems with authority, see Kuklick, *The Rise of American Philosophy*, pp. 123–126.

30. James mentions Müller's later book, *Science of Thought* (1887), in which Müller argued that "Thought and Speech are inseparable." See William James, *The Principles of Psychology*, 1:259.

31. See Donald Fleming, *John William Draper and the Religion of Science* (Philadelphia: University of Pennsylvania Press, 1950).

32. William James to his parents, [?] 1863; WJP.

33. William James to Katharine James Prince, September 12, 1863; *LWJ*, 1:43–44.

34. Depressed with theoretical struggles, James read Marcus Aurelius while in Germany in 1867. And much later, in 1900, when he was working on what would become *The Varieties of Religious Experience*, he once again turned to the stoic philosophers.

35. William James to Alice James, September 13, 1863; *LWJ*, 1:50.

36. William James to Mary Walsh James, November 2, 1863; WJP.

37. For the argument that vocational indecision was merely symptomatic of deeper oedipal conflict and unfulfilled aspirations to be a painter, see Howard M. Feinstein, *Becoming William James* (Ithaca, N.Y.: Cornell University Press, 1984), pp. 146–168.

38. On nineteenth-century materialism, see Maurice Mandelbaum, *History, Man, and Reason: A Study in Nineteenth-Century Thought* (Baltimore: Johns Hopkins University Press, 1961); John Theodore Merz, *A History of European Thought in the Nineteenth Century*, 4 vols. (Edinburgh: Blackwood, 1896–1914), especially vol. 2; and Everett Mendelsohn, "Physical Models and Physiological Concepts: Explanation of Nineteenth-Century Biology," *British Journal for the History of Science* (1965), 2:201–219.

39. For example, in 1905 in an essay titled "The Thing and Its Relations," later published in *Essays in Radical Empiricism*, James remarks, "The great continua of time, space and the self envelope everything, betwixt them, and flow together without interfering." This is typical of his

mature philosophy of experience, which denied dualism. See *Essays in Radical Empiricism and a Pluralistic Universe* (Gloucester, Mass.: Peter Smith, 1967), p. 94.

5. IN THE ORIGINAL SEAT OF THE GARDEN OF EDEN

1. For the development of professional medicine, see John S. Haller, Jr., *American Medicine in Transition, 1840–1910* (Urbana: University of Illinois Press, 1981), pp. 192–233; George Rosen, *The Structure of American Medical Practice, 1875–1941* (Philadelphia: University of Pennsylvania Press, 1983); and Paul Star, *The Social Transformation of American Medicine* (New York: Basic Books, 1982), pp. 79–144.

2. Mark D. Altschule, *Medicine at Harvard: The First Three Hundred Years* (Hanover, N.H.: University Press of New England, 1977), p. 14.

3. Moses King, *Harvard and Its Surroundings* (Cambridge: King, 1878), p. 59. On the Harvard Medical School in the mid- and late nineteenth century, see Samuel Eliot Morison, *The Development of Harvard University Since the Inauguration of President Eliot: 1869–1929* (Cambridge: Harvard University Press, 1930), p. 555–563.

4. In 1874 Bowditch, Putnam, and James purchased a primitive cabin near Keene Valley, New York, deep in the Adirondacks. James would frequent "Putnam camp" and Keene Valley every summer that he was not abroad. By the late sixties Bowditch and especially Putnam and James were deeply concerned with brain physiology and psychopathology. See Nathan Hale, Jr., *James Jackson Putnam and Psychoanalysis: Letters Between Putnam and Sigmund Freud, Ernest Jones, William James, Sandor Ferenczi, and Morton Prince, 1877–1917* (Cambridge: Harvard University Press, 1971).

5. For surviving microscope sketches, see William James, 1866 Notebook; item 4499, WJP. For cadaver drawings, see William James, Sketches; WJP.

6. William James to [?], February 21, 1864; quoted in Perry, 1:216.

7. William James to Charles Peirce, March 10, 1909; WJP.

8. Aunt Kate had gone to live with the Jameses in the 1840s after a failed marriage. She had accompanied the James family on the 1855–1858 trip to Europe and was a permanent family member until Henry Sr.'s death in December 1882.

9. See Boston Society of Natural History to Ralph Barton Perry, October 26, 1910; WJP. William was a member from 1864–1868. The society later published his essay, "The Feeling of Effort," in their 1880 Anniversary Memoirs.

10. William James to Mary Walsh James, November 2, 1863; WJP.

11. William James [no date], quoted in Louise Hall Tharp, *Adventurous Alliance: The Story of the Agassiz Family of Boston* (Boston: Little, Brown, 1959), pp. 302–303.

12. William James to his parents, April 21–25, 1865; quoted in Perry, 1:217–218.

13. William James to Mary Walsh James, August 23–25, 1865; quoted ibid.

14. William James, 1865 Brazilian Diary; item 4498, WJP.

15. Louis and Elizabeth Agassiz, *A Journey in Brazil* (Boston: Houghton Mifflin, 1867), p. 79.

16. William James to his parents, April 21–25, 1865; quoted in Perry, 1:217.

17. Travel and new locales seemed to put James in acute touch with bodily and mental sensations. In such settings illness became insight; he seemed to experience what one scholar has called a "creative illness." See Henri Ellenberger, "La Maladie Creative," *Canadian Philosophical Review* (1964), 3:25–41.

18. Louis and Elizabeth Agassiz, *Journey in Brazil,* pp. 8–10.

19. See Edward Lurie, *Louis Agassiz: A Life in Science* (Chicago: University of Chicago Press, 1960), pp. 345–346.

20. William James to Henry James, Sr., September 12–15, 1865; *LWJ,* 1:65.

21. William James to Mary Walsh James, August 23–25; quoted in Perry, 1:222.

22. William James to Henry James, Sr., September 12–15, 1865; *LWJ,* 1:66.

23. I do not mean to imply that James "used" Agassiz in a Machiavellian way. Yet he avoided adopting a slavish attitude toward a mentor: he was always able to criticize, even as he admired and learned. James never clung to an older man with the emotional dependency and obvious transference that the mentored usually exhibited. See Daniel Levinson, *The Seasons of a Man's Life* (New York: Knopf, 1978), pp. 97–101 and 333–335.

24. William James to Henry James Sr., June 3, 1865; *LWJ,* 1:61.

25. William James to his parents, October 22, 1865; ibid., 1:69.

26. William James to his family, May 3–10, 1865; quoted in Gay Wilson Allen, *William James: A Biography,* (New York: Viking, 1967), p. 106.

27. William James to his family, June 3, 1865; quoted ibid.

28. Ibid.

29. "Speculative lines" did not necessarily mean William foresaw that he would become a philosopher. Speculation meant to *risk* inquiring into the nature of things, a venture that could be artistic and scientific as well as philosophical.

30. William James to Henry James, Jr., July 15, 1865; WJP.

31. William's correspondence from Brazil is rich in descriptions of nature and local color, particularly his letters to Henry. His letters to his brother offer extended commentary on natural history and on the traits of native Brazilians. For a sample see Perry, 1:303–319.

32. William James to Henry James, July 15, 1865; quoted in Allen, *William James: A Biography*, p. 109.

33. William James, 1865 Brazilian Diary and Sketch Book; item 4498, WJP.

34. William James to his parents, October 21–22, 1865; quoted in Perry, 1:223.

35. William James to Mary Walsh James, December 9, 1865; quoted ibid., p. 225.

36. William James to Henry James, Jr., July 15, 1865; quoted ibid., pp. 220–221.

37. Particularly germane is William James, Notebook, The Self-Pure Experience, etc., book 2, 1897–1898; item 4505, WJP. Also see "A World of Pure Experience," *Journal of Philosophical, Psychological, and Scientific Methods* (1904), 1:533–543 and 561–570.

38. Perry simply includes letters home without analysis, as if to say there was nothing of essential importance a future great philosopher could find in Brazil; see Perry, 1:216–226. Allen gives an excellently detailed account of the trip, notes James' improved mood after his illness, but makes nothing of it; see his *William James: A Biography*, pp. 101–116. Howard M. Feinstein interprets the Brazilian experience mostly as a relaxing interlude—he calls his chapter about it "Vacation in Brazil." William, he suggests, was smitten with a tropical languor reflected in drawings, and was no longer consumed with oedipal rage. See his *Becoming William James* (Ithaca, N.Y.: Cornell University Press, 1984), pp. 169–181.

39. Elizabeth Agassiz to William James [?] 1885; William James to Elizabeth Agassiz [?] 1885; quoted in Tharp, *Adventurous Alliance*, pp. 302–303.

40. James, Brazilian Diary and Sketch Book.

41. James, *The Principles of Psychology*, 1:233.

42. William James, "Confidence of a Psychical Researcher," *American Magazine* (October 1909), 68:588–589.

43. James called his 1898 experience his "Walpurgis Nacht," and related it to the Gifford Lectures that became *The Varieties of Religious Experience*. See William James to Alice Gibbens James, July 9, 1898; WJP.

44. See Roderick Nash, *Wilderness and the American Mind* (New Haven: Yale University Press, 1982), pp. 67–121.

45. William James to Hugo Münsterberg, June 28, 1906; WJP. James' defense of wild nature was made in critical reaction to Münsterberg's book, *Science and Idealism*.

6. THE ANTINOMIAN ROOT

1. William James to Garth Wilkinson James, February 25, 1866; WJP.
2. William James to Garth Wilkinson James, March 21, 1866; WJP.
3. William James to Thomas Ward, March 27, 1866; WJP.

4. Ibid.

5. On the younger Holmes, see Francis Biddle, *Mr. Justice Holmes* (New York: Scribner's, 1942); Catherine Drinker Bowen, *Yankee from Olympus: Justice Holmes and His Family* (Boston: Little, Brown, 1945); and David H. Burton, *Oliver Wendell Holmes, Jr.* (Boston: Twayne Publishers, 1980).

6. William wrote Wilky that Fanny was "about as fine they make 'em. That villain Wendell Holmes has been keeping her all to himself . . . but I hope to make her acquaintance now." William James to Garth Wilkinson James, February 25, 1866; WJP.

7. Oliver Wendell Holmes, Sr., quoted in Bowen, *Yankee from Olympus,* p. 219.

8. See Burton, *Oliver Wendell Holmes, Jr.,* p. 27.

9. William James to Oliver Wendell Holmes, Jr., quoted in Bowen, *Yankee from Olympus,* p. 219.

10. Oliver Wendell Holmes, Jr., "The Soldier's Faith" (1895), in *Speeches* (Boston: Little, Brown, 1913), p. 59. For discussion of Holmes' tough-minded metaphysics and its relationship to the Civil War experience, see George Fredrickson, *The Inner Civil War: Northern Intellectuals and the Crisis of the Union* (New York: Harper and Row, 1965), pp. 219–222; and Paul Boller, *American Thought in Transition: The Impact of American Evolutionary Naturalism, 1865–1900* (New York: Rand McNally, 1969), pp. 148–174.

11. William James to Henry James, Jr., June 12, 1869; WJP.

12. Oliver Wendell Holmes, Jr. to William James, April 19, 1968; quoted in Perry, 1:509.

13. William James to Oliver Wendell Holmes, Jr., May 15, 1868; quoted ibid., 1:512.

14. William James to Oliver Wendell Holmes, Jr., May 18, 1868; quoted ibid., 1:516.

15. See William James to Oliver Wendell Holmes, Jr., May 15, 1868; ibid., 1:513.

16. Shaw was a Harvard-educated Bostonian who commanded the Fifty-fourth Massachusetts Regiment, the first force to use black militia from a free state. Shaw was killed and William James later gave the dedication speech for a commemorative statue in Boston. See Fredrickson, *The Inner Civil War,* pp. 152–155.

17. See Lawrence Powell's excellent *New Masters: Northern Planters During the Civil War and Reconstruction* (New Haven: Yale University Press, 1980), and William C. Childers, "Letters of Garth Wilkinson James and Robertson James: Abolitionists in Alachua County During Reconstruction," manuscript in the possession of William C. Childers, English Department, University of Florida, Gainesville.

18. Garth Wilkinson James to Henry James, Sr., March 2, 1866; quoted in Childers, "Letters." The original letter is in the possession of Robertson James' grandson, Henry Vaux, Berkeley, California.

19. Garth Wilkinson James to William James, September 29, 1866; quoted ibid; original in VC.

20. William James to Garth Wilkinson James, March 13, 1867; WJP.

21. Wilky continued to fail in business ventures in Wisconsin, and by the early 1880s was mortally ill. Robertson became alcoholic, could find no satisfying profession, and had marriage problems (see chapter 12).

22. See Jean Strouse's perhaps overly sympathetic *Alice James: A Biography* (New York: Houghton Mifflin, 1980), and Ruth Bernard Yeazell, *The Death and Letters of Alice James* (Berkeley and Los Angeles: University of California Press, 1981), which discusses Alice's propensity to welcome death. Alice James' enduring monument is her diary, now considered an American classic. See Leon Edel, ed., *The Diary of Alice James* (New York: Penguin American Library, 1982), and Anna Robeson Burr, *Alice James: Her Brothers—Her Journal* (New York: Dodd, Mead, 1934).

23. William James to Robertson James, January 27, 1868; WJP.

24. William James to Alice James, December 12, 1866; quoted in Perry, 1:231.

25. See William James to Henry James, Jr., June 12, 1869; WJP.

26. See Strouse, *Alice James: A Biography,* pp. 111–112.

27. William James to Henry James, Sr., May 27, 1867; WJP.

28. William James to Mary Walsh James, June 12, 1867; WJP.

29. Ibid.

30. William James to Henry James, Sr., July 24, 1867; WJP.

31. William James to Henry James, Sr., September 5, 1867; WJP.

32. William James to Thomas Ward, December [?], 1867; WJP.

33. William James to Thomas Ward, September 12, 1867; WJP.

34. William James to Oliver Wendell Holmes, Jr., September 17, 1867; WJP.

35. William James to Henry James, Jr., September 26, 1867; WJP.

36. William James to Thomas Ward, November [?], 1867; WJP.

37. William James to Henry P. Bowditch, December 12, 1867; WJP.

38. William James to Henry P. Bowditch, June 15, 1868; WJP.

39. Emile du Bois-Reymond, along with Ernst Brücke and Herman Helmholtz, formed a pact as youths in the 1840s to overthrow vitalist physiology. See Frank Sulloway, *Freud, Biologist of the Mind: Beyond the Psychoanalytic Legend* (New York: Basic Books, 1979), pp. 14–15. Bois-Reymond provided William with more scientific evidence for a wholly empirical physiological reality.

40. William James to Henry P. Bowditch, December 12, 1867; WJP.

41. William felt pressured to justify shifts in scene to his parents, since each one meant extra expense. Since health had not blossomed for long in Berlin, he anticipated parental reaction: "I hope you don't think from having me back here [in Teplitz] that my loudly trumpeted improvement in the autumn was fallacious—on the contrary, I feel more than ever now that I am back in the presence of my old measures of strength . . .

only it has not yet bridged the way to complete soundness." William James to Henry James, Sr., January 22, 1868; WJP.

42. William James to Robertson James, March 29, 1868; VC.

43. Mary Walsh James to Robertson James, March 29, 1868; VC.

44. For example, when his brother Robertson complained a few years later of depression, back problems, kidney trouble, constipation, and fears of impotency, William replied, "Your case as you describe it belongs to the most trifling class, amenable to treatment in some form; and if continuing in spite of medical treatment, ready to cease when sexual intercourse begins regularly." William James to Robertson James, June 22, 1872; VC.

45. William James to Henry James, Sr., September 26, 1867; WJP.

46. Henry James, Sr. to William James, September 27, 1867; WJP.

47. William James to Henry James, Sr., October 28, 1867; WJP.

48. Ibid.

49. See William James, 1868–1873 Diary; WJP. The reading list appears early in the diary. See entries for 1868–1869.

50. Ralph Barton Perry set the general tone for interpreting the period of the late sixties and early seventies as a period of "Depression and Recovery," as he titled one chapter of his James biography. Indeed, he expanded this duality into a picture of James as a bifurcated personality, assigning to him "morbid" and "benign" traits. See Perry, 1:320–332 and 2:670–698. In 1968 Cushing Strout used Erik Erikson's identity crisis psychology to interpret the period as one of depression arising from unresolved vocational and oedipal conflicts with Henry Sr. See "William James and the Twice-Born Sick Soul," *Daedalus* (1968), 97:1062–1082; and "The Pluralistic Identity of William James," *American Quarterly* (1971), 23:135–149. The oedipal sources of the crisis have been most fully developed in Howard M. Feinstein's *Becoming William James* (Ithaca, N.Y.: Cornell University Press, 1984), which traces William's depression back to Henry Sr.'s unresolved conflict with his own father, William James of Albany. In a recent article, James William Anderson has maintained that William's depressive collapse was the result of a "self-fragmentation experience" which in part stemmed from a strained relationship with his mother, Mary Robertson Walsh James. See " 'The Worst Kind of Melancholia': William James in 1869," in *A William James Renaissance: Four Essays by Young Scholars,* special issue, *Harvard Library Bulletin* (October 1982), 30(4):369–386. S. P. Fullinwider has departed from the mainstream Freudian or Eriksonian interpretation of the crisis, maintaining instead that James suffered an existential trauma, one that was solved by writing *The Principles of Psychology*—a primer on survival. See "William James's Spiritual Crisis," *The Historian* (1975), 13:39–57. Marian C. Madden and Edward H. Madden have argued that James' difficulties were not deep enough to warrant psychoanalysis. They place the blame for his crisis on environmental factors, especially the constantly shifting

educational scene that made it difficult for him to develop a long-term perspective. See "The Psychosomatic Illnesses of William James," *Thought* (December 1979), 54(215):376–392. Gerald E. Myers, James' most recent full-scale biographer, is generally skeptical of psychoanalytic interpretation of the crisis. Although appreciative of the literature, Myers argues that a constellation of physical, psychological, and philosophical factors caused James' difficulty. Myers also maintains—much in the manner of Perry—that Charles Renouvier led James toward a new definition of freedom and hence toward recovery. See *William James: His Life and Thought* (New Haven: Yale University Press, 1986), pp. 46–53.

51. Mary Walsh James to Robertson James, May 3, [1868?]; VC. In 1868 William published five reviews, including two on Darwin; see John M. McDermott, *The Writings of William James* (Chicago: University of Chicago Press, 1977), pp. 812–813.

52. William James to Henry James, Jr., April 5, 1868; WJP.

53. William James, 1868–1873 Diary, April 10, 1868 entry; WJP. Hereafter cited parenthetically in text.

54. In "The Transcendentalist," a lecture delivered at the Masonic Temple in Boston, January 1842, Emerson exclaimed, "Our American literature and spiritual history are, we confess, in the optative mood." See Edward L. Ericson, ed., *Emerson on Transcendentalism* (New York: Ungar, 1986), p. 99.

55. William James to Thomas Ward, May 24, 1868; WJP.

56. See Kate E. Havens to Alice Gibbens James, July 10, 1913; WJP. For a discussion of William and Kate's Dresden romance, see Anderson, " 'The Worst Kind of Melancholia,' " pp. 379–380.

57. William James to Thomas Ward, May 24, 1868, WJP.

58. James loathed his tendency have "Vorstellungen disproportionate to the object," or imagination disproportionate to any practical effect; see Diary, May 22, 1868 entry.

59. Even when desperately ill with angina late in life, William still often found time to write up to fifty letters a day. If he could not carry on sustained intellectual activity when seriously ill, he nearly always carried on social relationships through correspondence.

60. William James to Henry P. Bowditch, October 2, 1868; WJP.

61. William James to Henry James, Jr., September 22, 1868; WJP.

62. William James to Henry P. Bowditch, October 21, 1868; WJP.

63. William James to Henry P. Bowditch, November 30, 1868; WJP.

64. William James to Thomas Ward, December 16, 1868; WJP.

65. William James to Thomas Ward, March [?], 1869; WJP.

66. On the Metaphysical Club, see Perry, 1:534–535, and Philip Wiener, *Evolution and the Founders of Pragmatism* (Cambridge: Harvard University Press, 1949).

67. William James, quoted in Perry, 1:522.

68. Reproduced ibid., pp. 525–528.

69. William James, Notes on Idealism, ca. 1870–1871; item 4503, WJP. Hereafter cited parenthetically in text.

70. James noted "that in the immediate . . . moment of consciousness, the object is cognized as in itself, independent[,] coordinate with the state of consciousness, not incident to it[,] not enveloped by it" (Notes on Idealism).

71. Mary Temple to John Chipman Gray, January 7, 1869; WJP.

72. James' counterimage of insanity through failure to embrace the active life has been canonized biographically, even though there is no direct evidence that in it William referred to himself. James included it in the chapter on "sick-mindedness" in *The Varieties of Religious Experience,* written when he was almost sixty. The unnamed author from whom James quotes the passage had visited an insane asylum and encountered "a black-haired youth with greenish skin, entirely idiotic. . . . The image and my fear entered in a species of combination with each other. *That shape am I,* I felt potentially. . . ." The image so disturbed the observer that "something hitherto solid within my breast gave way entirely, and I became a mass of quivering fear." William James, *The Varieties of Religious Experience* (New York: Random House, 1929), pp. 157–158. The optimistic resolve of James' discovery of Renouvier and the will to believe has often been contrasted to the frightful depressive power of this passage, which also resembles Henry Sr.'s 1843 vastation upon gazing into the fire and finding "some fetid personality" radiating "influences fatal to life." But if James did lapse into existential terror, the meaning of that fright symbolized more than his fear of not being able to act. Beneath the fear of the madness that would result from inertia was the greater foreboding about his intellectual future. Without a creative destiny he would be a living dead man. Action itself could not save James from madness; but finding creative grace could.

7. TRAGICAL MARRIAGE

1. William Dean Howells, *The Wedding Journey* (New York: Harper and Brothers, 1896), p. 54.

2. See Beatrice Sweeney, *The Grand Union Hotel: A Memorial and a Lament* (Saratoga Springs, N.Y.: Historical Society of Saratoga Springs, 1982).

3. William H. Murray, *Adventures in the Wilderness; or, Camp-Life in the Adirondacks of 1869* (Boston, 1869), p. 22. For an account of revived post–Civil War interest in the Adirondacks as a wilderness area, see Roderick Nash, *Wilderness and the American Mind* (New Haven: Yale University Press, 1982), pp. 116–121.

4. On James' initial interest in Keene Valley see Elizabeth Putnam McIver (James Putnam's daughter), "Early Days at Putnam Camp," paper delivered to the Keene Valley Historical Society, September 1941, in

the Keene Valley Library, Keene Valley, New York; and Nancy Lee, "Putnam Camp," *Adirondack Life* (November–December 1980), pp. 42–46.

5. See Perry, 1:375.

6. On Davidson see William James, "A Knight-Errant of the Intellectual Life," *McClure's Magazine* (May 1905), 25:3–11; William Knight, *Memorials of Thomas Davidson, the Wandering Scholar* (Boston: Ginn and Co., 1907); and Charles M. Bakewell, "Glenmore of the Old Days," manuscript, Keene Valley Library.

7. Years later William's married daughter, Margaret ("Peggy") James Porter, remarked to her brother Henry that "by a sort of wild justice the Dr. Murphy, who dispossessed not only the Gibbens family but the Putnams and the Canfields as well and accumulated thousands of acres, leaves one granddaughter in poverty on an acre or two. Perhaps if it hadn't been land he grabbed he would have died a millionaire." Margaret James Porter to Henry James III, December 31, 1933; WJP.

8. The fullest account of the Gibbens family's troubled past is Gay Wilson Allen's in *William James: A Biography* (New York: Viking, 1967), pp. 214–219.

9. William James to Robertson James, November 15, 1869: quoted in Jean Strouse, *Alice James: A Biography* (Boston: Houghton Mifflin, 1980), p. 126.

10. For an interesting discussion of medical and cultural pressures that influenced the decision for or against marriage, see John S. Haller and Robin M. Haller, *The Physician and Sexuality in Victorian America* (New York: Norton, 1977), pp. 191–234. Other helpful discussions of Victorian marriage include Barbara Rees, *The Victorian Lady* (London: Gordon and Cremonesi Publishers, 1977); Sidney Ditzion, *Marriage, Morals, and Sex in America: A History of Ideas* (New York: Octagon Books, 1975), pp. 123–155; and Carl N. Delgler, *At Odds: Women and the Family in America from the Revolution to the Present* (New York: Oxford University Press, 1980).

11. William James to Alice Howe Gibbens, [summer] 1876; WJP.

12. Ibid.

13. Sadly, the other half of the correspondence, Alice's letters to William, is almost nonexistent. William himself destroyed many of the early letters. The William James to Alice Gibbens James correspondence was made available to scholars working on the James Papers at Harvard's Houghton Library in 1981. For an insightful analysis of Alice's crucial role in helping William pass beyond melancholia, see Mark R. Schwehn, "Making the World: William James and the Life of the Mind," in *A William James Renaissance: Four Essays by Young Scholars,* special issue, *Harvard Library Bulletin* (October 1982), 30(4):439–444.

14. I do not mean that William disliked Alice when they were together, but only that distance seemed to renew romantic enthusiasm.

15. William James to Alice Howe Gibbens, [?] summer, 1876; WJP.

16. See *The Varieties of Religious Experience: A Study in Human Na-*

ture, first published in New York by Longmans, Green and Company in 1902; and *Pragmatism: A New Name for Some Old Ways of Thinking,* first published by Longmans, Green in 1907.

17. William James to Alice Howe Gibbens, [?] summer, 1876; WJP.
18. Ibid.
19. Ibid.
20. Fragment of letter, William James to Alice Howe Gibbens, November 23, 1876; WJP.
21. William James, 1868–1873 Diary, entry for April 30, 1870; WJP.
22. Fragment of letter, William James to Alice Howe Gibbens, November 23, 1876.
23. William James to Alice Howe Gibbens, March 15, 1877; WJP.
24. William James to Alice Howe Gibbens, April 12, 1877; WJP.
25. William James to Alice Howe Gibbens, April 30, 1877; WJP.
26. Ibid.
27. William James to Alice Howe Gibbens, May 1877; WJP.
28. Ibid.
29. William James to Alice Howe Gibbens, June 7, 1877.
30. James' first truly scholarly article was published early in 1878, some six months before he married Alice. See "Remarks on Spencer's Definition of Mind as Correspondence," *Journal of Speculative Philosophy* (January 1878), 12:1–18.
31. See William James, "Brute and Human Intellect," *Journal of Speculative Philosophy* (July 1878), 12:236–276.
32. Notebook excerpt dated May 16, 1873, in William James to Alice Howe Gibbens, June 7, 1877; WJP. Hereafter excerpt cited parenthetically in text.
33. See William James, "The Sentiment of Rationality, *Mind* (1879), 4:317–346.
35. William James to Alice Howe Gibbens, June 7, 1877; WJP.
35. Ibid.
36. William James, "What is an Emotion?" *Mind* (April 1884), 9:188–205. An even earlier 1880 essay illustrated his conviction that one could physically feel the emergence of ideas. See "The Feeling of Effort," in *Essays in Philosophy* in *The Works of William James* (Cambridge: Harvard University Press, 1983), pp. 83–124.
37. William James to Alice Gibbens James, [?], 1888; WJP.

8. CREATING THE REALM OF CONSCIOUSNESS

1. James himseslf shares responsibility for the reactive tone; he did at times portray the life of the mind as an intellectual battlefield where one parried with religious, scientific, and philosophical enemies. But Ralph Barton Perry canonized the view of James' thought as essentially responsive to established "isms" in nineteenth-century science and metaphysics. See, for instance, the chapters "The Influence of the British School" and

"The Struggle with Kantian Idealism" in Perry, 1:543–554 and 711–720. Richard Hofstadter encouraged several generations of historians to view James as one of a group of American intellectuals important mainly for their responses to Darwinism. See *Social Darwinism in American Thought* (Boston: Beacon, 1955), pp. 123–142.

2. Of course James had read Darwin in the early 1860s while a student in the Lawrence Scientific School. And in 1868 he reviewed one of Darwin's books; see his unsigned review of Charles Darwin's *The Variation of Animals and Plants Under Domestication* in *Atlantic Monthly* (1868), 22:122–124.

3. See William James, "Remarks on Spencer's Definition of Mind as Correspondence," *Journal of Speculative Philosophy* (January 1878), 12:1–18; "Brute and Human Intellect," *Journal of Speculative Philosophy* (July 1878), 12:236–276; and "Are We Automata?" *Mind* (1879), 4:1–22. Discussion of discrimination and reasoning roughly similar to that of "Brute and Human Intellect" appears in *The Principles of Psychology,* 1:457–518 and 2:952–993. Likewise, material in "Are We Automata?" on automation theory and the stream of thought reappears in *The Principles of Psychology,* 1:132–147 and 219–278.

4. William Woodward contends that James directed criticism at Spencer rather than at Darwin because the former was more centrally concerned with mental development. Woodward also observes that in "Are We Automata?" James defines the functional relationship between stimulus and response. See Woodward's introduction to William James, *Essays in Psychology,* especially pp. xiv–xxiv.

5. William James to Henry James, Jr., November 31, 1872; WJP.

6. G. Stanley Hall, *Life and Confessions of a Psychologist* (New York: Appleton, 1923), p. 218.

7. For the broad intellectual milieu of nineteenth-century science, see John Theodore Merz, *A History of European Thought in the Nineteenth Century,* 4 vols. (Edinburgh: Blackwood, 1896–1914); John C. Green, *The Death of Adam: Evolution and Its Impact on Western Thought* (Ames, Iowa: Iowa State University Press, 1959); Maurice Mandelbaum, *History, Man, and Reason: A Study in Nineteenth-Century Thought* (Baltimore: Johns Hopkins University Press, 1961); and Frank J. Sulloway, *Freud, Biologist of the Mind* (New York: Basic Books, 1979).

For the development of early scientific psychology, see James McKeen Cattell, "The Psychological Laboratory at Leipzig," *Mind* (1888), 13:37–51; David Galaty, "The Philosophical Basis of Mid-Nineteenth Century German Reductionism," *Journal of the History of Medicine* (1974), 29:295–316; Irving Kirsch, "Psychology's First Paradigm," *Journal of the History of the Behavioral Sciences* (1977), 13:317–325; R. W. Rieber, ed., *Wilhelm Wundt and the Making of a Scientific Psychology* (New York: Plenum Press, 1980); and Michael M. Sokal, ed., *An Education in Psychology: James McKeen Cattell's Journal and Letters from Germany and England, 1880–1882* (Cambridge: MIT Press, 1981).

8. See William James, Notes on Psychology, the Baltimore and Lowell Lecture Notes, 1878–1879; item 4395, WJP. Unless otherwise indicated, all subsequent quotations in this chapter are from these notes.

9. See Lawrence Lowell Abbott to Henry James III, January 26, 1920; WJP. James wrote to Josiah Royce a year after the Baltimore Lectures that "a proposal from Gilman to teach in Baltimore three months yearly for the next three years had to be declined as incompatible with work here." William James to Josiah Royce, February 16, 1879; WJP.

10. The other two Lowell Lectures series were "Exceptional Mental States," delivered in 1896, and "Pragmatism," in 1906.

11. For the Lowell setting see Eugene Taylor, "William James on Psychopathology: The 1896 Lowell Lectures on 'Exceptional Mental States,' *A William James Renaissance: Four Essays by Young Scholars,* special issue, *Harvard Library Bulletin* (October 1982), 30(4):470–471.

12. James had begun to read Kant while studying the German Romantics in Dresden during the 1867–1868 trip.

13. Helmholtz's great pioneering work in physiological optics was *Treatise on Optics* (1867). He extended color theory and explained how the lens mechanism accommodated the eye. Earlier, in 1851, he had invented the ophthalmoscope.

14. See William James, "The Spatial Quale," *Journal of Speculative Philosophy* (1879), 13:64–87, and *The Principles of Psychology,* 2:776–912. As early as 1863 William had written, "Now my notion is this: Retina and skin give the spatial *quale* immediately . . . only this *sui generis* spacial sensation is yet vague in all its three dimensions. . . . Which explains why when we perceive space it appears to us far more in the shape of [a] room to float smoothly off in than of [a] room articulated and interrupted as it would be if the notion of muscular progression entered largely into [it]." William James, 1863 Notebook entitled "Theory of Space"; WJP.

15. But clearly this speculation was aimed at avoiding the skepticism of Darwinian empiricism: "Skepticism tried to show by the way in which our sensations vary that we have no true knowledge." But "those physiologists who study sensations of the eye" had shown that sensations are in fact "ingulfed, caught up, used, neglected or altered by the idea we gain of the *thing.*" James, the Baltimore and Lowell Lecture Notes.

16. See *The Principles of Psychology,* 1:15–147.

17. An interesting recent example is Robert J. Richards, "The Personal Equation in Science: William James' Psychological and Moral Uses of Darwinian Theory," in *A William James Renaissance: Four Essays by Young Scholars,* special issue, *Harvard Library Bulletin* (October 1982), 30(4):387–425. Richards argues imaginatively but I think unconvincingly that James found in Darwinism an objective theory that gave the mind creative power, a scientific support for Renouvier's free will. Rather, as the Lowell Lectures clearly show, James considered Darwinism a form of determinism and created his own mental theory as a scientific antidote

to the unconvincing notion of consciousness in evolutionary theory. Gerald E. Myers shows that Jamesian consciousness combined introspection with physiological psychology, two developments in nineteenth-century science that developed largely independent of the Darwinian model. Later James would locate the "stream of thought" in experience rather than in introspection. See Gerald E. Myers, *William James: His Life and Thought* (New Haven: Yale University Press, 1986), pp. 54–80.

18. Two excellent studies that examine the problematical nature of work in late nineteenth-century America are David A. Noble, *America by Design: Science, Technology, and the Rise of Corporate Capitalism* (New York: Knopf, 1977) and James B. Gilbert, *Work Without Salvation: America's Intellectuals and Industrial Alienation, 1880–1910* (Baltimore: Johns Hopkins University Press, 1977).

19. James, *The Principles of Psychology,* 1:233.

20. See Samuel Haber, *Efficiency and Uplift: Scientific Management in the Progressive Era* (Chicago: University of Chicago Press, 1960); and Matthew Hale, Jr., *Human Science and Social Order: Hugo Münsterberg and the Origins of Applied Psychology* (Philadelphia: Temple University Press, 1980), pp. 148–163.

21. After establishing his reputation, however, James became almost exclusively interested in such branches of psychology as psychical research and psychopathology. Indeed, his next appearance before Lowell audiences dealt exclusively with psychopathology. See Eugene Taylor, "Historical Introduction," in *William James on Exceptional Mental States: The 1896 Lowell Lectures* (New York: Scribner's, 1982), pp. 1–14, and his "William James on Psychopathology: The 1896 Lowell Lectures on "Exceptional Mental States," in *A William James Renaissance: Four Essays by Young Scholars,* special issue, *Harvard Library Bulletin* (October 1982), 30(4):455–497.

9. A WANDERING PIECE OF PROPERTY

1. Mary Walsh James to Robertson James, May 20, 1879; VC.

2. William James to Robertson James, May 26, 1879; VC.

3. William James to Robertson James, August 18, 1879; VC.

4. This contrasted with his father's practice of invariably being on hand during the weaning period.

5. William James to G. Stanley Hall, October 10, 1879; WJP.

6. William James to G. Stanley Hall, January 16, 1880; WJP. James did meet Hall in Heidelberg where they had "the highest and most instructive conversation." William James to Henry P. Bowditch, July 19, 1880; WJP.

7. On moral philosophy, especially at Harvard, see Bruce Kuklick, *The Rise of American Philosophy: Cambridge, Massachusetts, 1860–1930* (New Haven: Yale University Press, 1977), pp. 5–62. On fad psychology see Donald Meyer, *The Positive Thinkers: Religion as Pop Psychology*

from Mary Baker Eddy to Oral Roberts (New York: Pantheon Books, 1980), pp. 21–125; R. Laurence Moore, *In Search of White Crows: Spiritualism, Parapsychology, and American Culture* (New York: Oxford University Press, 1977); and Robert C. Fuller, *Mesmerism and the American Cure of Souls* (Philadelphia: University of Pennsylvania Press, 1982).

8. See Kuklick, *The Rise of American Philosophy,* p. 135. James' experimental psychology undermined Bowen's religiously oriented metaphysical priorities. William's courses came to signify a formal switch to the evolutionary perspective.

9. William James to Josiah Royce, February 3, 1980; *LWJ,* 1:204.

10. William James to Robertson James, May 26, 1879; VC.

11. See William James to Josiah Royce, February 3, 1880; *LWJ,* 1:204.

12. William James to Charles Renouvier, June 1, 1880; *LWJ,* 1:206.

13. William James, "The Sentiment of Rationality," *Mind* (1879), 4:344.

14. There is a good overview of Hodgson by George E. Davie in Paul Edwards, ed., *The Encyclopedia of Philosophy,* 8 vols. (New York: Macmillan, 1967), 3:47–48. The James-Hodgson philosophical relationship is treated at length in Perry, 2:611–627. Richard High claims that James was especially indebted to Hodgson for his conception of space. See his "Shadworth Hodgson and William James' Formulation of Space Perception: Phenomenology and Perceptual Realism," *Journal of the History of Behavioral Sciences* (1981), 17:466–485.

15. William James to Alice Gibbens James, June 27, 1880; WJP.

16. William James to Alice Gibbens James, July 10, 1880; WJP.

17. William James to Alice Gibbens James, July 15, 1880; WJP.

18. See William James, 1868–1873, Diary; WJP. For a widely accepted version of the sixties crisis, see Perry, 1:320–382.

19. William James to Charles Renouvier, July 29, 1876; *LWJ,* 1:187–188.

20. On Renouvier see Alexander J. Dunn, *Modern French Philosophy: A Study of the Development Since Comte* (London: Unwin, 1922), and see Perry, 1:654–710, for a review of James-Renouvier correspondence from the late sixties to the latter's death in 1904. Gerald E. Myers argues that Renouvier was second only to Henry Sr. in influence on William. No other thinker, Myers maintains, "had formulated a point of view so close to his own." See *William James: His Life and Thought* (New Haven: Yale University Press, 1986), pp. 46–47. For a brief overview of Renouvier's life and metaphysics, see George Boas' sketch in Edwards, ed., *The Encyclopedia of Philosophy,* vol. 7/8, pp. 180–182.

21. To James, *all* intellectual alternatives tended toward empiricism or idealism, thus oversimplifying and misrepresenting reality. On Scottish realism see J. David Hoeveler, Jr., *James McCosh and the Scottish Intellectual Tradition: From Glasgow to Princeton* (Princeton, N.J.: Princeton University Press, 1981).

22. William James to Alice Gibbens James, July 10, 1880; WJP.

23. William James to Alice Gibbens James, July 31, 1880; WJP.

24. Ibid.

25. See, for instance, what James called his "Walpurgis Nacht," in William James to Alice Gibbens James, July 9, 1898; WJP. In 1882 James admitted his attraction to the idea of a unified world in an article which nevertheless scathingly criticized a Hegelian world where all distinctions ultimately collapsed. He could not find such a world after sober intellectual reflection. After inhaling nitrous oxide and in a state of hallucination, however, James felt all distinctions as artificial. See "On Some Hegelism," *Mind* (April 1882), 7:186–208, especially pp. 206–208.

26. For an annotated list of James' publications during the late seventies and early eighties, see John J. McDermott, *The Writings of William James* (Chicago: University of Chicago Press, 1977), pp. 817–821. In 1880–1881, all of James' major articles appeared in popular or religious journals as opposed to professional philosophical ones.

27. William James to Thomas Davidson, May 29, 1881; WJP.

28. Catherine Walsh to Mrs. C. P. Cranch, March 22, 1882; JFP.

29. Henry James to Thomas Sergeant Perry, February 3, 1882; quoted in Virginia Harlow, *Thomas Sergeant Perry: A Biography* (Durham, N.C.: Duke University Press, 1950), p. 312.

30. See James William Anderson, "In Search of Mary James," *The Psychohistory Review* (1979), 8:63–70.

31. Alice Gibbens James to Mary Holton James, March 24, 1882; VC.

32. William James to Alice Gibbens James, September 24, 1882; *LWJ*, 1:211.

33. See William James to Alice Gibbens James, June 27, 1882; WJP.

34. William James to Thomas Davidson, April 16, 1882; WJP.

35. Henry James, Sr. to Henry James, May 9, 1882; WJP.

36. William James to Josiah Royce, April 23, 1882; WJP.

37. William James to Josiah Royce, June 13, 1882; WJP.

38. William James to Alice Gibbens James, June 26, 1882; WJP.

39. William James to Alice Gibbens James, June 25, 1882; WJP.

40. William James to Alice Gibbens James, September [?], 1882; WJP.

41. William James to Alice Gibbens James, September 4, 1882; WJP.

42. William James to Alice Gibbens James, September 14, 1882; WJP.

43. William James to Alice Gibbens James, September 25, 1882; WJP.

44 Ibid.

45. William James to Henry James, Jr., October 2, 1882; WJP.

46. William James to Henry James, Jr., October 16, 1882; WJP.

47. William James to Henry James, Sr., October 23, 1882; WJP. Also quoted in Perry, 1:385.

48. William James to Henry James, October 16, 1882. WJP.

49. William James to Alice Gibbens James, October 25, 1882. WJP.

50. William James to Alice Gibbens James, October 28, 1882.

51. On Stumpf see E. B. Titchener, "Prof. Stumpf's Affective Psychology," *American Journal of Psychology* (1917), 28:263–277, and Edwin

G. Boring, *History of Experimental Psychology* (New York: Appleton-Century-Crofts, 1950), pp. 362–371. Stumpf's correspondence to James (in German) appears in Perry, 2:738–744. For Stumpf's own views on his career, see Carl Murchison, ed., *History of Psychology in Autobiography*, 3 vols. (New York: Russell and Russell, 1961), 1:388–441.

52. In his 1879 article "The Spatial Quale," James reported that he agreed "with Stumpf, that we cannot feel plane space *as a plane* without in some way cognizing the cubic spaces which the plane separates." *Journal of Speculative Philosophy* (1879), p. 71.

53. William James, *The Principles of Psychology*, 2:282.

54. William James to Alice Gibbens James, November 2, 1882; WJP.

55. See Frank J. Sulloway, *Freud, Biologist of the Mind* (New York: Basic Books, 1979), p. 274.

56. On Mach see P. Carus, "Professor Mach and His Work," *The Monist* (1911), 16:21–25; Edwin G. Boring, *A History of Experimental Psychology*, pp. 392–395; and Peter Alexander's sketch in Edwards, *The Encyclopedia of Philosophy*, 5:115–119. There is an interesting discussion of the similarity of scientific imagination in Mach and James in Lewis Feurer, *Einstein and the Generations of Science* (New York: Basic Books, 1974), pp. 26–41. Gerald Myers points out, however, that James could not abide by Mach's attempts to put sensation in the frame of reference of physics rather than of physiology. See *William James: His Life and Thought*, pp. 569–570.

57. William James to Alice Gibbens James, November 2, 1882; WJP.

58. William James to Alice Gibbens James, November 3, 1882; WJP.

59. William James to Alice Gibbens James, November 6, 1882; WJP.

60. William James to Alice Gibbens James, November 9, 1882; WJP. The article in which James refers to Baginsky is "The Sense of Dizziness in Deaf Mutes," *American Journal of Otology* (1882), 4:239–254. Baginsky had attacked the theory "that the semicircular canals are unconnected with the sense of hearing, but serve to convey to us the feeling of movement of our head through space, a feeling which, when very intensely excited, passes into that of vertigo or dizziness." James wrote that "Baginski's experiments seem to me far from conclusive" (p. 239).

61. On James' scientific critics in America see Daniel W. Bjork, *The Compromised Scientist: William James in the Development of American Psychology* (New York: Columbia University Press, 1983), pp. 73–102 and 163–174.

62. There has been a recent revival of interest in Wundt. See Wolfgang G. Bringmann, William D. G. Balance, and Rand B. Evans, "Wilhelm Wundt 1832–1920: A Brief Biographical Sketch," *Journal of the History of the Behavioral Sciences* (1975), 11:287–297; Arthur L. Blumenthal, "A Reappraisal of Wilhelm Wundt," *American Psychologist* (1975), 30:1081–1088; Kurt Danziger, "The Positivist Repudiation of Wundt," *Journal of the History of the Behavioral Sciences* (1979), 16:205–

230; and David E. Leary, "Wundt and After: Psychology's Shifting Relations with the Natural Sciences, Social Sciences, and Philosophy," *Journal of the History of the Behavioral Sciences* (1979), 15:231–241.

63. William James to Alice Gibbens James, November 11, 1882; WJP.

64. William James to Alice Gibbens James, November 18, 1882; WJP.

65. See James' review of Wundt's *Grundzuge der Physiologischen Psychologie* in the *North American Review,* (1875), 121:195–201.

66. Carl Stumpf to William James, September 8, 1886; quoted in Perry, 2:67.

67. William James to Carl Stumpf, February 6, 1887; quoted in Perry, 2:68–69.

68. See for example the exchange between James and Cattell in *Science:* Letter, *Science* (May 6, 1898), 7:640–641 and Letter, *Science* (May 6, 1898), 7:641–642. Also note W. B. Pillsbury, "Titchener and James," *Psychological Review* (1943), 50:71–73.

69. Wilhelm Wundt, quoted in Arthur L. Blumenthal, *Language and Psychology: Historical Aspects of Psycholinguistics* (New York: Wiley, 1970), p. 238.

70. See Shadworth Hodgson to William James, January 26, 1882; quoted in Perry, 1:621.

71. On Delboeuf, see Sulloway, *Freud, Biologist of the Mind,* p. 40, and a scattered but valuable discussion in Perry, 1:677–679 and 686–690. James had mentioned in "The Spatial Quale" (p. 82) Delboeuf's "brilliant and original article on Vision" in *Revue Philosophique.* Unlike James, Renouvier found Delboeuf's ideas almost incomprehensible.

72. William James to Carl Stumpf, November 26, 1882; quoted in Perry, 2:61.

73. William James to Alice Gibbens James, November 20, 1882; WJP.

74. William James to Alice Gibbens James, November 29, 1882; WJP.

75. Alice Gibbens James to William James, November 30, 1882; WJP.

76. William James to Alice Gibbens James, December [?], 1882; WJP. William had received a telegram saying that his father was no longer lucid: "Brain softening. Possibly live months. All insist Wm. shall not come."

77. William James to Henry James, Sr., December 14, 1882; WJP.

78. Alice Gibbens James to William James, December 18, 1882; WJP.

79. Ibid.

80. Ibid.

81. See Gay Wilson Allen, *William James, A Biography,* (New York: Viking, 1967), pp. 258–259.

82. Correspondence relating to Wilky and the will is in the VC.

83. Alice Gibbens James to William James, December 21, 1882; WJP.

84. William James to Alice Gibbens James, December 19, 1882; WJP.

85. William James to Alice Gibbens James, December 20, 1882; WJP.

86. Ibid.

87. *New York Post,* December 20, 1882.

88. *New York Times,* December 20, 1882.
89. *New York Post,* December 19, 1882.

10. CHARTING THE REALM OF CONSCIOUSNESS

1. James' professional promotion was as follows: he was appointed instructor of anatomy and physiology in 1873; assistant professor of physiology in 1876; professor of philosophy in 1885; professor of psychology in 1889; and he resumed the title Professor of Philosophy in 1898.

2. Daniel Levinson has suggested that professional academicians in their forties often find the limits of their creative drives, a realization that can produce frustration and depression, or settling in and acceptance. James, however, expanded his creative endeavor spectacularly during middle age. See *The Seasons of a Man's Life* (New York: Knopf, 1978), pp. 191–208.

3. James published thirteen major articles in this six-year period, in addition to numerous book reviews and the 119-page introduction to *The Literary Remains of the Late Henry James.* See John J. McDermott's annotated bibliography, *The Writings of William James* (Chicago: University of Chicago Press, 1977), pp. 818–823.

4. Among the most illuminating studies interpreting the great transformation of American life in the late nineteenth century are Robert Wiebe, *The Search for Order: United States, 1877–1920* (New York: Wang, 1967); T. J. Jackon Lears, *No Place for Grace: Antimodernism and the Transformation of American Culture 1880–1920* (New York: Pantheon, 1981); and Alan Trachtenberg, *The Incorporation of America: Culture and Society in the Gilded Age* (New York: Hill and Wang, 1982).

5. Jacques Barzun, a lifelong student of James, has recently asked, "Who, or what, was this Protean creature William James?" See *A Stroll with William James* (New York: Harper and Row, 1983), p. 3. James was a creature of his age and a preview of our own. Paradoxically, he was historically placed and yet placeless; interpretable as both a man beyond his age and in it. Hence in contrast to Barzun and his protean James, Burton Bledstein has remarked, "Though William James was not a typical American, he was an acutely self-conscious one, the kind of individual who is torn by the deepest tensions in a culture." See *The Culture of Professionalism: The Middle Class and the Development of Higher Education in America* (New York: Norton, 1976), p. 319.

6. See William James, *Pragmatism,* p. 13.

7. Rand Evans has carefully traced the progress of *The Principles* in an essay on the history of the writing of the book included in William James, *The Principles of Psychology,* 3:1532–1579. As Evans notes, Holt's side of his correspondence with James is housed in the Holt Archives, Princeton University Library, but copies are included in the WJP at the Houghton Library, Harvard.

8. See Evans, in *The Principles of Psychology,* 3:1536–1540.

9. William James, *Psychology: Briefer Course* (New York: Henry Holt, 1892), p. v.

10. For instance, parts of "On Some Omissions of Introspective Psychology," *Mind* (1884), 9:1–26, were incorporated into *The Principles,* in chapter 7, "The Methods and Snares of Psychology," chapter 9, "The Stream of Thought," and chapter 12, "Conception." And "What is an Emotion?" *Mind* (April 1884), vol. 9, was integrated here and there into chapter 25, "The Emotions." For a detailed account of the merger of articles into *The Principles,* see Evans' historical essay in *The Principles of Psychology* 3:1542–1549.

11. In contrast to the thirteen major articles composed between 1879 and 1885, James wrote only six from 1886 through 1890—portions of which were also incorporated into various chapters of *The Principles.* See McDermott, *The Works of William James,* pp. 823–227. Moreover, James' most influential psychological ideas—on habit, the stream of consciousness, cognition, and the emotions—were developed before 1886.

12. James had considered reasoning as an associative process in "Brute and Human Intellect," *Journal of Speculative Philosophy* (July 1878), 12:236–276. But he had not yet argued for the primacy of feeling and relation in the building up of a knowing consciousness—ideas which would set his psychology apart. Gerald E. Myers has recently discussed James' developing concept of the process of thought in *William James: His Life and Thought* (New Haven: Yale University Press, 1986), pp. 242–271. Myers concludes that James' work in this area was a classic inquiry into the relationship between introspective and metaphysical problems. But in the seventies James was intent on understanding the relationship between physiology and consciousness, and was only beginning to concentrate on matters that would bring introspection and metaphysics to the forefront of his concerns.

13. Complex consciousness raised the question of how the mind comes to know an object, and it would eventually bring James toward a phenomenology of the mind that he called "radical empiricism" and toward a philosophy of "pure experience."

14. The stream of thought and feelings of relation were broached in "On Some Omissions of Introspective Psychology." "Knowledge of," or acquaintance, and "knowledge about" were distinguished in "On the Function of Cognition," *Mind* (1885), 10:27–44.

15. The members of the group were Croom Robertson, Shadworth Hodgson, Thomas Sully, Carveth Read, Frederick Pollock, F. W. Maitland, Leslie Stephen, and Edmund Gurney. "Scratch Eight" simply referred to the number of participants; James became the ninth. William remarked to Alice that "discussion carried on by Sully, Hodgson, and Robertson . . . seemed to show me that there was a great opening for my psychology." William James to Alice Gibbens James, December 16, 1882; WJP.

16. William James to Alice Gibbens James, February 10, 1883; WJP.

James may also have been catalyzed by Shadworth Hodgson's growing opposition to his emerging psychological perspective. In the same letter he says, "Hodgson is constitutionally incapable of understanding any thoughts but those that grow up in his own mind,—with all the desire in the world to do justice to them, he simply can't reproduce them in himself."

17. Perry suggests that James' moment of "psychological truth" was the germ for the essay "On Some Omissions of Introspective Psychology"; see Perry, 1:38–39. Perry was not certain, however, since that article deals mainly with feelings of relation, whereas in the February 10 letter to Alice, William says that he spoke at the Scratch Eight on "the difference between feeling and Thought." Actually, "Loose Notes on Cognition" is concerned both with feelings of relation and with distinguishing between feeling and thought. In all probability these notes provided the genesis for both the "Omissions" article and the essay "The Function of Cognition."

18. James had another creative moment in which various ideas rushed together; these ideas, he maintained, were an inspiration for the Gifford Lectures, which became *The Varieties of Religious Experience*. See William James to Alice Gibbens James, July 9, 1898; WJP.

19. William James, Loose Notes on Cognition, [1882–83?]; item 4394, folder 2, WJP. Hereafter cited parenthetically in text as "Notes on Cognition."

20. This tendency toward faith rather than skepticism in science was an early indication of James' famous defense of belief in "The Will to Believe," *New World* (1896), 5:327–347.

21. On Condillac see Georges LeRoy, *La Psychologie de Condillac* (Paris: Boivin, 1937). Once before James' thinking had been stimulated by considering statues, when he compared classical and modern sculpture in Dresden in 1868. Now, however, the statue existed only in the mind, perhaps indicating that James' creative field had become more abstract, less dependent on actual works of art to activate the imagination.

22. James borrowed "the fragrance of the rose" example from the Scottish commonsense philosopher Thomas Brown.

23. James argues in "The Sentiment of Rationality," *Mind* (1879), 4:317–346, that most thinkers have a craving either for clarity or for unity, depending on temperament.

24. Scottish philosophers such as Hamilton and his predecessors Thomas Reid and Dugald Stewart had attacked what they viewed as the skepticism and nihilism of Locke and especially of Hume. In this they were aligned with Kant and idealism, although they actually wanted a middle course between the scientifically inclined empiricism and romantic transcendentalism. Unlike Kant, the Scots believed the mind could know the thing directly, in an "intentional realism." But James believed that Hamilton assumed too much about the power of the mind to grasp objects. Also, the Scottish tradition retained an antiscientific bias that James

found unprogressive. And he may have found that this tradition competed with his own design to create an alternative to empiricism and idealism. For discussion of Scottish philosophy, see Bruce Kuklick, *The Rise of American Philosophy: Cambridge, Massachusetts, 1860–1930* (New Haven: Yale University Press, 1977), pp. 16–21, and J. David Hoeveler, Jr., *James McCosh and the Scottish Intellectual Tradition: From Glasgow to Princeton* (Princeton, N.J.: Princeton University Press, 1981), pp. 111–146.

25. For a general discussion of the marriage of physiology and psychology in Europe, see R. W. Rieber, *William Wundt and the Making of a Scientific Revolution* (New York: Plenum Press, 1980). For an outline of the major trends in late nineteenth-century American psychology, see Thomas M. Camfield, "The Professionalization of American Psychology, 1870–1917," *Journal of the History of the Behavioral Sciences* (1973), 9:66–75; R. W. Rieber and Kurt Salzinger, eds., *The Roots of American Psychology: Historical Influences and Implications for the Future* (New York: New York Academy of Sciences, 1977); and Donald S. Napoli, *Architects of Adjustment: The History of the Psychological Profession in the United States* (New York: Kennikat Press, 1981).

26. The linguistic sources of James' psychology have been slighted, but see William Woodward's introduction to *Essays in Psychology*, especially pp. xxvii–xxviii.

27. John J. McDermott has observed that the "feeling of relation" shows up time after time in *The Principles*, in chapter 7, "The Methods and Snares of Psychology; chapter 9, "The Stream of Thought"; chapter 10, "The Consciousness of Self"; and chapter 12, "Conception." See his introduction to James, *Essays in Radical Empiricism*, p. xviii. All of these chapters owe much to the Loose Notes on Cognition.

28. Only in positing a world of pure experience in a growing universe does James move metaphysically beyond dualism. And he would not realize this unity until the last few years of his life.

11. THE PROFESSIONAL COIL

1. See Samuel Eliot Morison, *The Development of Harvard Since the Inauguration of President Eliot, 1869–1929* (Cambridge: Harvard University Press, 1930), pp. xic–xc; and Bruce Kuklick, *The Rise of American Philosophy: Cambridge, Massachusetts, 1860–1930* (New Haven: Yale University Press, 1977), pp. 233–235.

2. See Lists of William James' Salaries at Harvard; item 4577, WJP.

3. See Ralph Barton Perry, memorandum, List of Courses Taught by William James at Harvard; item 4575, WJP.

4. For a list of doctorates in philosophy at Harvard from 1878 to 1930, see Kuklick, *The Rise of American Philosophy*, pp. 582–589. During the 1890s there were twenty Ph.Ds. But in the first decade of the twentieth century there were sixty-six.

5. See Rand B. Evans, "The Historical Context," in William James, *The Principles of Psychology,* vol. 1, pp. xli–lxviii.

6. See Gerald E. Myers, "The Intellectual Context," in William James, *The Principles of Psychology,* vol. 1, pp. xi–xl, and Thomas M. Camfield, "The Professionalization of American Psychology, 1870–1917," *Journal of the History of the Behavioral Sciences* (1973), 9:66–75.

7. William James, quoted in Morison, *The Development of Harvard University,* p. 5.

8. Kuklick, *The Rise of American Philosophy,* p. 28.

9. Bruce Kuklick, the historian of Harvard philosophy, believes that this "golden age" was replaced after World War I with a "professional mentality" which exchanged the great general questions of philosophy for specialized expertise in positivistic metaphysics, and that the shift indicated the decline of moral and religious problems as reasons for taking up the philosophic vocation. See *The Rise of American Philosophy,* pp. 451–480 and 565–572.

10. Palmer headed the department from 1891 to 1894, and from 1898 to 1900; Royce from 1894 to 1898; and Münsterberg from 1900 to 1906. See ibid., pp. 243–244.

11. On Royce's life and thought, see Bruce Kuklick, *Josiah Royce* (Indianapolis: Bobbs-Merrill, 1972).

12. George Herbert Palmer, quoted in Kuklick, *The Rise of American Philosophy,* p. 237. For Palmer's biography and approach to higher education, see pp. 215–277 of the same work, as well as George Herbert Palmer and Alice Freeman Palmer, *The Teacher: Essays and Addresses on Education* (Boston: Houghton Mifflin, 1908).

13. William James to Josiah Royce, September 26, 1900; WJP.

14. William James to Alice Gibbens James, May 13, 1888; WJP.

15. William James to Alice Gibbens James, October 22, 1889; WJP.

16. William James to Alice Gibbens James, November 3, 1889; WJP.

17. William James to Alice Gibbens James, November 5, 1889; WJP.

18. G. Stanley Hall to William James, April 7, 1890; WJP. For Hall's role in early experimental psychology, see Dorothy Ross, *G. Stanley Hall: The Psychologist as Prophet* (Chicago: University of Chicago Press, 1972), pp. 148–185. Hall helped found the first New World professional journal, the *American Journal of Psychology* (1887), as well as the first national professional organization, the American Psychological Association.

19. Form Letter Soliciting Funds for the Renovation of the Psychological Laboratory Under the Direction of William James, found in James Family Letters to Various Persons; WJP.

20. William James to Alice Gibbens James, May 17, 1890; WJP.

21. William James to Henry James, June 4, 1890; WJP.

22. James never viewed introspection as an analytical process that broke consciousness down into elements, but as an observational method that confirmed the unity of consciousness. See Rand B. Evans, "The Historical Context," in *The Principles of Psychology,* vol. 1, pp. lvix–lxi. For

incisive accounts of the different historical perceptions of introspection, including James', see Kurt Danziger, "The History of Introspection Reconsidered," *Journal of the History of the Behavioral Sciences* (1980), 16:241–262, 24; and Gerald E. Myers, *William James: His Life and Thought* (New Haven: Yale University Press, 1986), pp. 64–80.

23. On Münsterberg see Matthew Hale, Jr.'s biography, *Human Science and the Social Order: Hugo Münsterberg and the Origins of Applied Psychology* (Philadelphia: Temple University Press, 1980).

24. William James to Josiah Royce, June 22, 1892; WJP.

25. Münsterberg never became "Americanized," and James did not live to see the political fallout at Harvard occasioned by "M's" outspoken support of Germany before and during World War I. See Hale, *Human Science and the Social Order,* pp. 87–105 and 164–183.

26. For the context and the relationship between *Psychology: Briefer Course* and *The Principles,* see Michael M. Sokal, "Introduction," in William James, *Psychology: Briefer Course,* pp. xi–xli.

27. See for example James' exchange with James McKeen Cattell in *Science* (May 6, 1898), 7:640–642.

28. See Rand B. Evans and Frederick J. Down Scott, "The 1913 International Congress of Psychology: The American Congress that Wasn't," *American Psychologist* (1978), 33:334–341.

29. Two outstanding studies of the social ferment in America at the end of the century are Robert Wiebe, *The Search for Order, 1877–1900* (New York: Hill and Wang, 1967); and Alan Trachtenberg, *The Incorporation of America: Culture and Society in the Gilded Age* (New York: Hill and Wang, 1982).

30. Don S. Browning, *Pluralism and Personality* (Lewisburg, Pa.: Bucknell University Press, 1982).

31. For context see Gerald E. Myers, "Introduction," in James, *Talks to Teachers on Psychology,* pp. xi–xxvii.

32. William James to Alice Gibbens James, July [?], 1896; WJP.

33. Ibid.

34. Ibid.

35. William James, "What Makes a Life Significant," in *Talks to Teachers,* p. 155.

36. William James to Alice Gibbens James, July 6, 1905; WJP.

37. Perry, 1:442.

38. John Jay Chapman, *Memories and Milestones* (New York: Moffact, Yard, 1915), p. 26.

39. William James to Kate E. Havens, February 22, 1876; WJP.

40. William James to Alice Gibbens James, June 13, 1888; WJP.

41. William James to Charles W. Eliot, March 1, 1897; WJP.

42. William James to Henry Lee Higginson, December 14, 1900; WJP.

43. Henry Holt to William James, April 2, 1890; WJP.

44. William James to Henry Holt, April 5, 1890; WJP.

45. William James to Henry James, June 4, 1890; WJP.

46. William James to Pauline Goldmark, January 2, 1900; WJP.
47. William James to Henry Holt, May 9, 1890; WJP.
48. William James to Alice Gibbens James, May 24, 1890; WJP.
49. For the history of early printings, see James. *The Principles of Psychology,* 3:1576–1577.
50. For an exceptionally acute discussion of Victorian mores in America, see Nathan Hale, Jr., *Freud and the Americans: The Beginnings of Psychoanalysis in the United States, 1876–1917* (New York: Oxford University Press, 1971), pp. 24–46.
51. See George Santayana, *The Genteel Tradition at Bay* (New York: Scribner's, 1931), pp. 54–61.
52. See Ronald Hayman's fine *Nietzsche: A Critical Life* (New York: Penguin, 1982).
53. William James, "The Gospel of Relaxation," *Scribner's Magazine* (1899), 25:506–507.

12. SOLIDLY ESTABLISHED

1. Garth Wilkinson James to William James, September 25, 1883; WJP.
2. Robertson James to William James, November 26, 1883; VC.
3. Ibid.
4. In 1872, for instance, when Bob was suffering a variety of complaints that included fear of sexual impotence, William had advised, "It would probably do you much good to unburden your mind about it. Your case as you describe it belongs to the most trifling class, amenable to treatment in some form; and if continuing in spite of medical treatment, ready to cease when sexual intercourse begins regularly. I don't think you need be in the least degree anxious about impotence. That is an effect of a totally different class of cases from yours, and moreover very uncommon." William James to Robertson James, June 22, 1872; VC.
5. William James to Robertson James, April 20, 1883; WJP.
6. Myers, although recognizing some validity in the Jacob-Esau interpretation, believes that Edel goes too far in suggesting that William was jealous because Henry was elected to membership in the prestigious National Institute of Arts and Letters on an earlier ballot than himself in early 1905. This was a late life example, Edel has argued, of the theme of usurpation that had never been far from the surface in the brothers' relationship; see Leon Edel, *Henry James,* 5 vols. (Philadelphia: Lippincott, 1953–1972), 1:141. Myers, however, inclines to the view that William's refusal to become a member of the Institute after Henry's earlier invitation was simply "playfulness." See *William James: His Life and Thought* (New Haven: Yale University Press, 1986), pp. 21–28 and 494. For an earlier rejection of Edel's view, see Jacques Barzun's discussion in the *New York Times Book Review* of April 16, 1972.
7. Alice Gibbens James to Henry James, April 11, 1892; WJP.
8. William James to Mary Holton James, October 4, 1884; VC.

9. William James to Henry James, April 1, 1885; WJP.

10. Higginson was a partner in the Boston investment firm Lee, Higginson, and Company. He served as James' financial advisor from the time of Henry Sr.'s death.

11. William James to Alice Gibbens James, August 3, 1885; WJP.

12. William James to Henry James, September 7, 1885; WJP.

13. William James to Alice Gibbens James, February 3, 1888; WJP.

14. William James to Alice Gibbens James, February 5, 1888; WJP.

15. See William James to Alice Gibbens James, February 6, 1888; WJP.

16. William James to Alice Gibbens James, February 6, 1888; WJP.

17. William James to Henry James, April 11, 1892; WJP.

18. Alice Gibbens James to Henry James, 1895; WJP.

19. Ibid.

20. William James to Henry James, May 9, 1897; WJP.

21. William James to Henry James, February 7, 1897; WJP.

22. William James to Alice Gibbens James, July 10, 1897; WJP. Robertson did stop seeing Mrs. Schishkar, however, in early July 1897. See William James to Alice Gibbens James, July 5, 1897; WJP.

23. Alice Gibbens James to Henry James, February 14, 1898; WJP.

24. William James to Alice Gibbens James, July 5, 1898; WJP.

25. William James to Henry James, September 8, 1907; WJP.

26. By 1890 James' Harvard salary was $4,000.

27. William James to Henry James, June 9, 1885; WJP.

28. William James to Alice Gibbens James, February 7, 1888; WJP.

29. William James to Henry James, September 17, 1886; WJP.

30. William James to Alice Gibbens James, September 6, 1886; WJP.

31. William James to Alice Gibbens James, May 20, 1889; WJP.

32. William James to Alice Gibbens James, May 21, 1889; WJP.

33. William James to Alice Gibbens James, May 22, 1889; WJP. James made a balance sheet listing "outlays and income, and glory to be God, I find all the money we have borrowed this year has just about gone to investments and none to living expenses." By his calculation they had borrowed $8,688 dollars—$5,688 from two banks, $2,000 from Alice's mother, Elizabeth Gibbens, and $1,000 from Mary Salter, Alice's recently married sister. William James to Alice Gibbens James, May 23, 1889; WJP.

34. William James to Kitty James Prince, June 12, 1885; WJP.

35. William James to Kitty James Prince, July 12, 1885; WJP. Perhaps William must fully describe Herman's death to Kitty Prince because she had much experience with suffering. Kitty was the daughter of William's half-uncle, Reverend William James of Albany. She had spent years in mental institutions.

36. William James to Elizabeth Webb Gibbens, March 24, 1887; WJP.

37. William James to Alice Gibbens James, June 30, 1889; WJP.

38. William James to Henry James, January 20, 1892; WJP.

39. William James to Shadworth Hodgson, March 28, 1892; WJP.

40. See Jean Strouse, *Alice James: A Biography* (Boston: Houghton Mifflin, 1980), p. 184.
41. William James to Henry James, March 7, 1892; WJP.
42. William James to Henry James, March 24, 1894; WJP.
43. See William James to Henry Lee Higginson, March 26, 1892; WJP.
44 William James to Henry B. Bowditch, June 13, 1892; WJP.
45. Alice Gibbens James to Henry James, June 21, 1892; WJP.
46. William James to Henry James, October 11, 1892; WJP. This letter was handwritten by Alice Gibbens James.
47. William James to Alice Gibbens James, November 2, 1892; WJP.
48. William James to Alice Gibbens James, November 5, 1892; WJP.
49. William James to Alice Gibbens James, November [?], 1892; WJP.
50. William James to Josiah Royce, March 18, 1893; WJP.
51. Alice Gibbens James to Elizabeth Webb Gibbens, April 26, 1893; WJP.
52. Alice Gibbens James to Elizabeth Webb Gibbens, May, 9, 1893; WJP.
53. Alice Gibbens James, to Elizabeth Webb Gibbens, June 18, 1893; WJP.
54. William James to Shadworth Hodgson, June 23, 1893; WJP.
55. William James to Alice Gibbens James, July 19, 1893; WJP.
56. William James to F. W. H. Myers, December 17, 1893; WJP.
57. William James to Alice Gibbens James, June 1, 1894; WJP.
58. Note by Henry James III, no date, but probably written in 1918 or 1919 when he was compiling his father's letters for publication as *The Letters of William James* (1920); WJP.
59. William James to Henry James, August 6, 1895; WJP.
60. William James to Alice Gibbens James, August 9, 1895; WJP.
61. William James to Alice Gibbens James, September 3, 1895; WJP.
62. William James to Alice Gibbens James, September 5, 1895; WJP.
63. Alice James to Sara Darwin, August 9, 1879; JFP.
64. William James to Alice Gibbens, James, June 22, 1895; WJP.
65. William James to Alice Gibbens James, September 15, 1896; WJP.
66. William James to Alice Gibbens James, September 13, 1897; WJP.
67. William James to Alice Gibbens James, September 17, 1897; WJP.
68. Alice Gibbens James to Henry James, April 23, 1897; WJP.
69. Alice Gibbens James to Henry James, May 31, 1897; WJP.
70. William James, quoted in Bliss Perry, *Life and Letters of Henry Lee Higginson* (Boston: Atlantic Monthly Press, 1921), p. 414.
71. William James to Henry James, April 10, 1898; WJP.
72. Alice Gibbens James to Henry James, February 4, 1898; WJP.
73. William James to Alice Gibbens James, April 19, 1898; WJP.
74. William James to Henry James, February 21, 1898; WJP.
75. James Jackson Putnam to William James, March 9, 1898; WJP.
76. William James to James Jackson Putnam, March 2, 1898; WJP.

77. William James to James Jackson Putnam, March 3, 1898; WJP.

78. William James to Henry James, December 9, 1896. William did not cite the specific source for this description.

13. CONQUERING THE REALM OF CONSCIOUSNESS

1. William James, *A Pluralistic Universe,* p. 149.

2. Introductions to the various volumes of the definitive edition of James' publications, *The Works of William James* (Cambridge: Harvard University Press, 1977—), provide excellent context and analyses of the later philosophical books. See H. S. Thayer's essay in *Pragmatism,* pp. xi–xxxviii; Richard J. Bernstein's in *A Pluralistic Universe,* pp. xi–xxix; H. S. Thayer's in *The Meaning of Truth,* pp. xi–xivi; Peter H. Hare's in *Some Problems of Philosophy,* pp. xiii–xli; and John J. McDermott's in *Essays in Radical Empiricism,* pp. xi–xlviii.

3. For instance, James considered ten "proofs" for the existence of an unconscious mind and rejected them in the chapter "The Mind-Stuff Theory." He concluded that unconscious ideas did not exist in a separate mental compartment. Rather, unconscious states—which James fully admitted were in the mind's *experience*—were simply felt; "The *are* eternally as they feel when they exist, and can, neither actualy nor potentially be identified with anything else than their own faint selves." William James, *The Principles of Psychology,* 1:175.

4. It is interesting, however, that both James and Freud based their psychologies within a biological-physiological context and agreed that instincts and habit played a crucial role in determining the mind's activity. See Frank J. Sulloway, *Freud, Biologist of the Mind* (New York: Basic Books, 1979), pp. 290–291.

5. For a group of essays that explore regional mental illness, see George Gifford, ed., *Psychoanalysis, Psychotherapy, and the New England Medical Scene, 1894–1944* (New York: Science History Press, 1978). For a study on insanity in notable New England families—one which includes a thinly disguised reference to Henry Sr., Henry, and William—see Abraham Myerson and Rosalie D. Boyle, "The Incidence of Manic-Depressive Psychosis in Certain Socially Important Families," *American Journal of Psychiatry* (July 1941), 98:11–21. For a recent study on the medical context of neurasthenia, see F. G. Gosling, *Before Freud: Neurasthenia and the American Medical Community, 1870–1910* (Urbana, Ill.: University of Illinois Press, 1987).

6. For an excellent discussion of the proliferation of religious conversions in the mid-nineteenth century, see Whitney Cross, *The Burnt-Over District* (New York: Harper and Row, 1965). On spiritualism and James' role as both advocate and critic within the broader cultural context, see R. Laurence Moore, *In Search of White Crows: Spiritualism, Parapsychology, and American Culture* (New York: Oxford University Press, 1977), pp. 146–147, 166–168, 183–184, and 239–241, and Robert C.

Fuller, *Mesmerism and the American Cure of Souls* (Philadelphia: University of Pennsylvania Press, 1982), pp. 169–173.

7. See William James, "Moral Medication," *Nation* (1868), 7:50–52, an unsigned review of A. A. Liébeault's *Du Sommeil et des Etats Analogues.* Liébeault's book later inspired Hippolyte Bernheim, the leader of the Nancy School of hypnotism. William believed that hypnosis had therapeutic potential. Also note William James, unsigned review of Epes Sargent's *Planchette; or, The Despair of Science,* in *Boston Daily Advertiser,* March 10, 1869.

8. See Nathan G. Hale, Jr., "James Jackson Putnam and Boston Neurology: 1877–1891," in George E. Gifford, Jr., ed., *Psychoanalysis, Psychotherapy, and the New England Medical Scene, 1894–1944* (New York: Science History Publications, 1978), pp. 149–154.

9. See William James, 1868–1873 Diary; WJP.

10. William James to Kate E. Havens, June 14, 1874; WJP.

11. See for example his unsigned reviews of the following: Henry Maudsley's *Responsibility in Mental Disease,* in *Atlantic Monthly* (1874), 34:364–365; W. B. Carptenter's *Principles of Mental Physiology,* in *Atlantic Monthly* (1874), 34:495; A. Winter's *Borderlands of Insanity and Other Allied Papers,* in *Nation* (1875), 21:330; and R. L. Dugdale's "*The Jukes": A Study in Crime, Pauperism, Disease, and Heredity,* in *Atlantic Monthly* (1878), 41:405.

12. William James to Thomas Davidson, February 1, 1885; *LWJ,* 1:250.

13. James became vice president of the English Society for Psychical Research in 1890, served as president from 1894 to 1895, and remained vice president until his death.

14. William James, "Certain Phenomena of Trance," *Proceedings of the Society for Psychical Research* (1890), 6:652. On Mrs. Piper see M. Sage, *Mrs. Piper and the Society for Psychical Research* (New York: Scott-Thaw, 1904), and Anne Manning Robbins, *Past and Present with Mrs. Piper* (New York: Henry Holt, 1921).

15. The results of James' census on hallucinations were published in *Psychological Review* (1895), 2:69–75. For a sample of James' publications on psychical research, see William James (and G. M. Carnochan), "Report of the Committee on Hypnotism," *Proceedings of the American Society for Psychical Research* (July 1886), 1:95–102; William James, "Reaction-Time in the Hypnotic Trance," *Proceedings of the American Society for Psychical Research* (December 1887), 1(3):246–248; William James, "Notes on Automatic Writing," *Proceedings of the American Society for Psychical Research* (March 1889), 1(4):548–564; and William James, "Record of Observations of Certain Phenomena of Trance," *Proceedings of the Society for Psychical Research* (1890), 6:651–659.

16. William James to Carl Stumpf, January 1, 1886; *LWJ,* 1:248.

17. William James, Notes on Psychical Research, undated [1890?], item 4534, WJP.

18. Even James' replacement at Harvard, Hugo Münsterberg, a psy-

chologist with wide-ranging interests, felt James had gone too far off the legitimate scientific path. See Hugo Münsterberg, "Psychological Mysticism," *Atlantic Monthly* (1899), 81:67–85.

19. See Eugene Taylor, "James on Psychopathology: The 1896 Lowell Lectures on 'Exceptional Mental States' in *A William James Renaissance: Four Essays by Young Scholars,* special issue, *Harvard Library Bulletin* (October 1982), 30(4):455–479.

20. See Nathan Hale, Jr's introduction in *James Jackson Putnam and Psychoanalysis: Letters Between Putnam and Sigmund Freud, Ernest Jones, William James, Sandor Ferenczi, and Morton Prince, 1877–1917.* (Cambridge: Harvard University Press, 1971), pp. 1–63.

21. See *William James on Exceptional Mental States: The 1896 Lowell Lectures,* reconstructed by Eugene Taylor (New York: Scribner's 1982).

22. Henry James III to Mrs. William James (Alice Gibbens James), July 21, 1920; WJP. Apparently in concordance with the appearance of *The Letters of William James,* Henry was organizing a display of his father's publications for the Widener Library at Harvard.

23. Perry, 2:155.

24. For the context of the lectures, see Eugene Taylor, "Historical Introduction," in *William James on Exceptional Mental States,* pp. 1–14.

25. William James to George H. Howison, April 5, 1897; WJP.

26. William James, notes for "Dreams and Hypnotism," of the Lowell Lectures of 1896; WJP. Notes for all eight 1896 Lowell Lectures are located in WJP. Subsequent quotations from these lecture notes will be identified in the text.

27. For an extensive discussion of the problematical diagnostic climate in psychopathology in the nineteenth century, see Henri Ellenberger, *The Discovery of the Unconscious: The History and Evolution of Dynamic Psychiatry* (New York: Basic Books, 1970).

28. William James, Notebook, Seminary of 1895–1896, The Feelings, book 1; item 4504, WJP. Hereafter cited parenthetically in the text as "Feelings Notebook."

29. These expressions for reality were new in philosophy and were most fully developed by James in the Hibbert Lectures delivered at Manchester College, Oxford University, in 1909, and published as *A Pluralistic Universe* (1909) and *Essays in Radical Empiricism* (1912).

30. For the James-Bradley correspondence, see Perry, 2:637–657. Gerald E. Myers argues that James found sensations more real than did Bradley. See *William James: His Life and Thought* (New Haven: Yale University Press, 1986), pp. 110–112. H. S. Thayer maintains that Bradley's astute philosophical idealism helped James to articulate all facits of his later metaphysical development. See Thayer's introduction to *The Meaning of Truth,* especially p. xvii.

31. The address was published the same year, with the same title, in the *University of California Chronicle.* James gave too much credit to

Charles Peirce for originating pragmatism; see H. S. Thayer's introduction to William James, *Pragmatism,* pp. xx–xxvii.

32. See Ellen Kappy Suckiel, *The Pragmatic Philosophy of William James* (Notre Dame, Ind.: University of Notre Dame Press, 1982), p. 11.

33. William James, Notebook, The Self–Pure Experience, etc., book 2, 1897–1898; item 4505, WJP. Hereafter cited parenthetically in the text as "Self–Pure Experience Notebook."

34. William James, Notebook, Pure Experience Continued, book 3, 1897–1898; item 4506, WJP. Hereafter cited parenthetically in the text as "Pure Experience Cont'd. Notebook."

35. See Clyde Griffin, "The Professive Ethos," in Stanley Coben and Lorman Ratner, eds., *The Development of an American Culture* (Englewood Cliffs, N.J.: Prentice Hall, 1970), pp. 120–149, and R. Jackson Wilson, *In Quest of Community: Social Philosophy in the United States, 1860–1920* (New York: Wiley, 1968).

36. A stream of critical articles that fleshed out the shift toward a metaphysics of experience appeared in 1904 and 1905, primarily in the *Journal of Philosophy, Psychology, and Scientific Methods* (*JPPSM*). They included "Does Consciousness Exist?" *JPPSM* (1904), 1:477–491; "A World of Pure Experience," *JPPSM* (1904), 1:533–543 and 561–570; "Humanism and Truth," *Mind* (1904), 13:457–475; "The Pragmatic Method," *JPPSM* (1904), 1:673–687; "The Experience of Activity," *Psychological Review* (January 1905), 12:1–17; "The Thing and Its Relations," *JPPSM* (1905), 2:29–41; "How Two Minds Can Know One Thing," *JPPSM* (1905), 2:176–181; "Is Radical Empiricism Solipsistic?" *JPPSM* (1905), 2:235–238; and "The Place of Affectional Facts in a World of Pure Experience," *JPPSM* (1905), 2:281–287.

37. Ralph Barton Perry dates the World of Pure Experience notebook as being completed by June 1904 (Perry, 2:385). But references to pure experience in James' correspondence with Henri Bergson make it probable that he had begun composing the notebook as early as December 1902.

38. For complete publication information see n. 36 above.

39. William had actually opened correspondence with Bergson during the period when he was composing the World of Pure Experience notebook. In December 1902 James reread *Matière et Mémoire* (Matter and Memory) and wrote, "It fills *my* mind with all sorts of new questions and hypotheses and brings the old into a most agreeable liquefaction." James believed that Bergson had accomplished a "conclusive demolition of the dualism of object and subject in perception"—which was the major theoretical obstacle to creating a world of pure experience. "How good it is sometimes simply to *break away* from all old categories, deny worn out beliefs, and restate things *ab initio* making the lines of division fall into entirely new places!" he exclaimed to a fellow originator (William James to Henri Bergson, December 14, 1902; WJP). James appropriated

Bergson's perspective into his own and predicted new philosophical agreement: "I am convinced that a philosophy of *pure experience*, such as I conceive yours to be, can be made to work, and will reconcile many of the old inveterate oppositions of the schools." William James to Henry Bergson, February 25, 1903; WJP.

For Bergson's philosophy, see Jacques Chevalier, *Henri Bergson* (New York: Macmillan, 1928), and Thomas Hanna, ed., *The Bergsonian Heritage* (New York: Columbia University Press, 1962).

40. William James to Henri Bergson, June 13, 1907; WJP.

41. William James, Notebook, World of Pure Experience, ca. 1903; item 4471, WJP. All subsequent quotations from James in this chapter are from this notebook.

42. On his critics, also see William James, Notebook, Miller-Bode Criticism, 1905–1908; item 4515, WJP. This notebook runs to some one hundred closely argued pages. James never tired of answering critics such as former student Dickinson S. Miller, who maintained that a metaphysics of experience never closed the dualistic chasm. Even though he had found his final philosophical solution, James could not stop considering pure experience within a never-ending, multifacited intellectual context. If denied critics, William would probably have invented them and carried on a dialogue with himself.

14. GOD HELP THEE, OLD MAN

1. William James, Notebook, Miscellaneous Notes on Pluralistic Metaphysics, ca. 1903–1907; item 4513, WJP.

2. Even Gerald E. Myers, whose recent biography has challenged Perry's as the denfinitive study of James' intellectual life, accepts the conventional view that James was "twice born" and never again suffered the acute melancholia of the late sixties and early seventies. See *William James: His Life and Thought* (New Haven: Yale University Press, 1986), p. 46.

3. John Jay Chapman, *Memories and Milestones* (New York: Moffat, Yard, 1915), p. 26.

4. William James to Alice Gibbens James, July 9, 1898; WJP.

5. See Henry James III's note regarding the above letter, with family correspondence in WJP, as well as *LWJ*, 2:78–79. Perhaps Alice's dislike for Miss Goldmark stemmed from the latter's fondness for Keene Valley, which Alice may have seen as forging a bond with William that she could not share.

6. William James to Alice Gibbens James, August 23, 1898; WJP.

7. See John S. Haller, Jr., *American Medicine in Transition, 1840–1900* (Urbana, Ill.: University of Illinois Press, 1981), pp. 193–233, and John Duffy, *The Healers: A History of American Medicine* (Urbana, Ill.: University of Illinois Press, 1979), pp. 109–128 and 180–188.

8. William James to Alice Gibbens James, May 17, 1899; WJP.

9. William James to Alice Gibbens James, June 19, 1899; WJP.

10. Alice Gibbens James to Henry James, May 21, 1899; WJP.

11. William James to Henry James, June 21, 1899; WJP.

12. William James to Pauline Goldmark, August 12, 1899. WJP.

13. Henry James III, note in family correspondence, undated; WJP.

14. William James to Charles W. Eliot, August 7, 1899, Harvard Archives.

15. William James to Henry James, September 16, 1899; WJP.

16. William James to Henry James, September 18, 1899, WJP.

17. William James to Charles W. Eliot, August 7, 1899; Harvard Archives. Eliot apparently found something out of order in Baldwin, for James withdrew his endorsement for the professorship. He had not realized "how ill a man he is." William James to Charles W. Eliot, September 7, 1899; Harvard Archives.

18. September 15, 1899 entry in Alice Gibbens James' diary, quoted in Gay Wilson Allen, *William James: A Biography* (New York: Viking, 1967), p. 401.

19. William James to Charles W. Eliot, December 20, 1899; Harvard Archives.

20. For a description of the extract's ingredients see William James to F. W. H. Myers, December 6, 1900; WJP.

21. On the James-Flournoy relationship, see Robert C. Le Clair's introduction to their correspondence in Robert C. Le Clair, ed., *The Letters of William James and Theodore Flournoy* (Madison: University of Wisconsin Press, 1966), pp. xiii–xix.

22. William James to Henry James, May 14, 1900; WJP.

23. Ibid.

24. "I had a pretty bad spell, and know now a new kind of melancholy. It is barely possible that the recovery may be due to a mind-curer with whom I tried 18 sittings," James had written to F. W. H. Myers, December 17, 1893; WJP.

25. William James to Henry James, May 7, 1900; WJP.

26. William James to Henry James, August 23, 1900; WJP.

27. See Duffy, *The Healers,* 109–128; 180–188.

28. James spoke out against licensing in the *Boston Transcript.* See William James to James Jackson Putnam, March 2, 1898; WJP.

29. William James to Alice Gibbens James, August 23, 1900; WJP.

30. William James to Alice Gibbens James, August 25, 1900; WJP.

31. William James to Alice Gibbens James, September 8, 1900; WJP.

31. William James to Theodore Flournoy, August [?], 1894; quoted in Le Clair, *The Letters of William James and Theodore Flournoy,* p. 35.

33. William James to Alice Gibbens James, September 8, 1900; WJP.

34. Ibid.

35. Ibid.

36. William James to Alice Gibbens James, September 9, 1900, WJP.

37. Ibid.

38. William James to Henry James, November 13, 1900; WJP.

39. Ibid.

40. For a discussion of electrical theory and practice in late nineteenth-century medicine, see John S. Haller and Robin M. Haller, *The Physician and Sexuality in Victorian America* (New York: Norton, 1977), pp. 9–23.

41. William James to Josiah Royce, May 17, 1901; WJP.

42. William James to Henry James, June 17, 1901; WJP.

43. William James to Shadworth Hodgson, August 22, 1901; WJP.

44. William James to Alice Gibbens James, September 16, 1901; WJP.

45. William James to Henry James, December 7, 1901, WJP.

46. William James, Notebook, Memorandum for the Gifford Lectures, and Original Plan for a Philosophy, 2nd vol., ca. 1900; item 4509, WJP. All subsequent quotations in this chapter are from this notebook.

47. William Barrett, *The Illusion of Technique* (Garden City, N.Y.: Doubleday, Anchor Press, 1978), pp. 298–322.

48. William James, quoted ibid., p. 298.

15. GIVE ME A WORLD ALREADY CREATED

1. William James to Henry James, January 31, 1902; WJP.

2. William James to Shadworth Hodgson, June 25, 1902; WJP.

3. William James to Henry James, October 25, 1902; WJP.

4. William James to Henry James, December 22, 1902; WJP.

5. Ibid.

6. William James to Alice Gibbens James, January 24, 1903; WJP.

7. See William James to Horace Fletcher, October [?] 1903; WJP.

8. William James to Horace Fletcher, June 5, 1904; WJP.

9. William James to Charles W. Eliot, December 12, 1903; Harvard Archives.

10. See Charles W. Eliot to William James, December 17, 1903; Harvard Archives.

11. William James to Pauline Goldmark, February 23, 1904; WJP.

12. William James to F. C. S. Schiller, February 1, 1904; WJP.

13. See Darrell Rucker, *The Chicago Pragmatists* (Minneapolis: University of Minnesota Press, 1969).

14. Charles W. Eliot to William James, January 13, 1905; Harvard Archives.

15. On Papini see Vittoria Vettori, *Giovanni Papini* (Torino; Italy: Borla, 1967), and Antonio Santucci's sketch in Paul Edwards, ed., *The Encyclopedia of Philosophy* (New York: Macmillan, 1967), 6:37–38.

16. William James to Alice Gibbens James, April 25, 1905; WJP.

17. William James to Alice Gibbens James, April 30, 1905, WJP.

18. William James to Alice Gibbens James, May 13, 1905, WJP.

19. William James to Alice Gibbens James, May 22, 1905; WJP.

20. William James to Alice Gibbens James, July 6, 1905; WJP.

21. William James to Mary Margaret James, July 11, 1905. WJP.

22. William James to Henry James, October 22, 1905; WJP.
23. William James to Charles W. Eliot, January 25, 1906; Harvard Archives.
24. William James to Alice Gibbens James, January 13, 1906; WJP.
25. William James to Alice Gibbens James, January 17, 1906; WJP.
26. William James to Mary Margaret James, February 7, 1906. WJP.
27. Alice Gibbens James, typescript regarding the San Francisco earthquake, dated Wednesday, April 18, 1906; VC.
28. William James to Charles W. Eliot, September 23, 1906; Harvard Archives.
29. William James to Alice Gibbens James, August 28, 1906; WJP. William wrote to Peggy, "Your feeling words about Keene Valley speak to my heart. There is something unequally peculiar about that place, and I want never to be cut off from it. I felt it just as much this time as I ever did; and I should like my children to keep some connexion with it forever." William James to Mary Margaret James, September 11, 1906; WJP. James wanted to buy a share in the "Reserve" at one end of Keene Valley especially for the children. However, his hope for perpetual James ownership of East Hill was not to be. After Alice's death the children sold the property.
30. William James to Charles W. Eliot, September 23, 1906; Harvard Archives. Eliot arranged with the Carnegie Foundation to provide James with an annual "retirement allowance" of $3,000. See Charles Eliot to Alice Gibbens James, July 16, 1907; Harvard Archives.
31. William James to Mary Margaret James, December 9, 1906; WJP.
32. William James to Henry James, December 4, 1906; WJP.
33. Horace M. Kallen to Ralph Barton Perry, August 3, 1920; WJP. Kallen went on to gain a considerable philosophical reputation at the University of Wisconsin.
34. David A. Pfromn to Alice Gibbens James, August 27, 1913; WJP.
35. William James to Mary Margaret James, January 25, 1907; WJP.
36. William James to Mary Margaret James, February 11, 1907; WJP.
37. William James to Henry James, February 14, 1907; WJP.
38. William James to Henry James, November 30, 1907; WJP.
39. William James to Henry James, April 29, 1908; WJP.
40. William James to Mary Margaret James, May 28, 1908; WJP.
41. William James to Alice Gibbens James, July 7, 1908; WJP.
42. William James to William James, Jr., June 28, 1908; WJP.
43. William James to Henry P. Bowditch, September 1908; WJP.
44. William James to Henry James, October 21, 1908; WJP.
45. William James to Henry James, January 24; 1909; WJP.
46. William James to Henry James, May 31, 1909; WJP.
47. William James to W. Lutoslawski, May 6, 1906; *LWJ*, 2:255.
48. William James, *The Energies of Men*, Religion and Science Series (New York: Moffat, Yard, 1908). The material that became the book—or rather, pamphlet—was first delivered as the presidential address to

the American Philosophical Association in 1906. The next year the address was published as "The Energies of Men," *Philosophical Review* (January 1907), 16:1–20.

49. William James, Notebook, Miller-Bode Criticism, 1905–1908; item 4515, WJP.

50. William James to Pauline Goldmark, September 5, 1909; WJP.

51. For a discussion of calomel and purgative medicine in general see Haller, *American Medicine in Transition,* pp. 37–99. William's first reference to using calomel was in early September. See William James to Alice Gibbens James, September 5, 1909; WJP.

52. See Haller, *American Medicine in Transition,* p. 36.

53. William James to Henry James, October 6, 1909; WJP.

54. William James to Robertson James, October 30; 1909; VC.

55. For a perceptive discussion of the Christian Scientists and the mind cure culture in general, see Donald Meyer, *The Positive Thinkers: Religion as Pop Psychology from Mary Baker Eddy to Oral Roberts* (New York: Pantheon Books, 1980), pp. 21–125.

56. William James to Alice Runnells, December 15, 1909; WJP.

57. William James to Henry P. Bowditch, November 6, 1909; WJP.

58. See William James to Henry James, February 4, 1910; WJP.

59. Henry James III, Note by Henry James, Jr., 1910; WJP.

60. William James to Mary Margaret James, May 7, 1910; WJP.

61. William James to Henry P. Bowditch, June 4, 1910; WJP.

62. William James to Alice Gibbens James, May 11, 1910; WJP.

63. William James to Alice Gibbens James, May 16; 1910; WJP.

64. Quoted in Note by Henry James, Jr., 1910; WJP.

65. William James to Alice Gibbens James, May 18, 1910; WJP.

66. William James to Alice Gibbens James, May 20, 1910; WJP.

67. William James to Henry James, May 28, 1910; WJP.

68. On Blood see William James, "A Pluralistic Mystic," *Hibbert Journal* (1910), 8:739–759.

69. Alice Gibbens James to Elizabeth Webb Gibbens, July 15, 1910; WJP.

70. Alice Gibbens James to Elizabeth Webb Gibbens, August 7, 1910; WJP.

71. William James to Theodore Flournoy, July 9, 1910; in Le Clair, ed., *The Letters of William James and Theodore Flournoy,* p. 239.

72. Alice Gibbens James to Frances R. Morse, August 26, 1910; WJP.

73. Gay Wilson Allen's description of William's suffering and Alice's compassion during the last days can scarcely be improved. See *William James: A Biography* (New York: Viking, 1967), pp. 469–494.

74. John Jay Chapman to Elizabeth Chanler Chapman, November 17, 1910; quoted in M. S. DeWolfe Howe, *John Jay Chapman and His Letters* (Boston: Houghton Mifflin, 1937), p. 249.

75. James did, however, continue to elaborate his perspective. See "Pragmatism's Conception of Truth," *Journal of Philosophy, Psychol-*

ogy, and Scientific Methods (March 14, 1907), 4:141–155; "Controversy about Truth," (with John E. Russell), *Journal of Philosophy, Psychology, and Scientific Methods* (May 23, 1907) 4:289–296; "A Word More About Truth," *Journal of Philosophy, Psychology, and Scientific Methods* (July 18, 1907), 4:396–406; "A Case of Clairvoyance," *Proceedings of the American Society for Psychical Research* (January 1907), vol. 1, part 2, pp. 221–236; "Psychical Phenomena at a Private Circle," *Journal of the American Society for Psychical Research* (1909), 3:109–113; "Report on Mrs. Piper's Hodgson-Control," *Proceedings of the Society for Psychical Research* (1909) 23:1–121, also published in *Journal of the American Society for Psychical Research* (1909), 3:470–589 (*Proceedings* was the journal of the English chapter of the society); and "The Confidences of a Psychical Researcher," *American Magazine* (October 1909), 68:580–589.

EPILOGUE

1. William James, 1868–1873 Diary; WJP.

2. John J. McDermott has suggested that James' mature philosophy anticipated a view of technology as a matter of "relatings." The idea that neither matter nor mind make up the world but give the world back in a continuum of unending, manipulative forms—the self-less expressions of the electronic media—is the contemporary mode. This was profoundly prophetic, and of more significance to modern culture than anything in James' moral legacy, which was backward-looking and less original. See John J. McDermott, "A Metaphysics of Relations: James' Anticipation of Contemporary Experience," in Walter Robert Corti, ed., *The Philosophy of William James* (Hamburg: Felix Meiner, 1976), pp. 81–100.

3. William James, Notebook, Pure Experience Continued, book 3, 1897–1898; item 4505, WJP.

Bibliographical Essay

I HAVE NOTED only works most pertinent to the development and expression of James' biography. For the most complete annotated bibliography, see Ignas K. Skrupskelis, *William James: A Reference Guide* (Boston: G. K. Hall, 1977). Still useful is Ralph Barton Perry's *Annotated Bibliography of the Writings of William James* (New York: Longmans, Green, 1920).

The definitive edition of James' writings is *The Works of William James*, edited by Frederick Burkhardt et al. (Cambridge: Harvard University Press, 1975—). Through 1987 the following have appeared: *Pragmatism, The Meaning of Truth,* and *Essays in Radical Empiricism*, introduced by H. S. Thayer; *Essays in Philosophy* and *Essays in Religion and Morality*, introduced by John J. McDermott; *A Pluralistic Universe,* Richard J. Bernstein; *The Will to Believe,* Edward H. Madden; *Some Problems of Philosophy,* Peter H. Hare; *The Principles of Psychology,* 3 vols., Rand B. Evans and Gerald E. Myers; *Psychology: Briefer Course,* Michael M. Sokal; *Talks to Teachers on Psychology,* Gerald E. Myers; *Essays in Psychology,* William R. Woodward; *The Varieties of Religious Experience,* John E. Smith; *Essays in Psychical Research,* Robert A. McDermott; and *Essays, Comments, and Reviews,* Ignas K. Skrupskelis. For a convenient anthology of selections, particularly from writings with philosophical import, see John J. McDermott, ed., *The Writings of William James: A Comprehensive Edition* (Chicago: University of Chicago Press, 1977).

Interpretation of James began even before his death. In 1907 two of his colleagues at Harvard, psychologist Hugo Münsterberg and philosopher Ralph Barton Perry, wrote short sketches marking James' retirement from the Harvard Philosophy Department. See Münsterberg, "Professor James as a Psychologist," *The Harvard Illustrated Magazine* (February 1907), 7:97–98; and Perry, "Professor James as a Philosopher," *The Harvard Illustrated Magazine* (February 1907), 7:96–97. Münsterberg described James as a "descriptive" rather than an "explanatory" psychologist, while Perry placed James' philosophy within the British empiricist tradition.

Upon James' death in 1910 other friends and colleagues eulogized him with sketches that touched on one or another of his personal and intellectual qualities. John Jay Chapman portrayed James as a serious and tragic thinker despite his renowned wit. Chapman was the first to suggest that James actually loathed philosophy and was primarily interested in religious and moral life. See Chapman, "William James: A Portrait," *Harvard Graduates' Magazine* (December 1910), 19:233–238. Walter Lippmann, a former student, popularized a James who was tolerant of differing points of view even when they clashed with prevailing scientific opinion, as did psychical research. See Lippmann, "An Open Mind: William James," *Everybody's Magazine* (December 1910), 23:800–801. Dickinson S. Miller, another student and close friend, maintained that James' psychology was essentially the study of human nature. James came to philosophy with human nature in mind and never let his metaphysics "ignore the way life feels." See Miller, "Some of the Tendencies of Professor James's Work," *The Journal of Philosophy* (November 1910), 7:645–664. Still another Harvard student who continued there as a philosopher, Arthur O. Lovejoy, defended James from detractors who claimed he was too impressionistic and less than scrupulous in logical analysis. See Lovejoy, "William James as Philosopher," *The International Journal of Ethics* (January 1911), 21:125–153. Josiah Royce, James' beloved philosophical opponent, gave a Harvard Phi Beta Kappa oration on James that was widely circulated. Royce was the first to portray James as the third great "representative American philosopher," giving him a hallowed place alongside Jonathan Edwards and Ralph Waldo Emerson. See Royce, "William James and the Philosophy of Life," in his *William James and Other Essays on the Philosophy of Life* (New York: Macmillian, 1911), pp. 3–45. The address was also published in *Science* (July 14, 1911), 34:33–45 and in *Harvard Graduates' Magazine* (1911–1912), 20:1–18.

Only one Harvard colleague and former student partially dissented from the widespread and varied praise for James' intellectual contribution. George Santayana objected to James' being identified as a philosopher, not so much because he felt, as had Chapman, that James was basically interested in religion, but rather because in his view James only made raids against prevailing metaphysics and was not interested in building philosophical edifices at all; see George Santayana, *Character and Opinion in the United States* (New York: Scribner's, 1920), pp. 64–96. Moreover, for all of James' criticism of what Santayana called "the genteel tradition," in Santayana's eyes James still embraced the genteel perspective. Whereas Royce praised James' cultural authenticity and representativeness, Santayana saw James as too culturally bound to the prevailing Victorian ethic. See Santayana, "The Genteel Tradition in American Philosophy," *The University of California Chronicle* (October 1911), 13:357–

380. That the Spanish-born Santayana dared to belittle the philosophical contribution and cultural instincts of the recently deceased James did not pass unnoticed. William's former student and friend Dickinson S. Miller came to his former professor's defense by characterizing Santayana as a snob who simply did not grasp James' vastly superior humanity, let alone his philosophy. See Miller, "Mr. Santayana and William James," *Harvard Graduates' Magazine* (March 1921), 29:348–364. For Santayana's rejoinder see "Professor Miller and Mr. Santayana," *Harvard Graduates' Magazine* (September 1921), 30:32–36.

Little advance toward a major James biography was made between the early eulogies and the publication of Henry James, ed., *The Letters of William James,* 2 vols. (New York: Longmans, Green, 1920). As James' eldest surviving son, Henry James assembled portions of his father's personal correspondence. The letters are interesting chiefly for their insights into James' social circle rather than for light shed on his psychological and philosophical friends. Henry James the novelist had earlier written *Notes of a Son and Brother* (New York: Scribner's, 1914), an interesting but highly protective account of their brotherly childhood and youth.

During the 1920s and early 1930s several writers fashioned a pragmatic James. In 1922 John Dewey, already thought of by some as James' major philosophical successor, maintained that James had developed pragmatism essentially as a theory of truth. See his "Le Development du Pragmatisme Americain," *Revue de Metaphysique et de Morale* (October–December, 1922), vol. 29; available in English as "The Development of American Pragmatism," in vol. 2 of *Studies in the History of Ideas,* 2 vols. (New York: Columbia University Press, 1925). Dewey also reemphasized the connection between James' pragmatism and the "American spirit," making James and pragmatism the authentic national philosopher and philosophy; see "William James in Nineteen Twenty-Six," *New Republic* (June 30, 1926), 47:163–165. And Horace Kallen, a former student and later University of Wisconsin philosopher, called attention to James' pluralistic pragmatism, his ability to look at truth as instrumental in our "struggle to live." See especially Kallen's introduction to *The Philosophy of William James* (New York: Random House, 1925), pp. 1–55.

In *Religion in the Philosophy of William James* (Boston: Marshall Jones, 1926, Julius Seeyle Bixler attempted to interpret James in terms other than pragmatic, stressing the tension between passive and active religious strains in James. But F. C. S. Schiller, who acted as James' pragmatic bulldog, argued that religion was vital to James' pragmatism; James' will to believe was not a personal strategy for survival but an epistemology more valuable to science than to religion. See Schiller, review of Bixler's *Religion in the Philosophy of William James,* in *Mind* (October 1926), 36:509–511.

Thus well before the publication of Ralph Barton Perry's landmark biography in 1935, James had been interpreted primarily as a pragmatist. Dewey and others saw James as the philosophical father of what by the twenties and thirties had become a broad progressive movement that advocated pragmatic social change in American institutions, from schools to political parties. See Sidney Hook, "Our Philosophers," *Current History,* (March 1935), 41:698–704, and his "William James," *The Nation* (December 11, 1935), 141:684 and 686–687. In the latter article Hook, a well-known pragmatist, finds James more important as an intellectual force than as a systematic thinker. He argues that James initiated the pragmatic frame of mind rather than definitively solving metaphysical puzzles.

The only pre-Perry study aside from Santayana's that dissents from the chorus of scholarly voices lauding James as America's pioneering early twentieth-century philosopher is that of Clinton Hartley Grattan. Grattan examines the father, Henry Sr., and the two famous sons, Henry and William, and concludes that William was essentially a shallow thinker because as a pragmatist he offered too-easy solutions to immovable problems of human nature. It is interesting that Grattan evaluates James only in terms of the prevailing view: Grattan did not like pragmatism; James was a pragmatist; therefore, James was a shallow thinker. See Grattan, *The Three Jameses: A Family of Minds* (New York: Longmans, Green, 1932).

From the time of James' death, Ralph Barton Perry had been gathering materials for the first full-scale biography. His *The Thought and Character of William James,* 2 vols. (Boston: Little, Brown, 1935) became the indispensible, comprehensive account of James' personal and intellectual development. Despite its lack of any substantial criticism of either James' life or mind, until very recently it remained the definitive biography; perhaps because of its long tenure as such, it is still the most influential. In it Perry argues that James was first and foremost a philosopher, and one who effectively provided a pragmatic alternative to the prevailing British empiricism and German idealism; James' scientific psychology was but a prologue to pragmatism and a philosophy of pure experience. Perry's portrait of James as a sensitive, troubled young man who moved from art, to natural science, to psychology, and finally to his true calling, philosophy, is still widely accepted.

The reviews of *The Thought and Character of William James* were positive, for the most part. See for example Edward S. Robinson, *Yale Review* (March, 1936), 25:616–620. Robinson's review considers Perry's James book not only a contribution to the art of biography, but to the history of American thought. Dickinson S. Miller's "James's Philosophical Development; Professor Perry's Biography," *The Journal of Philosophy* (June 1936), 33:309–318 observes that James' faith in human nature was so high that it required him to defend positions (like the will to believe) not practical for most men. But Miller nonetheless concludes that Perry's biography is "indisputable." Only Horace M. Kallen takes fundamental issue with Perry's James. Kallen's objection is suggestive; he notes that Perry in effect deintellectualizes James by dividing his creative life into four separate components—art, natural science, psychology, and philosophy. For Kallen, James was an "essentially integrated personality"; further, James was bookish and did not assume that achievement was superior to contemplation. See Kallen, "Remarks on R. B. Perry's Portrait of William James," *The Philosophical Review* (January 1937) 46:68–78. Kallen's view was soon forgotten, however, and Perry continued to develop a picture of James as not only a pragmatic philosopher who fashioned an empiricist theory of knowledge and a metaphysics of experience, but also as a committed, militant progressive; see Perry, *In the Spirit of William James* (New Haven: Yale University Press, 1938).

In 1942 the centenary of James' birth brought a great harvest of Jamesian scholarship. Two anthologies of articles appeared, emphasizing facets of James' intellectual interests from philosophy to religion to depth psychology: Brand Blanshard and Herbert W. Schneider, eds., *In Commemoration of William James: 1842–1942* (New York: Columbia University Press, 1942), and Max C. Otto et al., *William James: The Man and the Thinker* (Madison: University of Wisconsin Press, 1942). John Dewey was a contributor to both, continuing to extol James' pragmatism but now emphasizing the "pluralistic" dimension of his thinking. See his "William James and the World Today," in *William James: The Man and the Thinker,* pp. 89–97, and "William James as Empiricist," in *In Commemoration of William James,* pp. 48–57. Boyd H. Bode and Justice Buchler continued to connect James' pragmatism to indigenous American values. See Bode's "William James in the American Tradition," in *William James: The Man and the Thinker,* and Buchler's "The Philosopher, the Common Man, and William James," *The American Scholar* (October 1942), 11:416–426. Harvard psychologist A. A. Roback wrote an odd book, an attempt to analyze James' personality by studying his handwriting. James' graceful, regular letters reveal to Roback an artistic bent, while the broken spaces between his letters indicate exceptional intuition. Roback proposes that James was much more intellectually influential through his personality than through the objective power of his psychology or philosophy. See *William James: His Marginalia, Personality, and Contribution* (Cambridge, Mass.: Sci-Art Publishers, 1942).

The centenary adoration sanctified the by now conventional interpretation that James was less an artist or a scientist than a philosopher who advocated a broad, morally responsible pragmatism. Nonetheless, several voices presaged different emphases that were still decades away from full expression. In "William James as Artist," *The New Republic* (February 15, 1943), 108:218–220, the intellectual historian Jacques Barzun argues that James was more artistic than philosopher. His whole life was a work of art. Arthur F. Bentley, a psychologist, states categorically that the centenary writers had missed the point about James' mind. Pragmatism or a philosophy of action was not James' great discovery; his insight into "pure" or "immediate" experience was far more significant. See "The Jamesian Datum," *Journal of Psychology* (July 1943), 16:35–79, and "Truth, Reality, and Behavioral Fact," *The Journal of Philosophy* (April 1943), 40:169–187. Aron Gurwitsch notes that James' interest in the "immediately experienced continuity of consciousness" was directly related to the concerns of phenomenology, a point that had been suggested two years earlier by phenomenologist Alfred Schuetz. See Gurwitsch, "William James' Theory of the 'Transitive Parts' of the Stream of Consciousness," *Philosophy and Phenomenological Research* (June 1943), 3:449–477, reprinted in *Studies in Phenomenology and Psychology* (Evanston, Ill.: Northwestern University Press, 1966); and Schuetz, "William James' Concept of the Stream of Thought Phenomenologically Interpreted," *Philosophy and Phenomenological Research* (June 1941), 1:442–443, reprinted in *Collected Papers,* vol. 3 (The Hague: Nijhoff, 1966).

Bentley, Schuetz, and Gurwitsch were aware that Edmund Husserl, the founder of phenomenology, was indebted to James' stream of thought psychology, even though James never developed "intentionality," the notion that consciousness cannot exist without intending its object. Thus, from the time of the centenary, a few scholars ventured that James' intellectual significance was to be found more in those aspects of his psychology and philosophy that presaged phenomenology (stream of consciousness, pure experience, and radical empiricism) than in pragmatism. It was not until the 1960s, however, that a phenomenological James became a viable alternative to Perry's pragmatic one, although by no means replacing him. See for example Herbert Spiegelberg, *The Phenomenological Movement: A Historical Introduction* (The Hague: Nijhoff, 1960); James M. Edie, "Notes on the Philosophical Anthropology of William James," in James M. Edie, ed., *An Invitation to Phenomenology* (Chicago: Quadrangle Books, 1965) 110–132; Johannes Linschoten, *On the Way Toward a Phenomenological Psychology: The Psychology of William James* (Pittsburgh: Duquesne University Press, 1968); Bruce Wilshire, *William James and Phenomenology: A Study of "The Principles of Psychology"* (Bloomington: Indiana University Press, 1968); John Daniel Wild, *The Radical Empiricism of William James* (Garden City, N.Y.: Doubleday, 1969); Fred Kersten, "Franz Brentano and William James," *Journal of the History of Philosophy* (April 1969), 7:171–191; and in James M. Edie, ed., *New Essays in Phenomenology* (Chicago: Quadrangle Books, 1969), see James M. Edie, "Necessary Truth and Perception: William James on the Structure of Experience," pp. 233–255; Robert R. Ehman, "William James and the Structure of the Self," pp. 256–270; and John Wild, "William James and the Phenomenology of Belief," pp. 271–289.

Although substantial biographical revision did not emerge until 1968, several noteworthy studies did appear in the late forties and fifties, including *The James Family: Including Selections from the Writings of Henry James, Senior, William, Henry, and Alice James* (New York: Knopf, 1948), by F. O. Matthiessen, an English professor at Harvard. The book includes little fresh material on William, however, concentrating most heavily on the novelist Henry. Two years later Morris Lloyd's *William James: The Message of a Modern Mind* (New York: Scribner's, 1950) was published. Lloyd characterizes James as an ancestral liberal, the American philosopher who most influenced Oliver Wendell Holmes, Jr.'s legal thinking, as well as Franklin D. Roosevelt's New Deal.

In 1953 the first volume of Leon Edel's monumental five-volume biography of Henry James appeared, *Henry James: The Untried Years (1843–1870)* (Philadelphia: Lippincott). *Henry James: The Conquest of London (1870–1881)* and *Henry James: The Middle Years (1882–1895)* were published in 1962; *Henry James: The Treacherous Years (1895–1901)* in 1969; and *Henry James: The Master (1901–1916)* in 1972. Edel was the first scholar to seriously explore—although without a psychoanalytic framework—psychosomatic illness among the Jameses. He builds an interesting narrative centered on the James family domestic scene, and he uses many family letters that Perry had ignored. He presents William and Henry as companion-rivals, a modern reenactment of the biblical Jacob-Esau relationship. William T. Stafford also looks at the brotherly competition, in "William James as Critic of His Brother, Henry," *The Personalist* (Autumn 1959), 40:341–343.

In "William James and the Clue to Art," in *The Energies of Art: Studies of Authors Classic and Modern* (New York: Harper & Brothers, 1956), pp. 325–355, Jacques Barzun resumes his interpretation of James as an artist. Here Barzun maintains that *The Principles of Psychology* is closely associated with the neo-Romantic movement. He notes that James reestablished the mind as "myth-maker and artist." In a 1958 article, Saul Rosenzweig traces James' stream of consciousness back to a mid-nineteenth century Englishman, J. Garth Wilkinson, a close friend of Henry James, Sr., who had developed what he called a "method of impression" that was remarkably similar to the stream of consciousness idea. See "The Jameses' Stream of Consciousness," *Contemporary Psychology* (1958), 3:250–257.

In *The Ethics of William James* (New York: Bookman Associates, 1961), Bernard Brennan argues suggestively that there was no real center to James' work, but that his pragmatism, voluntarism, and radical empiricism must be seen in relation to one another—in itself a pragmatic judgment. And in *American Pragmatism: Peirce, James, and Dewey* (New York: Columbia University Press, 1961) and *William James* (New York: Washington Square Press, 1965), Edward C. Moore discovers that for James, pragmatism had from the beginning been a method to settle the troublesome personal question of God's existence.

Most original is English professor Frederick J. Hoffman's insight that James was the first to portray the self as a process rather than as a content. Hoffman's article suggests that this new way of thinking about the individual enormously influenced modern intellectual history. See "William James and the Modern Literary Consciousness," *Criticism* (Winter 1962), 4:1–13.

With the publication of Gay Wilson Allen's *William James: A Biography* (New York: Viking, 1967) a major alternative to Perry's *The Thought and Character of William James* finally appeared. Allen's limited access to hitherto unavailable family correspondence enabled him to construct for the first time some sense of William's relationship with his wife, Alice Howe Gibbens James. The book also provides a much more detailed presentation of James' travels and health problems than anything preceding it. Allen, however, does not adequately discuss James' ideas; Perry's work is far superior as an intellectual biography. And Allen, like Perry, sees William as a philosopher in the making who endured a protracted vocational crisis before finding his natural niche.

Biographical interpretation advanced markedly in 1968 with the appearance of intellectual historian Cushing Strout's psychoanalytic article, which presented James for the first time in the context of a crushing oedipal conflict. The article draws heavily from Erik Erikson's identity crisis concept and portrays William as a troubled young man whose vocational crises stemmed from Henry Sr.'s role as an "inner court of tribunal" for his son; William was not free to develop his own original philosophy, pluralism, until after his father's death in 1882. Strout was successful in refocusing attention on William's psychic or spiritual troubles during the late 1860s and early 1870s. Indeed, since 1968, biographical interest in James has almost exclusively centered on the young man and his oedipal-vocational anxieties. See Strout, "William James and the Twice-Born Sick Soul," *Daedalus* (1968), 97:1062–1082, reprinted in his *The Veracious Imagination: Essays on American History, Literature, and Biography*, pp. 199–222 (Middletown, Conn.: Wesleyan University Press, 1981); and "The Pluralistic Identity of William James," *American Quarterly* (1971), 23: 135–149.

S. P. Fullinwider, also an intellectual historian, shifted interpretive ground in the mid-1970s with an important essay that portrays James' melancholia and neurasthenia as symptomatic of an existential struggle. Fullinwider maintains that James regained mental health not so much by resolving oedipal conflict as by working out a creative psychology that put the sting of objectivity back into his dangerously solipsistic personal world. James' scientific psychology is best viewed, according to Fullinwider, as an existential map which recorded James' journey out of psychic sickness, rather than as an introductory chapter in the development of his pragmatic and pluralistic metaphysics. See "William James's Spiritual Crisis," *The Historian* (1975), 13:39–57.

The oedipal theme was reemphasized in 1977, however, when Howard Feinstein discovered an intergenerational father-son conflict that originated not with Henry Sr. and William, but with the first paternal American James, William James of Albany, and his son Henry. See "Fathers and Sons: Work and the Inner World of William James, an Intergenerational Inquiry," Ph.D. dissertation, Cornell University, 1977. Feinstein shows that Henry Sr. had never resolved his oedipal conflict with William of Albany and had projected these lingering anxieties onto William. Henry Sr. disapproved of any vocational choice William made because his own father, William James of Albany, had never accepted Henry's antinomian

Protestantism. Hence, when William chose to be a painter, Henry Sr. opposed him; when William leaned toward science and then metaphysics, Henry Sr. disagreed with those choices as well. Feinstein was also the first scholar to offer a psychoanalytic interpretation of William as thwarted from his true calling as an artist. A revised version of the dissertation appeared as *Becoming William James* (Ithaca, N.Y.: Cornell University Press, 1984). My own initial approach to James was indebted to Feinstein's dissertation, but emphasized how James' natural artistry was transformed into an original psychological language that clashed with the laboratory-oriented scientific psychologies of his American contemporaries. See Daniel W. Bjork, *The Compromised Scientist: William James in the Development of American Psychology* (New York: Columbia University Press, 1983).

In the 1980s biographical interest in James has, if anything, increased. Jacques Barzun has recently portrayed James as a sophisticated artist-thinker of international caliber who added a unique dimension to modern western thought. Barzun observes that James' stream of consciousness psychology and open-universe metaphysics provided a vital alternative to the overwhelmingly deterministic perspective of late nineteenth-century science and philosophy. But Barzun so completely sympathizes with James as his favorite genius and hero that the book offers too few insights critical of its subject. See *A Stroll with William James* (New York: Harper and Row, 1983).

Intellectual historian John Owen King III depicts James as an archtypical American who struggled to transform an old American experience with spiritual conversion into the language of Victorian neurosis. Unlike the Puritans who saw a fall into melancholia as a sign that one's soul was close to salvation, James viewed melancholia negatively and sought a "twice born" road to healthy-mindedness. King's attempt to construct a historical lineage of conversion experiences, however, is fundamentally dependent on James' relationship with Henry James, Sr., and therefore does not depart far from the Strout-Feinstein oedipal context. See *The Iron of Melancholy: Structures of Spiritual Conversion in America from the Puritan Conscience to Victorian Neurosis.* (Middletown, Conn.: Wesleyan University Press, 1983).

Penetrating interpretation has recently come from those who have worked extensively with ignored or newly available James materials. Eugene Taylor, an independent scholar associated with the Francis A. Countway Library of Medicine, has tracked down hundreds of books that were originally in James' personal library but were later donated to the Widener and Houghton Libraries at Harvard. Many of these volumes contain James' own marginalia and underlinings. Using this reconstructed library to interpret references in James' notes for the 1896 Lowell Lectures on psychopathology, Taylor has in effect assembled a new book *by* William James. His reconstruction provides a missing link between publication of *The Principles of Psychology* in 1890 and the appearance of *The Varieties of Religious Experience* in 1902. Taylor shows clearly that James did not, as has been widely assumed, simply turn away from science toward "mysticism" after 1890. See *William James on Exceptional Mental States: The 1896 Lowell Lectures* (New York: Scribner's, 1982), and "William James on Psychopathology: The 1896 Lowell Lectures on Exceptional Mental States," in *A William James Renaissance: Four Essays by Young Scholars,* special issue, *Harvard Library Bulletin* (October 1982), 30(4):455–479.

The three other essays in the special October 1982 issue of *Harvard Library Bulletin,* by James William Anderson, Robert J. Richards, and Mark R. Schwehn, have encouraged biographical rethinking. Anderson's contribution once again examines William's youthful psychic crisis, and maintains that James suffered from the fear of intimacy and found close contact with people bitterly painful. He suffered a self-fragmentation experience which brought him close to suicide in 1869, but he was saved by embracing a maternal image of God, one that Anderson implies owed something to his actual mother, Mary James. Anderson is the first to suggest that Mary James as well as Henry Sr. needs to be considered in any psychoanalytic reconstruction of William's psychic crisis. See his " 'The Worst Kind of Melancholia': William James in 1869," in the above-cited *A William James Renaissance,* pp. 369–386, and his "In Search of Mary James," *The Psychohistory Review* (1979), 8:63–70.

The essay by Robert J. Richards, a historian of science, is challenging if not totally convincing, arguing that constructive interpretation of James' psychic or spiritual crisis must adequately relate his emotional travail to his unfolding views of science, particularly of Darwinism. Richards suggests that by emphasizing the creative rather than the determining side of evolution, James found the intellectual strength to overcome an emotional crisis. It was his study of Darwinism, Richards argues, that stimulated James' belief in man's creative conscious life. This view challenges the prevailing opinion that the deterministic implications of evolution brought on James' melancholy—a view that dates back to the Perry biography. See Richards, "The Personal Equation in Science: William James's Psychological and Moral Uses of Darwinian Theory," in *A William James Renaissance,* pp. 387–425. Perhaps even more significantly, Richards' article moves the biographical foundation away from oedipal conflict toward a reexamination of James' intellectual life—something S. P. Fullinwider had begun to do in the mid-1970s. Richards, like Taylor and Anderson, has also underscored the necessity for delving into the riches of the James papers and not relying on Perry and other oft-quoted James publications as authoritative guides to James' intellectual development.

Mark R. Schwehn is the first scholar to recognize what Allen mostly ignored in the William James to Alice Gibbens James correspondence. Schwehn shows that these letters bear directly on the development of James' intellectual as well as his emotional life. He finds that James not only revealed to Alice his deepest convictions about the relationship of morality and marriage, but that he connected these convictions to his emerging view of Victorian science. As science became more positivistic, James told Alice, it reduced human experience to the brute existence that he loathed. Just as he felt he had a duty to keep his marriage based in morality, so too did he firmly resolve to Alice to provide science and metaphysics with a moral foundation. Although Schwehn does not pursue the correspondence beyond the late 1870s, he nonetheless illustrates its potential in biographical revision. See "Making the World: William James and the Life of the Mind," in *A William James Renaissance,* pp. 426–454.

The most recent James book to benefit from post-Allen biographical revision is Gerald E. Myers' *William James: His Life and Thought* (New Haven: Yale University Press, 1986). Myers, a philosopher at the City University of New York and the author of introductions to two volumes of *The Works of William James,* has produced the most comprehensive analysis of James' intellectual contribution since Perry. In thirteen chapters he explicates both the historical and intellectual context of James' major ideas, and he manages this with respect for the profundity of the problems that James tackled, while at the same time offering critical commentary. Moreover, Myers does not slight James' popular contributions in the social and moral areas. No one after reading this work can ever again simply describe James as a pragmatist. His scientific and philosophical contributions were far too interwoven within the web of the intellectual development of the modern western world.

Yet in one sense Myers' book is not a traditional biography. The first chapter summarizes James' life. The work does not balance James' life and thought in chronological progression—one advantage, perhaps, of Perry's approach. Although Myers uses many neglected and newly available sources, his book is a lengthy explication of James' thought, rather than an integrated narrative of life and thought.

It is unlikely that interest in either James' life or mind will diminish in the twenty-first century. Indeed, the generous attention devoted to James within the last twenty years points to an even greater interpretive harvest. Something will compel us to continue to place him in the center of our vision.

Index